Abiotic Stress in Crop Plants - Ecophysiological Responses and Molecular Approaches

Edited by Mirza Hasanuzzaman and Kamrun Nahar

Published in London, United Kingdom

Abiotic Stress in Crop Plants - Ecophysiological Responses and Molecular Approaches
http://dx.doi.org/10.5772/intechopen.114576
Edited by Mirza Hasanuzzaman and Kamrun Nahar

Contributors
A. Amaregouda A, Abida Sultana, Amal Morsy, Amena Khatun, Angela Sibanda Makuvise, Angeles Calatayud, Augustine Antwi-Boasiako, Ayyanagouda Patil, B.K. Prasad, Bakht Afroze, Beyazit abdurrahman Sanli, Bushra Rashid, Daniel Bárcenas-Santana, Daouda Coulibaly, David Kawadza, Elizabeth García-León, Enetia Bobo, Fabián Pérez-Labrada, Faomida Sinthi, Fatima Batool, Gladys Edith Jiménez Nopala, Grace Mabadahanye, Hafiza Fizzah Zulfiqar, Hamidreza Ghorbani, Hany M. Almotairy, Jacinta Adoma Opoku, José Luis Espinoza-Acosta, Juan Gerardo Ortiz-Montiel, Katlego Sehlabane, Manuel Mandujano Piña, Maria del Socorro Sánchez Correa, Mari Carmen López-Pérez, María del Rocío Reyero-Saavedra, Mirza Hasanuzzaman, Muhammad Ahmed, Muhammad Saad Bhutta, Mutsa Takundwa, Nahla El-Sherif, Neo Mametja, Orkun Gencer, Oziniel Ruzvidzo, Parastoo Majidian, Patience Chatukuta, Paul Mintah, Prakash H. Kuchanur, Prem Sagar S P, Priscilla Amponsah, R. Yashaswini, Ramón Gisbert-Mullor, Salvador López-Galarza, Sameera Hassan, Samiha Rummana, Samiul Alam, Sana Shakoor, Suvarna Veeresh, Tshegofatso Dikobe, V.C. Raghavendra, Yaiza G. Padilla, Zahide Neslihan Öztürk

First published in London, United Kingdom, 2024 by IntechOpen
IntechOpen is the global imprint of INTECHOPEN LIMITED, registered in England and Wales, registration number: 11086078, 167-169 Great Portland Street, London, W1W 5PF, United Kingdom

British Library Cataloguing-in-Publication Data
A catalogue record for this book is available from the British Library

Additional hard and PDF copies can be obtained from orders@intechopen.com

Abiotic Stress in Crop Plants - Ecophysiological Responses and Molecular Approaches
Edited by Mirza Hasanuzzaman and Kamrun Nahar
p. cm.
Print ISBN 978-0-85014-006-4
Online ISBN 978-0-85014-007-1
eBook (PDF) ISBN 978-0-85014-008-8

For EU product safety concerns:
IN TECH d.o.o., Prolaz Marije Krucifikse Kozulić 3, 51000 Rijeka, Croatia,
info@intechopen.com or visit our website at intechopen.com.

Meet the editors

Dr. Mirza Hasanuzzaman is a professor of agronomy at Sher-e-Bangla Agricultural University, Dhaka, Bangladesh. He received his Ph.D. from Ehime University, Japan. As a recipient of the Japan Society for the Promotion of Science (JSPS) postdoctoral fellowship, he completed his postdoctoral research at the Center of Molecular Biosciences, University of the Ryukyus, Japan. He was also the recipient of the Australian Government Endeavour Research Fellowship for postdoctoral research as an adjunct senior researcher at the University of Tasmania. Dr Hasanuzzaman's current research focus is the physiological and molecular mechanisms of environmental stress tolerance. He has published over 300 articles in peer-reviewed journals, edited 20 books, and written over 40 book chapters on important aspects of plant physiology, plant stress tolerance, and crop production. According to Scopus®, Dr Hasanuzzaman's publications have received over 24,000 citations with an h-index of 80. He is an editor and reviewer for more than 50 peer-reviewed international journals and was a recipient of the Publons Peer Review Award in 2017, 2018, and 2019. He received the World Academy of Science (TWAS) Young Scientist Award 2014; University Grants Commission (UGC) Gold Medal 2018; Bangladesh Academy of Sciences (BAS) Gold Medal Award 2022 (Senior Group); Global Network of Bangladeshi Biotechnologists (GNOBB) Award 2021; Society for Plant Research Young Scientist Award (Agriculture) 2023; Eminent Scientist Award 2022 from the Society for Science of Climate Change and Sustainable Environment; and Distinguished Scientists Award (2020) and Global Scientist Award-2023 from AETDS, India. He is a fellow of the Bangladesh Academy of Sciences (BAS), the World Academy of Sciences (TWAS), the Royal Society of Biology, and the International Society of Environmental Botanists, and a foreign fellow of the Society for Science of Climate Change and Sustainable Environment.

Dr. Kamrun Nahar is a professor of agricultural botany at Sher-e-Bangla Agricultural University, Dhaka, Bangladesh. She received her Ph.D. in the environmental stress physiology of plants from the United Graduate School of Agricultural Sciences, Ehime University, Japan, with a Japanese government (MEXT) scholarship. Dr. Nahar has been involved in research with field crops, with a focus on stress physiology, since 2006. Several of her research work and continuing research projects have been funded by the Sher-e-Bangla Agricultural University Research System and the Ministry of Science and Technology, Bangladesh. Dr. Nahar has published over 100 articles and chapters on plant physiology and environmental stresses. Her publications have achieved over 12,000 citations with an h-index of 60 according to Google Scholar. She is involved in editorial activities and is a reviewer of international journals. She is an active member of about 20 professional societies. Dr. Nahar has presented 20 papers and posters at national and international conferences. She is an associate fellow of the Bangladesh Academy of Sciences.

Contents

Preface

In the era of rapid climate change, the resilience of our agricultural systems is more critical than ever. Crop plants, the cornerstone of global food security, are increasingly subjected to various abiotic stressors. The responses of crop plants to abiotic stress are complex and multifaceted, governed by a range of factors that vary across species and varieties. Unlike mobile organisms, plants are sessile and cannot escape these harsh environmental conditions. However, they have evolved remarkable mechanisms to adapt and survive. Understanding these mechanisms and enhancing stress tolerance in crop plants has become a primary focus for crop scientists.

Over the past few decades, substantial progress has been made in the field of plant research, particularly in understanding the responses and tolerance mechanisms of plants under abiotic stress conditions. This book, *Abiotic Stress in Crop Plants – Ecophysiological Responses and Molecular Approaches*, aims to compile and present these recent advancements comprehensively.

This book is structured into 13 chapters, each meticulously crafted to delve into different aspects of abiotic stress in crop plants. It covers a wide array of topics, including the ecophysiology of plant responses, the adaptation mechanisms employed by various species, and the molecular biology underlying stress tolerance. By bringing together these diverse yet interconnected areas of study, the book provides a holistic view of the strategies that plants use to cope with environmental challenges. From ecophysiological responses to molecular approaches, each chapter offers valuable insights into the adaptive mechanisms of plants. By bringing together cutting-edge research and developments, this book serves as a crucial resource for scientists, researchers, and students dedicated to enhancing crop resilience and ensuring sustainable agricultural practices.

We would like to thank all the authors for their outstanding and timely contributions. We are very grateful to Marica Novakovic, publishing process manager, and Lucija Tomicic-Dromgool, senior commissioning editor, at IntechOpen for their prompt responses during the production of this book. We also thank Faomida Sinthi, Abida Sultana, and Samiul Alam, Department of Agronomy, Sher-e-Bangla Agricultural University for their reviews and formatting.

We hope that the knowledge and insights presented in this book will inspire further research and innovation in the field of plant stress biology, ultimately contributing to the development of crops that can withstand the challenges posed by a changing climate.

Mirza Hasanuzzaman and Kamrun Nahar
Department of Agronomy,
Sher-e-Bangla Agricultural University,
Dhaka, Bangladesh

Section 1

Plant Responses and Ecophysiology

Chapter 1

Perspective Chapter: Enhancing Plant Resilience to Salinity Induced Oxidative Stress – Role of Exogenous Elicitors

Mirza Hasanuzzaman, Faomida Sinthi, Samiul Alam, Abida Sultana, Samiha Rummana and Amena Khatun

Abstract

Plants face various abiotic stresses, among which soil salinity is a significant threat. It reduces plants' growth and development remarkably due to its detrimental effects. Salt stress occurs when soluble salts accumulate in the soil solution. A considerable portion of the world's soil is negatively impacted by salinity, even at low concentrations. Salinity can hinder plant physiological processes by inducing osmotic stress and ionic toxicity, generating excessive reactive oxygen species (ROS) and leading to oxidative stress within plant cells. The formation of ROS is a normal plant metabolic phenomenon, but excessive ROS-induced oxidative stress can disrupt membrane activities, enzymes, and cell walls, cause cell damage, and, in severe cases, plant death. Various strategies, such as chemical substances and plant growth-promoting substances, are applied exogenously to diminish ROS-induced oxidative stress. Additionally, the use of rhizobacteria that have plant growth-promoting traits, organic amendments, rhizospheric fungi, and various genetic approaches are considered when addressing salt stress in plants. These mechanisms for scavenging ROS enhance plants' tolerance to saline stress by developing an antioxidant defense system, reducing oxidative damage at the cellular level, and maintaining ion homeostasis. This chapter focuses on the latest research regarding the alleviation of salinity-induced oxidative stress in several crops through the exogenous application of stress elicitors.

Keywords: agronomic management, climate-smart agriculture, environmental stress, phytohormones, plant-environment interaction, sustainable agriculture

1. Introduction

The global community has become increasingly concerned about climate change, driven by global warming, and its adverse effects on the environment and agroecosystems. It is increasingly recognized that the objective of environmental sustainability is seriously threatened by global warming. Weather fluctuations due to climate change profoundly affect agriculture by interrupting planting and harvesting schedules, changing crop growth patterns, and increasing the prevalence of pests and pathogens, thereby jeopardizing crop yields and the food security of the world. Numerous environmental factors, including salinity, high temperatures, drought, heat waves, frost, cool temperatures, and high levels of carbon dioxide (CO_2), impact both annual and perennial plants [1].

Among these factors, salinity stress affects hectares of cultivable land worldwide, and its severity is expected to escalate further as a consequence of climate change. Salinity adversely affects over 1 billion hectares of land, damaging 20% of irrigated cultivable land [2, 3]. The excessive accumulation of salts has a detrimental consequence on plant growth, development, seed performance, and other plant properties due to stress caused by salt. Sources of salt in the soil include seawater intrusion, inherent soil salinity, poor drainage, plant transpiration, seawater sprays, wind-borne salts, surface evaporation, excessive fertilizer use, soil additives, sewage sludge, and industrial brine dumping. According to the Food and Agriculture Organization [4], around 833 million hectares of subsoil (30–100 cm) and 424 million hectares of topsoil (0–30 cm) are disrupted by salt in 118 nations or 85% of the world's land surface. Of the topsoil damaged by salt, 10% are sodic, 5% are saline-sodic, and 85% are saline.

Generally, soil salinity refers to the existence of an abundant amount of soluble salts within a particular area. When the soil's electrical conductivity (EC) value is 4 dS m^{-1} or more, it is considered saline soil [5]. Soil solutions usually contain soluble salts such as sodium sulfate (Na_2SO_4) and sodium chloride (NaCl), although there are also partially soluble salts, including calcium sulfate ($CaSO_4$), and even magnesium sulfate ($MgSO_4$) with low solubility, sodium bicarbonate ($NaHCO_3$), potassium nitrate (KNO_3), and many others. The primary increase in salinity essentially involves Na^+ and Cl^- ions. Although both Na^+ and Cl^- ions are toxic to plants, the Cl^- ion poses a more significant threat [6, 7]. These ionic imbalances in soil pose a burgeoning problem for agriculture that needs irrigation, significantly constraining crop productivity and dictating their spatial distribution.

Elevated soil salinity exerts adverse effects on plant physiological processes, encompassing growth and development. The accumulation of excessive salt in the soil disrupts cellular functions and crucial metabolic processes, including seed germination and photosynthesis, leading to physiological abnormalities and even potential death of the plant. Salt stress also detrimentally impacts various physiological traits of plants, such as respiration rates, transpiration, stomatal density, and count. Plants experience salt-related damage through two primary mechanisms: Firstly, osmotic stress, which impairs water absorption and suppresses cell growth and division; Secondly, ionic toxicity, which disrupts ionic balance and cellular processes, ultimately culminating in premature senescence [7]. Salinity-exposed plants have an abundance of Na^+ and Cl^- ions in the soil, which are absorbed by the plant more quickly and subsequent depletion of other essential ions crucial for its growth and development. For example, it causes ionic stress by preventing K^+ accumulation and upsetting the balance of nutrients in plants. When this occurs, it affects the relationship between plants and water by bringing about physiological drought conditions. This causes osmotic stress, which thus causes a deterioration in stomatal conductance (g_s) and photosynthetic enzyme activity, in turn causing plants to produce reactive oxygen species (ROS) [7]. As a cumulative result, salinity disrupts redox homeostasis, inhibiting electron transport chains' ability to move electrons to oxygen reduction pathways in different locations of the cell, causing excess ROS in plants [8, 9].

Though the synthesis of ROS is a natural aspect of plant metabolism, their excessive accumulation can negatively interact with different components of cells like proteins, nucleic acids, lipids, and chlorophyll (Chl). Elevated ROS levels can harm the organelles, cells, and tissues of both the shoot and root systems, affecting enzymes, cell walls, and membrane functions. The role of ROS in plants swings between acting as signaling molecules and causing oxidative damage, depending on the balance between their production and removal. The equilibrium between the ROS and the antioxidant defense mechanism is disturbed by oxidative stress which is brought on by salinity.

Antioxidants are essential for neutralizing ROS and play a significant role in counteracting the stress conditions imposed by abiotic factors [10]. Generally, plants use their oxidative stress defense mechanism to delay their cell death by protecting cells from ROS-induced oxidative stress, protein degradation, lipid peroxidation, and damage to DNA and nucleic acids. This system maintains cellular integrity and enhances stress resilience in plants. To counteract the surge in ROS, plants have developed various antioxidant defense strategies against environmental stressors [11]. Recent studies primarily concentrate on physiological, biochemical, and molecular strategies to improve a plant's resilience to salt stress by mitigating oxidative stress. Given the discussed elements, employing an external stress defense system emerges as a critical method for alleviating oxidative stress due to salinity. Furthermore, these external elicitors have been found to boost photosynthetic activity, gas exchange, photosynthetic pigments, antioxidant responses, ion balance, and hormone signaling pathways, thereby enhancing ROS metabolism and the ability to tolerate salinity in plants. This chapter offers a detailed examination and recent insights into the mechanisms and responses of various external strategies to confer salt-induced oxidative stress tolerance.

2. Overview and mechanisms of salt stress

Salt stress, particularly prevalent in coastal regions, poses a major challenge for plants. From the beginning of agriculture, 10,000 years ago, the increase in salinity in soil has been considered a major environmental issue. After soil erosion, salinity is the leading reason for global land degradation, affecting about 33% of irrigated land and 20% of cultivated land [12]. Salinization can severely inhibit cropping, reduce yields by 10–25%, and lead to desertification [13]. Due to the buildup of excessive salt, stress can halt crop growth and lead to crop death [7].

Plants respond to salt stress differently based on their resistance. Glycophytes cannot grow in high Na salt concentrations [14], whereas halophytes tolerate high salt levels [15]. Halophytes are further categorized based on their salt tolerance into severe euhalophytes and intermediate oligohalophytes [16]. Facultative halophytes thrive in non-saline environments, while obligate halophytes do not [17]. Many crop species are sensitive to salt, and their growth response to salinity varies significantly [18]. Even NaCl at lower concentrations than seawater can disrupt crop development [19]. Halobacteria are the only organisms that tolerate salt at the molecular level [20]. High levels of soluble salts affect plant development through osmotic inhibition, toxicity, and nutritional impacts mediated by specific ion effects or direct toxicity. Salinity effects can be categorized as osmotic, nutritional, and toxic, with the first two being secondary salt-induced stresses and the latter a primary salt injury [21]. Plants may defend against osmotic dehydration in secondary stress injuries by accumulating salt until the osmotic concentration exceeds that of the root medium. In contrast, primary salt injury worsens with salt accumulation.

Various physiological traits are related to a plant's capacity to adapt and yield in highly saline conditions. For example, most halophytes are succulent, diluting internal salt concentrations [15]. Some species have salt-excreting glands, while others control turgor through organic solute synthesis or adapt by developing small leaves, water-storage hairs, and aerenchyma [15]. Salinity remains a critical challenge in global agriculture and is expected to intensify. Addressing plant salt tolerance remains complex, necessitating innovative solutions and continued research. Differences in compartmentation, ion selectivity, and water usage between halophytes and glycophytes are notable, with some plant species showing greater adaptability to salty conditions.

3. Reactive oxygen species generation and its consequence on plants under salinity

Reactive oxygen species, which act as either signaling molecules or cell-damaging agents, increase under high salinity, potentially leading to cell and plant death. Plants generate abundant amounts of ROS, including free radicals like superoxide ($-O_2^{\bullet-}$), hydroxyl radical ($^{\bullet}OH$), and non-radicals like hydrogen peroxide (H_2O_2) and singlet oxygen ($^1O_2$) [22, 23]. Additional forms are the peroxy radical ($ROO^{\bullet}$), alkoxy radical ($RO^{\bullet}$), excited carbonyl ($RO^{\bullet}$), and H_2O_2 organic hydroperoxide (ROOH) [22, 24]. The balance between ROS generation and their removal regulates whether they function as stressors or signaling molecules. Oxidative stress occurs when this balance is disrupted under abiotic stress, leading to damage that is counteracted by the plant's antioxidant protection system, comprising enzymatic and non-enzymatic components [22, 25].

Salinity induces osmotic stress within cells, quickly hindering water absorption and cell growth, causing ionic toxicity that leads to long-term ionic imbalance and cellular disruption [7]. Excessive salt reduces plant yield and can lead to death by disrupting mineral distribution and damaging cell membranes.

Higher accumulation of Na^+ and Cl^- ions by greenery exposed to saline soil reduces essential ions and disrupts redox balance, leading to increased ROS production (**Figure 1**). This results in osmotic and ionic stresses, affecting g_s and photosynthetic activity [7]. Studies show that salinization increases malondialdehyde (MDA) and H_2O_2 levels, causing more damage to the root than the shoot due to direct exposure to saline conditions [26, 27]. In sorghum, salt stress led to higher MDA levels in roots along with increased H_2O_2 in shoots [28]. Salinity also influences plant development by affecting gene expression related to ROS metabolism and interacting with stress-related pathways, although the exact mechanisms remain under study [29].

Cell organelles for instance cell walls, plasma membranes, chloroplasts, mitochondria, peroxisomes, endoplasmic reticulum, glyoxysomes, cytosol, and apoplast produce ROS through various pathways [24, 30]. Stress conditions lead to increased formation of rigid cell walls, as peroxidase (POD) and ROS polymerize phenolic compounds and glycoproteins. Malate dehydrogenase from nicotinamide adenine dinucleotide (NADH) takes part in catalyzing the formation of H_2O_2 [31].

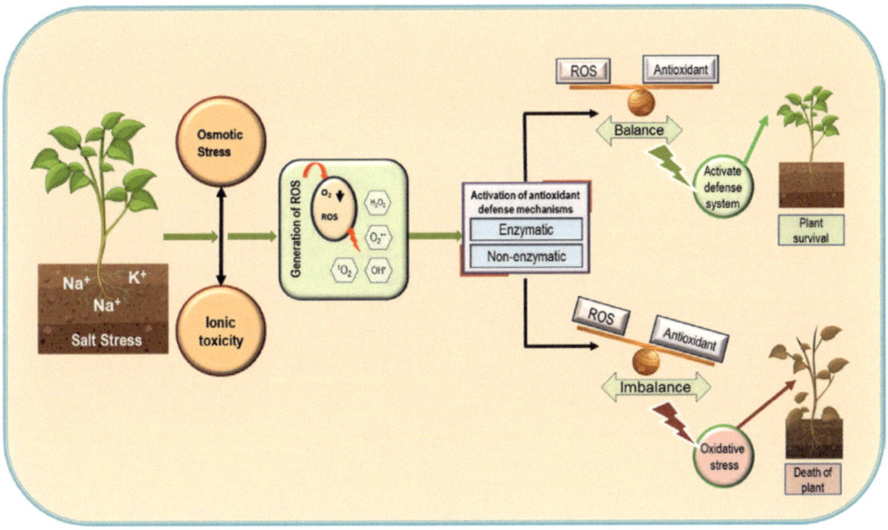

Figure 1.
Oxidative stress and antioxidant defense in plants under salinity. This illustration shows that a balance between ROS and antioxidant defense is necessary for plant survival under salt stress [22].

The plasma membrane produces a significant amount of $O_2^{\bullet-}$ due to the action of nicotinamide adenine dinucleotide phosphate (NADPH) reconciled by quinone reductase and NADPH oxidase. With the latter, an electron is transferred to create $O_2^{\bullet-}$ out of cytoplasmic NADPH. This reactive radical is later transformed into H_2O_2 [32]. Chloroplasts are crucial organelles for ROS production, especially under reduced g_s and CO_2 assimilation rates. Excited triplet chlorophyll ($^3Chl^*$) accelerates photo-oxidation, leading to overproduction of ROS and hindrance of the photosynthetic electron transport chain (ETC) [33, 34]. Photosystems I and II (PS I and PS II) are sites within the chloroplast where singlet oxygen ($^1O_2$) and $O_2^{\bullet-}$ are abundantly generated [23].

In non-green plant components, mitochondria are primary sites for ROS formation, with ubiquinone (complex I) and the ETC (complex III) being common sources of $O_2^{\bullet-}$. Superoxide dismutase (SOD) viz. Mn-SOD and Cu-Zn-SOD induced the transfiguration of $O_2^{\bullet-}$ into H_2O_2 [34, 35]. Peroxisomes are believed to primarily produce ROS through glycolate oxidase (GOX), leading to stomatal closure and reduced CO_2 availability for RuBisCO, thereby causing photorespiration and H_2O_2 production [36, 37].

In the endoplasmic reticulum, reduced NADPH levels lead to activities by cytochrome P450 and cytochrome P540 reductase, resulting in minor production of $O_2^{\bullet-}$ [38]. Glyoxysomes, derived from peroxisomes, engage in fatty acid β-oxidation and the glyoxylate cycle, where GOX along with urate oxidase activities, produce both $O_2^{\bullet-}$ and H_2O_2 [39].

In the cytosol, the degree of ROS formation is lower than in other compartments. However, the cytosol plays a vital participation in redox signaling and ROS generation, influencing gene expression in the cell nucleus [40]. Quinone reductase, NADPH oxidase, and lipoxygenases (LOX) are key enzymes contributing to ROS production in apoplasts [41, 42].

Excessive production of ROS damages proteins, lipids, carbohydrates, and DNA, leading to nucleic acid damage. Under stress, oxidative bursts damage cell lipid

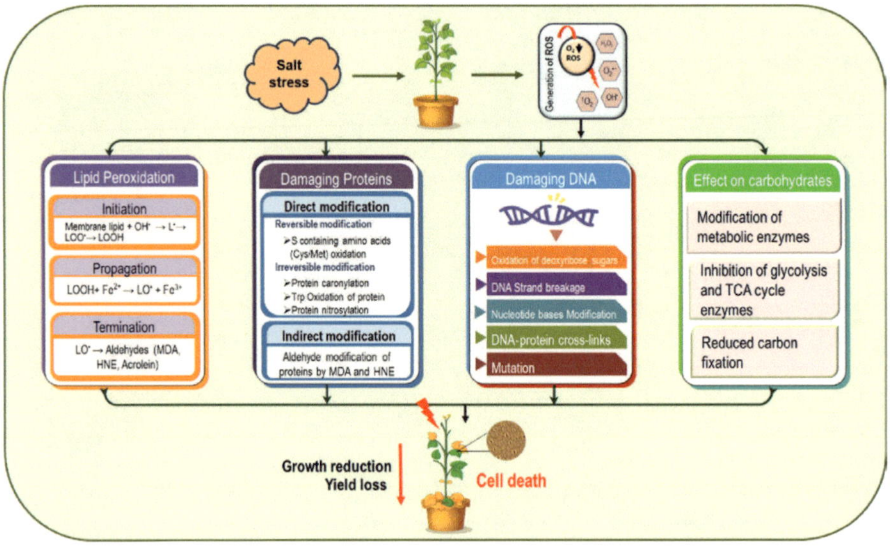

Figure 2.
Cellular impairment caused by reactive oxygen species and its consequences in plants under salinity. Salt stress causes overgeneration of ROS, which leads to damage to lipids, proteins, DNA, and carbohydrates. Various damage subsequently results in cell death and reduction of plant growth and yield reduction [43].

membranes and cause lipid peroxidation, targeting the unsaturated C-C double bonds in membrane phospholipids (**Figure 2**). The phases of lipid peroxidation include initiation, propagation/progression, and termination, with MDA serving as a marker for oxidative damage [25, 43, 44].

Protein oxidation is the process of chemically modifying proteins either by ROS directly or by attaching them to breakdown products of fatty acid peroxidation [45]. Several ROS selectively cause protein denaturation, among which the $^{\bullet}OH$ is notably aggressive, indiscriminately damaging protein molecules [22]. Non-radical oxidants typically cause less damage due to their slower oxidation rates, whereas rapid radicals significantly damage the protein structure, mainly by extracting hydrogen atoms from the α-carbon, resulting in stabilized carbon-centered radicals. This damage alters protein function through degradation of the peptide backbone, leading to alterations such as glutathionylation, carbonylation, nitrosylation, and disulfide bond generation.

Among free radicals, $^{\bullet}OH$ is particularly harmful as it oxidizes sugar residues, alters nucleotide bases (purine and pyrimidine), and induces DNA strand breaks, facilitating cross-linking between DNA and proteins [44]. In contrast, H_2O_2 and $O_2^{\bullet-}$ do not typically engage with nucleotide bases, while $^1O_2$ primarily responds to guanine [25]. Hydroxyl radicals also contribute to DNA damage by removing hydrogen from the C–H bonds of 2-deoxyribose and methyl groups, leading to the formation of deoxyribose radicals, hydroxymethyl urea, and thymine glycol. This can result in the oxidation of nucleotide bases, forming damaging compounds such as 8-hydroxyquinine and dehydro-2-deoxyguanosine, potentially leading to plant destruction [44].

In plants, the first pathways affected by free radical damage are the tricarboxylic acid (TCA) cycle and glycolysis. For instance, oxidative stress gives rise to the discontinuation of key enzymes in the pentose phosphate pathway, such as glyceraldehyde 3-phosphate dehydrogenase and fructose-1,6-bisphosphate aldolase, resulting in increased levels of cycle metabolites like ribose 5-phosphate and ribulose 5-phosphate. These changes facilitate NADPH synthesis and accelerate carbon flow to counterbalance ROS levels [22, 46]. Additionally, oxidative stress restricts the action of aconitase in the TCA cycle, leading to increased citrate biosynthesis, and impacts plant health by decreasing glycolysis, TCA cycle metabolism, and amino acid transport [46].

Salinity directly induces ionic and osmotic stress as well as secondary stresses including ROS-induced oxidative damage. At a certain low concentration that varies based on several conditions, these ROS perform a signaling role. However, the increased formation of ROS alters the equilibrium between ROS buildup and removal, leading to the destruction of various biomolecules in response to varied environmental stresses. In extreme cases, this damage to biomolecules includes the destruction of nucleic acids, protein oxidation, lipid peroxidation, enzyme deactivation, and Chl deterioration [25].

Young and mature leaves are the least impacted by salt stress, whereas root tissues suffer the most. A dose-dependent reduction was noted in root-shoot length, stem diameter, and plant biomass including root and shoot dry weight. The photosynthetic pigments Chl *a*, Chl *b*, Chl (*a* + *b*), Chl *a*/*b* ratio, and carotenoid (Car) content were reduced in salt stress. Due to salt exposure, the degree of Na^+ increased, and K^+ decreased; ultimately, the Na^+/K^+ ratio increased. On the other hand, Ca^{2+} content is reduced in a dose-dependent manner. Physiological attributes like relative water content showed a declining trend, and proline (Pro) content was uplifted because of salt accumulation [47].

In our experiment, we used soybean (*Glycine max*) plants to induce oxidative stress tolerance to salt, which was later alleviated by *Bacillus subtilis* (**Figure 3**). Oxidative stress indicators MDA and H_2O_2 were elevated for ROS generation. For analyzing the production of ROS (H_2O_2 and $O_2^{\bullet-}$) spot localization using staining (using diaminobenzene and nitroblue tetrazolium) was carried out. All of the salt treatments showed notable alterations. The brown H_2O_2 area and the dark blue $O_2^{\bullet-}$ areas were more noticeable in comparison to the control as the salt concentration increased [47].

In the ascorbate-glutathione (AsA-GSH) pool, AsA and GSH content was lessened, whereas dehydroascorbate (DHA) and oxidized glutathione (GSSG) level was escalated, which ultimately alleviated the ratio of AsA/DHA and GSH/GSSG. In the case of enzymatic antioxidants, ascorbate peroxidase (APX), glutathione reductase (GR), POD, glutathione *S*-transferase (GST), and glutathione peroxidase (GPX)

Figure 3.
Localization of the spots of H_2O_2 (A) and $O_2^{\bullet-}$ radicals (B) on the leaves of soybean as affected by bacillus sp. in response to different concentrations of stress. Here, C, BP, and BS indicate the control, seed inoculation, and soil with B. subtilis, respectively [47].

exhibited elevation on the contrary monodehydroascorbate reductase (MDHAR), dehydroascorbate reductase (DHAR), SOD, and catalase (CAT) contents were scaled down under the salt exposure at different doses. The reduction was unveiled in glyoxalase I (Gly I) and Gly II contrarily methylglyoxal (MG), which indicated an increasing trend [47].

4. Approaches to mitigate ROS toxicity in plants under salinity

In order to protect themselves from oxidative stress, plants need an efficient antioxidant defense mechanism consisting of non-enzymatic (such as AsA, GSH, α-tocopherol, non-protein amino acids, phenolic compounds, alkaloids, carotenoids, and flavonoids), and enzymatic (such as SOD, POD, APX, CAT, MDHAR, DHAR, GR, GST, and GPX) antioxidants [43]. Researchers found various ways to enhance antioxidant defense in pants to protect them from oxidative stress caused by salt.

4.1 Exogenous application of chemical substances

Chemical substances applied exogenously, through priming, irrigation, soil addition, or foliar spraying in small amounts, can mitigate ROS-induced oxidative damage under salt stress (**Table 1**). Abiotic stress tolerance, like salinity tolerance, can be enhanced by these chemicals by boosting the antioxidant defense mechanism and reducing oxidative damage at the cellular level [59].

4.1.1 Amino acids and their derivatives

The external application of various amino acids has emerged as a promising strategy to counteract ROS toxicity caused by salinity. Proline (Pro) is noteworthy as it acts as a molecular chaperone because of its ability to forage ROS and maintain cellular redox balance [60]. In a saline media where maize seedlings were grown, the addition of Pro and glycine betaine (GB) separately led to enhanced antioxidant systems and osmolyte metabolism, significantly reducing the generation of $O_2^{\bullet-}$ and H_2O_2. Furthermore, Pro and GB applications also bolstered the AsA-GSH pool and the glyoxalase system, providing cellular protection against excessive ROS, MG, and other toxic metabolites [61]. Wani et al. [62] demonstrated that spraying 20 mM Pro on *Brassica juncea* under varying salt concentrations (78, 117, and 176 mM NaCl) boosted endogenous Pro levels and the actions of antioxidant enzymes like CAT, POD, and SOD. Application of GB (25 and 50 mM) on *Phaseolus vulgaris* at the vegetative and flowering stages under salt stress improved shoot K^+ content, lessened Na^+ buildup, and enhanced the K^+/Na^+ ratio. This treatment also increased Pro and GSH levels, contributing to effective ROS scavenging and a noticeable reduction in oxidative stress [55]. Treating Glycine max plants with cysteine, a thiol-group containing α-amino acid, counteracted salinity effects through Pro accumulation and decreased H_2O_2 and lipid peroxidation levels, alongside increased action of SOD and CAT [63]. The addition of γ-aminobutyric acid along with salt to the nutrient solution reduced Na^+ uptake in *Solanum lycopersicum* and mitigated the excessive accumulation of $O_2^{\bullet-}$ and H_2O_2 [64]. Exogenous melatonin (150 μM) protected the

Species	Dose of NaCl	Chemical substances and dose	Method of application	Major effects	References
Triticum aestivum	NaCl (75, 150 and 225 mM)	Trehalose (10 mM)	Seed priming	Reduced lipid peroxidation and H_2O_2 content; reduced Pro accumulation and soluble sugar content	[48]
Solanum lycopersicum	NaCl (150 mM)	Melatonin (150 μM)	Added to nutrient solution	Reduced $O_2^{\bullet-}$ and H_2O_2 generation; increased Chl content; promoted APX, MDHAR, DHAR, GR, CAT, APX, SOD, and POD activities	[49]
S. lycopersicum	NaCl (100 mM)	Spermine (50 and 100 μM)	Foliar application	Decreased $O_2^{\bullet-}$ and H_2O_2 content, lipid peroxidation, and EL; increased Pro, soluble sugar, GB content; enhanced AsA-GSH cycle enzymes, SOD, and CAT activities	[50]
Capsicum annuum	NaCl (2000 and 4000 ppm)	Salicylic acid (1 mM) and Pro (10 mM)	Foliar application	Enhanced Chl *a* and *b* content, leaf RWC; decreased $O_2^{\bullet-}$ and H_2O_2 contents and lipid peroxidation; decreased CAT and POX activities	[51]
Cucumis sativus	NaCl (50 mM)	Sodium silicate (0.3 mM)	Added to nutrient solution	Decreased $O_2^{\bullet-}$ and H_2O_2 content; promoted Pro, SA, IAA content, and decreased ABA content	[52]
Luffa acutangula	NaCl (100 mM)	Putrescine (100 μM) and kinetin (100 μM)	Foliar application	Reduced $O_2^{\bullet-}$ and H_2O_2 content, lipid peroxidation, EL; promoted the activities of APX, SOD, CAT, and GR	[53]
Oryza sativa	NaCl (150 mM)	Melatonin (50, 100, 150 and 200 μM)	Foliar application	Enhanced the Na^+ efflux and K^+ influx, decreased H_2O_2 content	[54]
Phaseolus vulgaris	NaCl (50 and 100 mM)	Glycine betaine (25 and 50 mM)	Foliar application	Reduced MDA content, EL, and Na^+ accumulation; increased the synthesis of Chl and upregulation of SOD, POD, and CAT activities	[55]
Helianthus annus	NaCl (120 mM)	α-tocopherol (100, 200, and 300 mg L^{-1})	Seed priming	Reduced H_2O_2 and EL; enhanced AsA and total phenolics content; upregulated the activities of SOD, CAT, POD, and GR	[56]

Species	Dose of NaCl	Chemical substances and dose	Method of application	Major effects	References
O. sativa	NaCl (150 mM)	Spermidine (0.5, 1 and 1.5 mM)	Seed priming	Declined $O_2^{\bullet-}$ generation, lipid peroxidation, and EL	[57]
O. sativa	NaCl (50 mM)	Thiourea (7.5 μM) and H_2O_2 (1 μM)	Foliar application	Promoted the activities of antioxidant enzymes like SOD, CAT, APX, and GR	[58]

Table 1.
Exogenous application of different chemical substances to mitigate ROS toxicity under NaCl-induced salinity.

photosystem by enhancing antioxidant enzymes and the AsA-GSH pool, aiding in the detoxification of ROS caused by salinity [49]. Additionally, melatonin application reduced the root Na^+/K^+ ratio under stress induced by salinity, aiding in maintaining ion homeostasis [54]. Pre-treating young *B. napus* seedlings with exogenous β-aminobutyric acid (BABA, 150 μM) for 48 h before exposing them to salt environments (100 and 150 mM NaCl) for another 48 h showed higher tolerance to salt stress by upregulating antioxidants to neutralize toxic ROS [65].

4.1.2 Phytohormones

Exogenously applied plant growth regulators, including indole-3-acetic acid (IAA), jasmonic acid (JA), salicylic acid (SA), silicon (Si), gibberellic acid (GA), cytokinin (CK), and brassinosteroids, is a cost-effective method to alleviate the harmful consequences of abiotic stresses such as salinity. The combined application of kinetin (10 μM) and 24-epibrassinolide (EBL, 1 μM) has been shown to enhance photoprotection under salinity conditions (100 mM NaCl) by improving tissue water potential, mineral assimilation, Na^+ ion restriction, metabolite accumulation, and the antioxidant system. This is evidenced by reductions in H_2O_2 content, lipid peroxidation, and activities of protease and LOX, indicating ROS detoxification in kinetin and EBL-treated plants [66].

Raju et al. [67] showed that supplementation individually with IAA (10^{-5} M) or in combination with 28-homobrassinolide (10–8 M) reduced Na^+ buildup in the roots and shoots of *S. melongena* under various concentrations of NaCl (5 and 15 mM). Additionally, applications of IAA and 28-homobrassinolide improved the efficiency of PS II and the metabolism of nitrogen. A decrease in $O_2^{\bullet-}$ and H_2O_2 levels due to enhanced antioxidant activity such as increased SOD, CAT, POD, and GST was observed, highlighting the positive effects of IAA and homobrassinolide.

Combined seed priming and foliar spraying of JA (60 μM) increased salt stress (100 mM NaCl) tolerance by boosting the endogenous levels of GA, JA, and abscisic acid (ABA). Exogenous application of JA under salt stress modified physiological processes like the rate of photosynthesis, transpiration, and intercellular concentration of CO_2, significantly reducing H_2O_2 and lipid peroxidation in both salt-sensitive and tolerant cultivars of *G. max* [68].

In *S. lycopersicum*, seed priming and simultaneous application of SA (10 μM) mitigated salinity-induced ion toxicity and osmotic stress (125 mM NaCl). Exogenous SA application was found to elevate the endogenous SA level, inducing salt tolerance through the stimulation of the endogenous hormone signaling pathways [69].

Salicylic acid application (0.75 mM) under 125 mM NaCl salinity stress significantly activated antioxidant enzymes like CAT (by 2.25-fold), APX (by 2.22-fold), and SOD (by 1.76-fold), leading to reduced H_2O_2 accumulation and protection against peroxidation of lipid and leakage of electrolyte [70].

Silicon (SiO_2) application also ameliorated various physiological and biochemical attributes, safeguarding plants from the adverse consequence of salt stress by downregulating genes associated with ROS and reactive nitrogen species and enhancing antioxidants, both enzymatic and non-enzymatic [71]. Pre-treating seeds with EBL (10^{-1} M) improved leaf relative water content (RWC), Pro, GB, and flavonoid levels in *S. lycopersicum* seedlings under NaCl stress, reducing ROS toxicity alongside upregulated actions of antioxidants [72].

4.2 Application of organic amendments

Organic amendments like biochar (BC), lignite, and compost have been shown to lessen the damaging effects of oxidative stress by regulating the synthesis of plant enzymes endowed with antioxidant properties. Under stress induced by salinity, adding compost to soil increases the ratio of K^+/Na^+ in *S. lycopersicum* seedlings. Additionally, the upregulation of enzymes connected to the AsA-GSH cycle and activities of other enzymatic antioxidants was observed in compost-treated, salinity-stressed plants [73]. Biochar employment in saline soils has been shown to enhance soil physicochemical properties, including increases in organic matter content and cation exchange capacity, along with a decrease in exchangeable Na ions. In salinity stress, the beneficial effects of BC were highlighted by Farhangi-Abriz and Torabian [74], who stated that it reduced $O_2^{\bullet-}$, H_2O_2 content, and lipid peroxidation in *P. vulgaris*, leading to decreased action of CAT, APX, SOD, polyphenol oxidase (PPO), and POD, and also reduced concentrations of osmolytes like Pro, GB, and soluble sugar. Furthermore, Torabian et al. [75] demonstrated that the growth of *Vigna radiata* under control and salt stress improved with BC and lignite applications at 50 and 100 g kg^{-1} soil, attributed to the enhancement of potassium (K), calcium (Ca), magnesium (Mg), iron (Fe), and zinc (Zn) and diminution of Na and ROS production. In saline soil, compost supplementation improved endurance of salt stress in *Chenopodium quinoa* by improving actions of SOD, CAT, and APX, particularly at the time of foliar spraying of plants with Pro (0.5 and 1 mM) and/or trehalose (2.5 and 5 mM; [76]). Additionally, polysaccharide-enriched seaweed liquid extract, used as a biostimulant or biofertilizer, was seen to reduce salinity-induced ROS toxicity in plants. Hashem et al. [77] reported the beneficial effects of three seaweeds on *B. napus*, by documenting increased growth and yield attributes, such as higher total Chl content, total carbohydrate accumulation, and enhanced endogenous phytohormone levels (IAA, GB, GA, and CK) when grown in seaweed-incorporated soil under normal or salinity stress conditions (75 and 150 mM NaCl).

4.3 Use of microorganisms

4.3.1 Plant growth-promoting rhizobacteria

The restoration of soil health heavily relies on microbial communities in soil [78]. These microbial communities can be adjusted to improve the rejuvenation of agricultural lands that have been degraded due to different types of abiotic imbalances, especially salinity problems. It is essential to comprehend the role of beneficial microbes to maximize soil restoration endeavors, with an emphasis on groups such as rhizobacteria and mycorrhizal fungi that promote plant growth and development. The utilization of plant growth-promoting rhizobacteria (PGPR) emerges as a productive tactic for mitigating the adverse consequences of salinity stress on plant vitality [79–81]. These bacteria thriving in the rhizosphere promote plant health through various mechanisms, including hormonal stimulation, siderophore synthesis, exopolysaccharides secretion, osmoprotectant accumulation, ion exchange, and the stimulation of antioxidant enzymes (**Figure 4**) [82]. Improved antioxidant system by PGPR plays a critical role in detoxifying ROS molecules in plants to deal with ROS-induced oxidative stress [83]. Plant growth-promoting rhizobacteria can decrease the stress-induced activities of ROS in plants via two mechanisms: The generation of antioxidant molecules and the alteration of gene expression of plants that synthesize ROS scavengers. For example, administering halotolerant *Enterobacter* sp. UPMR18 to *Abelmoschus esculentus* plants resulted in a notable augmentation of antioxidant enzymes, including CAT and SOD, within the treated plants [84]. Research findings indicate that certain microorganisms, including *Bacillus licheniformis* and *Pseudomonas plecoglossicida*, which have the capability to form biofilms, may positively influence the growth of *Helianthus Annuus* when subjected to salt stress. This influence is attributed to their ability to increase the production of 1-aminocyclopropane-1-carboxylic acid (ACC) deaminase, phytohormones, and a variety of antioxidant enzymes, such as CAT, SOD, and GPX [85]. In a separate study, a notable reduction in H_2O_2 content within plant nodules was observed in *Vigna unguiculata* when seeds were simultaneously inoculated with *Bradyrhizobium* and PGPR (either *Actinomadura* sp. or *Paenibacillus graminis*). This reduction is attributed to changes in the collective amounts of reduced and oxidized forms of AsA and GSH, coupled with increased SOD and CAT activities. Specifically, plants subjected to coinoculation with *Bradyrhizobium* and *Streptomyces* sp. exhibited a 55% increase in CAT activity compared to plants inoculated with *Bradyrhizobium* alone [86]. Similarly, Yasmin et al. [87] reported that rhizobacteria belonging to *Pseudomonas* sp. and *Bacillus* sp. enhanced salt tolerance in *G. max* L. by adjusting the concentrations of MDA, CAT, PPO, SOD, POD, APX, and Phenylalanine ammonia-lyase (PAL) within plant system. The advantageous

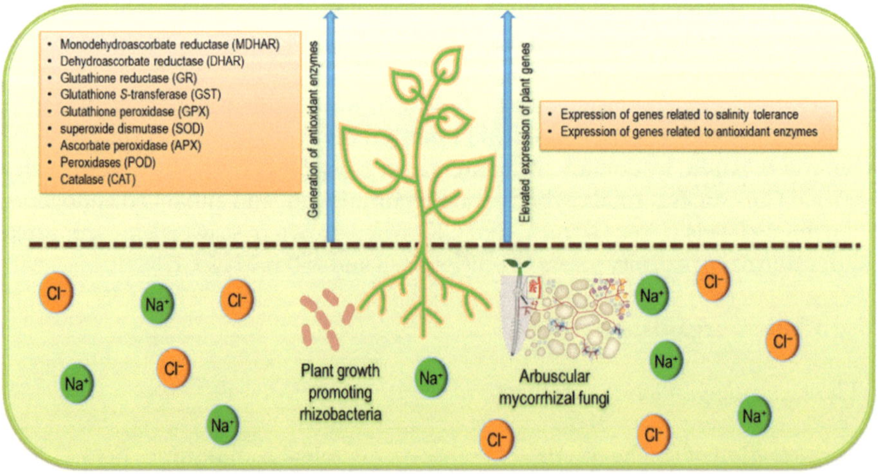

Figure 4.
Plant growth-promoting bacteria and rhizospheric fungi induce the production of antioxidant enzymes and induce their gene expressions in plants related to salt stress and antioxidant enzymes when exposed to saline conditions, thereby mitigating the stress.

impacts of *Rhodopseudomonas palustris* G5 in improving salt stress resistance in *Cucumis sativus* were demonstrated through increased levels of phytohormones and soluble sugars alongside elevated activities of SOD and POD enzymes. Strain G5 is capable of fixing nitrogen and solubilizing K and P [88]. Additionally, in a separate experiment, the combination of bacterial inoculation (*Pseudomonas* sp. RS-198 and *Azospirillum brasilense* RS-SP7) and treatments with phytohormones including SA and JA were employed to alleviate the detrimental consequences of salt stress caused by 50 and 100 mM NaCl in *Brassica napus*. Results indicated that the simultaneous application of *Pseudomonas* and SA yielded the most favorable outcome, reducing the generation of $O_2^{\bullet -}$, H_2O_2, and lipid peroxidation by increasing endogenous SA levels and antioxidant enzyme activities [89].

Furthermore, El-Esawi and coworkers [90] revealed that introducing *Bacillus firmus* SW5 into soybean plants enhanced the activation of genes linked to salinity tolerance and antioxidant enzymes (APX, SOD, POD, and CAT) particularly while exposed to high salt concentrations (80 mM NaCl). Additionally, PGPB-mediated upregulation of genes such as *GmFLD19* and *GmNARK* has been found to alleviate oxidative stress in soybeans, leading to reduced Na^+ and MDA and enhanced effectiveness of antioxidant enzymes in plants [80]. Moreover, the synergistic action of *Piriformospora indica* and *Azotobacter chroococcum* exhibited the ability to promote the growth of *Artemisia annua* L. by easing MDA content and fostering the levels of antioxidants [91]. These findings underscore the significant role of potential PGPR in alleviating ROS in plants subjected to salinity stress, wherein the heightened levels of PGPR-induced antioxidants contribute to enhanced growth and yield to mitigate the stress.

4.3.2 Rhizospheric fungi

The judicious application of arbuscular mycorrhizal fungi (AMF) has been shown to improve various physiological and biochemical characteristics of crop plants facing different types of abiotic stresses, including salinity (**Figure 4**). When subjected to salt stress, AMF can bolster the development and productivity of plants by boosting water and nutrient absorption, preserving ion balance, mitigating ROS toxicity, controlling lipid peroxidation, and accumulating osmolytes in inoculated plants [92]. The symbiotic interactions between plants and AMF also augment the synthesis of SA, JA, and other inorganic nutrients. It has been observed that under salt stress conditions, AMF-treated *Cucumis sativus* exhibited higher levels of Ca, N, Mg, K, and P compared to non-inoculated plants [93]. Furthermore, mycorrhizal treatment in *Lactuca sativa* has led to increased Pro accumulation, enhanced N uptake, and reduced Na^+ accumulation in contrast to plants without mycorrhizal treatment under saline conditions [94]. Although one of the most important anticipated facets of this symbiosis is the stimulation of the antioxidant activities in plants, investigations into the role of AMF in amplifying antioxidant activity in plants under saline stress have yielded conflicting results. Some studies indicate amplified action of antioxidant enzymes, like SOD, CAT, POD, and APX [95], while others report nonsignificant changes or reduced activity. For instance, tomato plants inoculated with mycorrhizae exposed to salinity stress exhibited varying responses in the activation or involvement of these ROS scavengers [96, 97].

5. Conclusion

Environmental sustainability is disrupted by climate change, primarily driven by global warming. This results in weather fluctuations that negatively impact agricultural practices and, hereafter, production declines. Salinity is one of the major destructive environmental factors affecting agriculture, which is exacerbated by climate change. Intensification of salt in the soil creates this stress condition, disrupting cellular functions and metabolic processes and leading to plant death. Salt stress exacerbates plant damage through increased osmotic stress and ionic toxicity. Affected plants exhibit an abundance of Na^+ and Cl^- ions, which are absorbed more rapidly than vital ions, leading to physiological drought conditions and osmotic stress, ultimately originating ROS, resulting in membrane damage and metabolic abnormalities. In this regard, various antioxidants, phytohormones, signaling molecules, and plant nutrients have been identified as crucial for the detoxification of ROS in situations of abiotic stress, in turn enhancing the plant's defense system. Different approaches have been taken under consideration to lessen this adversity, including the exogenous application of amino acids, polyamines, vitamins, plant growth-promoting substances, organic amendments, and exploiting the microorganisms that are beneficial for plants, as well as genetic approaches to scavenge ROS and improve plant physiological and metabolic functions. These interventions offer a significant opportunity for both soil and plants to alleviate salt stress, particularly in coastal regions where salinity is a major concern.

Acknowledgements

We acknowledge Naznin Ahmed for her help in collecting some important literature and critical reading of the manuscript.

Conflict of interest

The authors declare no conflict of interest.

Author details

Mirza Hasanuzzaman[1*], Faomida Sinthi[1], Samiul Alam[1], Abida Sultana[1], Samiha Rummana[2] and Amena Khatun[3]

1 Department of Agronomy, Sher-e-Bangla Agricultural University, Dhaka, Bangladesh

2 Independent Researcher, Dhaka, Bangladesh

3 Department of Agriculture, Noakhali Science and Technology University, Noakhali, Bangladesh

*Address all correspondence to: mhzsauag@yahoo.com

References

[1] Li YP, Ye W, Wang M, Yan XD. Climate change and drought: A risk assessment of crop-yield impacts. Climate Research. 2009;**39**:31-46

[2] Ghassemi F, Jakeman AJ, Nix HA. Salinisation of Land and Water Resources: Human Causes, Extent, Management and Case Studies. Wallingford, UK: CABI International; 1995

[3] Qadir M, Quillerou E, Nangia V, Murtaza G, Singh M, Thomas RJ, et al. Economics of salt-induced land degradation and restoration. Natural Resources Forum. 2014;**38**:282-295

[4] FAO. Global Map of Salt Affected Soils Version 1.0. Rome: FAO; 2021. Available from: https://www.fao.org/soils-portal/data-hub/soil-maps-and-databases/global-map-of-salt-affected-soils/en/ [Accessed: February 29, 2024]

[5] Yildiz M, Poyraz İ, Çavdar A, Özgen Y, Beyaz R. Plant responses to salt stress. In: Abdurakhmonov YI, editor. Plant Breeding-Current and Future Views. London, UK: IntechOpen; 2020

[6] Choudhury S, Panda P, Sahoo L, Panda SK. Reactive oxygen species signaling in plants under abiotic stress. Plant Signaling and Behavior. 2013;**8**:e23681

[7] Hasanuzzaman M, Raihan MRH, Masud AAC, Rahman K, Nowroz F, Rahman M, et al. Regulation of reactive oxygen species and antioxidant defense in plants under salinity. International Journal of Molecular Sciences. 2021;**22**:9326. DOI: 10.3390/ijms22179326

[8] Munns R, Tester M. Mechanisms of salinity tolerance. Annual Review of Plant Biology. 2008;**59**:651-681

[9] Ozgur R, Uzilday B, Sekmen AH, Turkan I. Reactive oxygen species regulation and antioxidant defence in halophytes. Functional Plant Biology. 2013;**40**:832-847

[10] Qados AMS. Effect of ascorbic acid antioxidant on soybean (*Glycine max* L.) plants grown under water stress conditions. International Journal of Advanced Research in Biological Sciences. 2014;**1**:189-205

[11] Berni R, Luyckx M, Xu X, Legay S, Sergeant K, Hausman JF, et al. Reactive oxygen species and heavy metal stress in plants: Impact on the cell wall and secondary metabolism. Environmental and Experimental Botany. 2019;**161**:98-106

[12] Machado RMA, Serralheiro RP. Soil salinity: Effect on vegetable crop growth. Management practices to prevent and mitigate soil salinization. Horticulturae. 2017;**3**:30. DOI: 10.3390/horticulturae3020030

[13] Shahid SA, Zaman M, Heng L. Introduction to soil salinity, sodicity and diagnostics techniques. In: Zaman M, Shahid SA, Heng L, editors. Guideline for Salinity Assessment, Mitigation and Adaptation Using Nuclear and Related Techniques. Cham: Springer; 2018. pp. 1-42

[14] Green TGA, Sancho LG, Pintado A, Saco D, Martín S, Arróniz-Crespo M, et al. Sodium chloride accumulation in glycophyte plants with cyanobacterial symbionts. AoB Plants. 2017;**9**:plx053. DOI: 10.1093/aobpla/plx053

[15] Meng X, Zhou J, Sui N. Mechanisms of salt tolerance in halophytes: Current

understanding and recent advances. Open Life Sciences. 2018;**13**:149-154

[16] Taia WK. Wide scene on halophytes. International Journal of Contemporary Research and Review. 2020;**11**:20219-20232. DOI: 10.15520/ijcrr.v11i01.777

[17] Nazish M, Zafar M, Ahmad M, Sultana S. Halophytes diversity as an indicator to saline environment in Pakistan. In: Fahad S, Adnan M, Saud S, Nie L, editors. Climate Change and Ecosystems: Challenges to Sustainable Development. Florida, USA: CRC Press; 2022. pp. 191-202

[18] Hameed A, Ahmed MZ, Hussain T, Aziz I, Ahmad N, Gul B, et al. Effects of salinity stress on chloroplast structure and function. Cells. 2021;**10**:2023. DOI: 10.3390/cells10082023

[19] Atzori G, de Vos AC, van Rijsselberghe M, Vignolini P, Rozema J, Mancuso S, et al. Effects of increased seawater salinity irrigation on growth and quality of the edible halophyte *Mesembryanthemum crystallinum* L. under field conditions. Agricultural Water Management. 2017;**187**:37-46

[20] Vauclare P, Natali F, Kleman JP, Zaccai G, Franzetti B. Surviving salt fluctuations: Stress and recovery in *Halobacterium salinarum*, an extreme halophilic archaeon. Scientific Reports. 2020;**10**:3298

[21] Safdar H, Amin A, Shafiq Y, Ali A, Yasin R, Shoukat A, et al. A review: Impact of salinity on plant growth. Nature and Science. 2019;**17**:34-40

[22] Hasanuzzaman M, Bhuyan MHMB, Zulfiqar F, Raza A, Mohsin SM, Mahmud JA, et al. Reactive oxygen species and antioxidant defense in plants under abiotic stress: Revisiting the crucial role of a universal defense regulator. Antioxidants. 2020a;**9**:681

[23] Gill SS, Tuteja N. Reactive oxygen species and antioxidant machinery in abiotic stress tolerance in crop plants. Plant Physiology and Biochemistry. 2010;**48**:909-930

[24] Bhattacharjee S. ROS and oxidative stress: Origin and implication. In: Bhattacharjee S, editor. Reactive Oxygen Species in Plant Biology. New Delhi: Springer; 2019. pp. 1-31

[25] Sachdev S, Ansari SA, Ansari MI, Fujita M, Hasanuzzaman M. Abiotic stress and reactive oxygen species: Generation, signaling, and defense mechanisms. Antioxidants. 2021;**10**:277

[26] Mohsin SM, Hasanuzzaman M, Parvin K, Fujita M. Pretreatment of wheat (*Triticum aestivum* L.) seedlings with 2,4-D improves tolerance to salinity-induced oxidative stress and methylglyoxal toxicity by modulating ion homeostasis, antioxidant defenses, and glyoxalase systems. Plant Physiology and Biochemistry. 2020;**152**:221-231

[27] Ghosh S, Mitra S, Paul A. Physiochemical studies of sodium chloride on mungbean (*Vigna radiata* L. Wilczek) and its possible recovery with spermine and gibberellic acid. The Scientific World Journal. 2015;**2015**:858016

[28] Nexele X, Klein A, Ndimba BK. Drought and salinity stress alters ROS accumulation, water retention, and osmolyte content in sorghum plants. South African Journal of Botany. 2017;**108**:261-266

[29] Luo X, Dai Y, Zheng C, Yang Y, Chen W, Wang Q, et al. The ABI4-RbohD/VTC2 regulatory module promotes reactive oxygen species

(ROS) accumulation to decrease seed germination under salinity stress. New Phytologist. 2021;**229**:950-962

[30] Dmitrieva VA, Tyutereva EV, Voitsekhovskaja OV. Singlet oxygen in plants: Generation, detection, and signaling roles. International Journal of Molecular Sciences. 2020;**21**:3237

[31] Raja V, Majeed U, Kang H, Andrabi KI, John R. Abiotic stress: Interplay between ROS, hormones and MAPKs. Environmental and Experimental Botany. 2017;**137**:142-157

[32] Heyno E, Mary V, Schopfer P, Krieger-Liszkay A. Oxygen activation at the plasma membrane: Relation between superoxide and hydroxyl radical production by isolated membranes. Planta. 2011;**234**:35-45

[33] Shakirova FM, Allagulova CR, Maslennikova DR, Klyuchnikova EO, Avalbaev AM, Bezrukova MV. Salicylic acid induced protection against cadmium toxicity in wheat plants. Environmental and Experimental Botany. 2016;**122**:19-28

[34] Singh A, Kumar A, Yadav S, Singh IK. Reactive oxygen species-mediated signaling during abiotic stress. Plant Gene. 2019;**18**:100-173

[35] Huang S, VanAken O, Schwarzländer M, Belt K, Millar AH. The roles of mitochondrial reactive oxygen species in cellular signaling and stress responses in plants. Plant Physiology. 2016;**171**:1551-1559

[36] Kerchev P, Waszczak C, Lewandowska A, Willems P, Shapiguzov A, Li Z. Lack of glycolate oxidase1, but not glycolate oxidase2, attenuates the photorespiratory phenotype of catalase2-deficient arabidopsis. Plant Physiology. 2016;**171**:1704-1719

[37] Foyer CH, Noctor G. Redox regulation in photosynthetic organisms: Signaling, acclimation, and practical implications. Antioxidants & Redox Signaling. 2009;**11**:861-905

[38] Janků M, Luhová L, Petˇrivalský M. On the origin and fate of reactive oxygen species in plant cell compartments. Antioxidants. 2019;**8**:105

[39] Hu J, Baker A, Bartel B, Linka N, Mullen RT, Reumann S, et al. Plant peroxisomes: Biogenesis and function. The Plant Cell. 2012;**24**:2279-2303

[40] Van Breusegem F, Bailey-Serres J, Mittler R. Unraveling the tapestry of networks involving reactive oxygen species in plants. Plant Physiology. 2008;**147**:978-984

[41] Mittler R. ROS are good. Trends in Plant Science. 2017;**22**:11-19

[42] Choudhary A, Kumar A, Kaur N. ROS and oxidative burst: Roots in plant development. Plant Diversity. 2020;**42**:33-43

[43] Hasanuzzaman M, Bhuyan MB, Parvin K, Bhuiyan TF, Anee TI, Nahar K, et al. Regulation of ROS metabolism in plants under environmental stress: A review of recent experimental evidence. International Journal of Molecular Sciences. 2020;**21**:8695

[44] Das K, Roychoudhury A. Reactive oxygen species (ROS) and response of antioxidants as ROS-scavengers during environmental stress in plants. Frontiers in Environmental Science. 2014;**2**:53. DOI: 10.3389/fenvs.2014.00053

[45] Zhang W, Xiao S, Ahn DU. Protein oxidation: Basic principles and implications for meat quality. Critical Reviews in Food Science and Nutrition. 2013;**53**:1191-1201. DOI: 10.1080/10408398.2011.577540

[46] Lehmann M, Laxa M, Sweetlove LJ, Fernie AR, Obata T. Metabolic recovery of *Arabidopsis thaliana* roots following cessation of oxidative stress. Metabolomics. 2012;**8**:143-153

[47] Hasanuzzaman M, Raihan MRH, Nowroz F, Fujita M. Insight into the mechanism of salt-induced oxidative stress tolerance in soybean by the application of *Bacillus subtilis*: Coordinated actions of osmoregulation, ion homeostasis, antioxidant defense, and methylglyoxal detoxification. Antioxidants. 2022;**11**:1856. DOI: 10.3390/antiox11101856

[48] Alla MN, Badran E, Mohammed F. Exogenous trehalose alleviates the adverse effects of salinity stress in wheat. Turkish Journal of Botany. 2019;**43**:48-57

[49] Yin Z, Lu J, Meng S, Liu Y, Mostafa I, Qi M, et al. Exogenous melatonin improves salt tolerance in tomato by regulating photosynthetic electron flux and the ascorbate–glutathione cycle. Journal of Plant Interactions. 2019;**14**:453-463

[50] Ahanger MA, Qin C, Maodong Q, Dong XX, Ahmad P, Abd-Allah EF, et al. Spermine application alleviates salinity induced growth and photosynthetic inhibition in *Solanum lycopersicum* by modulating osmolyte and secondary metabolite accumulation and differentially regulating antioxidant metabolism. Plant Physiology and Biochemistry. 2019;**144**:1-13

[51] Abdelaal KA, El-Maghraby LM, Elansary H, Hafez YM, Ibrahim EI, El-Banna M, et al. Treatment of sweet pepper with stress tolerance-inducing compounds alleviates salinity stress oxidative damage by mediating the physio-biochemical activities and antioxidant systems. Agronomy. 2020;**10**:26

[52] Zhu Y, Jiang X, Zhang J, He Y, Zhu X, Zhou X, et al. Silicon confers cucumber resistance to salinity stress through regulation of proline and cytokinins. Plant Physiology and Biochemistry. 2020;**156**:209-220

[53] Kapoor RT, Hasanuzzaman M. Exogenous kinetin and putrescine synergistically mitigate salt stress in *Luffa acutangula* by modulating physiology and antioxidant defense. Physiology and Molecular Biology of Plants. 2020;**26**:2125-2137

[54] Yan F, Wei H, Ding Y, Li W, Chen L, Ding C, et al. Melatonin enhances Na^+/K^+ homeostasis in rice seedlings under salt stress through increasing the root H^+-pump activity and Na^+/K^+ transporters sensitivity to ROS/RNS. Environmental and Experimental Botany. 2021;**182**:104328

[55] Sofy MR, Elhawat N, Alshaal T. Glycine betaine counters salinity stress by maintaining high K^+/Na^+ ratio and antioxidant defense via limiting Na^+ uptake in common bean (*Phaseolus vulgaris* L.). Ecotoxicology and Environmental Safety. 2020;**200**:110732. DOI: 10.1016/j.ecoenv.2020.110732

[56] Lalarukh I, Shahbaz M. Response of antioxidants and lipid peroxidation to exogenous application of alpha-tocopherol in sunflower (*Helianthus annuus* L.) under salt stress. Pakistan Journal of Botany. 2020;**52**:75-83

[57] Nounjan N, Kumon-SA N, Theerakulpisut P. Spermidine priming promotes germination of deteriorated seeds and reduced salt stressed damage in rice seedlings. Notulae Botanicae Horti Agrobotanici Cluj-Napoca. 2021;**49**:12130-12130

[58] Pandey M, Paladi RK, Srivastava AK, Suprasanna P. Thiourea and hydrogen

peroxide priming improved K^+ retention and source-sink relationship for mitigating salt stress in rice. Scientific Reports. 2021;**11**:3000. DOI: 10.1038/s41598-020-80419-6

[59] Costa SF, Martins D, Agacka-Mołdoch M, Czubacka A, de Sousa AS. Strategies to alleviate salinity stress in plants. In: Kumar V, Wani SH, Suprasanna P, Tran LP, editors. Salinity Responses and Tolerance in Plants. Vol. 1. Cham: Springer; 2018. pp. 307-337

[60] Ben Rejeb K, Abdelly C, Savouré A. How reactive oxygen species and proline face stress together. Plant Physiology and Biochemistry. 2014;**80**:278-284

[61] Rohman MM, Talukder MZA, Hossain MG, Uddin MS, Amiruzzaman M, Biswas A, et al. Saline sensitivity leads to oxidative stress and increases the antioxidants in presence of proline and betaine in maize (*Zea mays* L.) inbred. Plant Omics. 2016;**9**:35-47

[62] Wani AS, Ahmad A, Hayat S, Tahir I. Epibrassinolide and proline alleviate the photosynthetic and yield inhibition under salt stress by acting on antioxidant system in mustard. Plant Physiology and Biochemistry. 2019;**135**:385-394. DOI: 10.1016/j.plaphy.2019.01.002

[63] Sadak MS, Abd El-Hameid AR, Zaki FS, Dawood MG, El-Awadi ME. Physiological and biochemical responses of soybean (*Glycine max* L.) to cysteine application under sea salt stress. Bulletin of the National Research Centre. 2020;**44**:1-10

[64] Wu X, Jia Q, Ji S, Gong B, Li J, Lü G, et al. Gamma-aminobutyric acid (GABA) alleviates salt damage in tomato by modulating Na^+ uptake, the GAD gene, amino acid synthesis and reactive oxygen species metabolism. BMC Plant Biology. 2020;**20**:465. DOI: 10.1186/s12870-020-02669-w

[65] Mahmud JA, Hasanuzzaman M, Khan MIR, Nahar K, Fujita M. β-aminobutyric acid pretreatment confers salt stress tolerance in *Brassica napus* L. by modulating reactive oxygen species metabolism and methylglyoxal detoxification. Plants. 2020;**9**:241

[66] Ahanger MA, Mir RA, Alyemeni MN, Ahmad P. Combined effects of brassinosteroid and kinetin mitigates salinity stress in tomato through the modulation of antioxidant and osmolyte metabolism. Plant Physiology and Biochemistry. 2020;**147**:31-42

[67] Raju AD, Parihar P, Singh R, Kumar J, Prasad SM. Synergistic action of indole acetic acid with homobrassinolide in easing the NaCl-induced toxicity in *Solanum melongena* L. seedlings. Acta Physiologiae Plantarum. 2020;**42**:1-18

[68] Sheteiwy MS, Shao H, Qi W, Daly P, Sharma A, Shaghaleh H, et al. Seed priming and foliar application with jasmonic acid enhance salinity stress tolerance of soybean (*Glycine max* L.) seedlings. Journal of the Science of Food and Agriculture. 2021;**101**:2027-2041

[69] Gharbi E, Lutts S, Dailly H, Quinet M. Comparison between the impacts of two different modes of salicylic acid application on tomato (*Solanum lycopersicum*) responses to salinity. Plant Signaling & Behavior. 2018;**13**:e1469361

[70] Alsahli A, Mohamed AK, Alaraidh I, Al-Ghamdi A, Al-Watban A, El-Zaidy M, et al. Salicylic acid alleviates salinity stress through the modulation of biochemical attributes and some key antioxidants in wheat seedlings. Pakistan Journal of Botany. 2019;**51**:1551-1559

[71] Chung YS, Kim KS, Hamayun M, Kim Y. Silicon confers soybean resistance to salinity stress through regulation of reactive oxygen and reactive nitrogen species. Frontiers in Plant Science. 2020;**10**:1725

[72] Ahmad P, Abd-Allah EF, Alyemeni MN, Wijaya L, Alam P, Bhardwaj R, et al. Exogenous application of calcium to 24-epibrassinosteroid pre-treated tomato seedlings mitigates NaCl toxicity by modifying ascorbate–glutathione cycle and secondary metabolites. Scientific Reports. 2018;**8**:13515. DOI: 10.1038/s41598-018-31917-1

[73] Tartoura KA, Youssef SA, Tartoura ESA. Compost alleviates the negative effects of salinity via up-regulation of antioxidants in *Solanum lycopersicum* L. plants. Plant Growth Regulation. 2014;**74**:299-310

[74] Farhangi-Abriz S, Torabian S. Antioxidant enzyme and osmotic adjustment changes in bean seedlings as affected by biochar under salt stress. Ecotoxicology and Environmental Safety. 2017;**137**:64-70

[75] Torabian S, Farhangi-Abriz S, Rathjen J. Biochar and lignite affect H^+-ATPase and H^+-PPase activities in root tonoplast and nutrient contents of mung bean under salt stress. Plant Physiology and Biochemistry. 2018;**129**:141-149

[76] Abdallah MMS, El Sebai TN, Ramadan AAEM, El-Bassiouny HMS. Physiological and biochemical role of proline, trehalose, and compost on enhancing salinity tolerance of quinoa plant. Bulletin of the National Research Centre. 2020;**44**:96. DOI: 10.1186/s42269-020-00354-4

[77] Hashem HA, Mansour HA, El-Khawas SA, Hassanein RA. The potentiality of marine macro-algae as bio-fertilizers to improve the productivity and salt stress tolerance of canola (*Brassica napus* L.) plants. Agronomy. 2019;**9**:146

[78] Coban O, de Deyn GB, van der Ploeg M. Soil microbiota as game-changers in restoration of degraded lands. Science. 2022;**375**:abe0725

[79] Ahmed S, Heo TY, Roy Choudhury A, Walitang DI, Choi J, Sa T. Accumulation of compatible solutes in rice (*Oryza sativa* L.) cultivars by inoculation of endophytic plant growth promoting bacteria to alleviate salt stress. Applied Biolological Chemistry. 2021;**64**:1-14

[80] Khan MA, Hamayun M, Asaf S, Khan M, Yun BW, Kang SM, et al. Rhizospheric *bacillus* spp. rescues plant growth under salinity stress via regulating gene expression, endogenous hormones, and antioxidant system of *Oryza sativa* L. Frontiers in Plant Science. 2021;**12**:1145

[81] Khumairah FH, Setiawati MR, Fitriatin BN, Simarmata T, Alfaraj S, Ansari MJ, et al. Halotolerant plant growth-promoting rhizobacteria isolated from saline soil improve nitrogen fixation and alleviate salt stress in rice plants. Frontiers in Microbiology. 2022;**13**:905210

[82] Giannelli G, Potestio S, Visioli G. The contribution of PGPR in salt stress tolerance in crops: Unravelling the molecular mechanisms of cross-talk between plant and bacteria. Plants. 2023;**12**:2197

[83] Numan M, Bashir S, Khan Y, Mumtaz R, Shinwari ZK, Khan AL, et al. Plant growth promoting bacteria as an alternative strategy for salt tolerance

in plants: A review. Microbiological Research. 2018;**209**:21-32

[84] Habib SH, Kausar H, Saud HM. Plant growth–promoting rhizobacteria enhance salinity stress tolerance in okra through ROS-scavenging enzymes. Biomed Research International. 2016;**2016**:6284547. DOI: 10.1155/2016/6284547

[85] Yasmeen T, Ahmad A, Arif MS, Mubin M, Rehman K, Shahzad SM, et al. Biofilm forming rhizobacteria enhance growth and salt tolerance in sunflower plants by stimulating antioxidant enzymes activity. Plant Physiology and Biochemistry. 2020;**156**:242-256

[86] Santos ADA, Silveira JAGD, Bonifacio A, Rodrigues AC, Figueiredo MDVB. Antioxidant response of cowpea co-inoculated with plant growth-promoting bacteria under salt stress. Brazilian Journal of Microbiology. 2018;**49**:513-521

[87] Yasmin H, Naeem S, Bakhtawar M, Jabeen Z, Nosheen A, Naz R, et al. Halotolerant rhizobacteria *Pseudomonas pseudoalcaligenes* and *Bacillus subtilis* mediate systemic tolerance in hydroponically grown soybean (*Glycine max* L.) against salinity stress. PLoS One. 2020;**15**:e0231348

[88] Ge H, Zhang F. Growth-promoting ability of *Rhodopseudomonas palustris* G5 and its effect on induced resistance in cucumber against salt stress. Journal of Plant Growth Regulation. 2019;**38**:180-188

[89] Farhangi-Abriz S, Tavasolee A, Ghassemi-Golezani K, Torabian S, Monirifar H, Rahmani HA. Growth-promoting bacteria and natural regulators mitigate salt toxicity and improve rapeseed plant performance. Protoplasma. 2020;**257**:1035-1047

[90] El-Esawi MA, Alaraidh IA, Alsahli AA, Alamri SA, Ali HM, Alayafi AA. *Bacillus firmus* (SW5) augments salt tolerance in soybean (*Glycine max* L.) by modulating root system architecture, antioxidant defense systems and stress-responsive genes expression. Plant Physiology and Biochemistry. 2018;**132**:375-384

[91] Arora M, Saxena P, Abdin MZ, Varma A. Interaction between Piriformospora indica and *Azotobacter chroococcum* diminish the effect of salt stress in *Artemisia annua* L. by enhancing enzymatic and non-enzymatic antioxidants. Symbiosis. 2020;**80**:61-73

[92] Abdelhamid MT, El-Masry RR, Darwish DS, Abdalla MM, Oba S, Ragab R. The mechanisms involved in improving the tolerance of plants to salt stress using arbuscular mycorrhizal fungi. In: Giri B, Varma A, editors. Microorganisms in Saline Environments: Strategies and Functions. Cham: Springer; 2019. pp. 303-327

[93] Hashem A, Alqarawi AA, Radhakrishnan R, Al-Arjani AF, Aldehaish HA, Egamberdieva D. Arbuscular mycorrhizal fungi regulate the oxidative system, hormones and ionic equilibrium to trigger salt stress tolerance in *Cucumis sativus* L. Saudi Journal of Biological Sciences. 2018;**25**:1102-1114

[94] Santander C, Sanhueza M, Olave J, Borie F, Valentine C, Cornejo P. Arbuscular mycorrhizal colonization promotes the tolerance to salt stress in lettuce plants through an efficient modification of ionic balance. Journal of Soil Science and Plant Nutrition. 2019;**19**:321-331

[95] Latef AAHA, Chaoxing H. Effect of arbuscular mycorrhizal fungi on growth, mineral nutrition, antioxidant enzymes activity and fruit yield of tomato

grown under salinity stress. Scientia Horticulturae. 2011;**127**:228-233

[96] He Z, He C, Zhang Z, Zou Z, Wang H. Changes of antioxidative enzymes and cell membrane osmosis in tomato colonized by arbuscular mycorrhizae under NaCl stress. Colloids and Surfaces B: Biointerfaces. 2007;**59**:128-133

[97] Hajiboland R, Joudmand A. The K/Na replacement and function of antioxidant defence system in sugar beet (*Beta vulgaris* L.) cultivars. Acta Agriculturae Scandinavica Section B–Soil and Plant Science. 2009;**59**:246-259

Chapter 2

Salinity Stress in Plants: Challenges in View of Physiological Aspects

Parastoo Majidian and Hamidreza Ghorbani

Abstract

Increasing the worldwide population, the food supply has become a global crisis due to the existence of various environmental stresses. Salinity after drought is one of the devastating environmental stresses that affects about 50% of the world's agricultural lands. It is considered as one of the important abiotic stresses that cause plant growth restriction in different stages such as seed germination, photosynthesis, hormonal regulation, nutrient uptake, and seed quality and quantity. Under salinity conditions, plants undergo numerous changes as morphological (early flowering, prevention of lateral shoot development, and root adaptations), physiological (Na^+/K^+ discrimination, osmotic adjustment, ion homeostasis, and stomatal responses), and biochemical (accumulation of polyamines, antioxidant activity, proline, and change the hormone level). With the ever-increasing expansion of saline lands and highly costs spending for their rehabilitation, the preparation of high-yielding lines/genotypes tolerant to salinity will be of particular importance. Being aware of various pathways involved in plant resistance to salinity stress can be an effective tool to increase crop production and cultivated area in different parts of the world.

Keywords: antioxidant, climate variation, hormone, limiting factors, plant adaptability, saline lands

1. Introduction

Abiotic stress is one of the dreadful environmental stresses that affect plant growth and crop yield, even in irrigated areas throughout the world. Salinity stress is one of the major abiotic stresses that limits crop production and agricultural sustainability in many areas of the world because of the increased use of low-quality water for irrigation. An important issue for agricultural science and food production in the world is the study of salt stress mechanisms in crops. The plant response to high NaCl concentrations is complex and comprehensive; it contains numerous processes that should be correctly coordinated. The extra salt concentrations affect plants, leading to ionic imbalance and osmotic stress because of the accumulation of ions like Na^+ and Cl^-. Moreover, salt stress *leads to* mineral homeostasis of K^+ and Ca^{2+}. Recently, a new gene family involved in salt stress response in crops has been reported based on achievements in genomics, molecular biology, and transcriptomic procedures [1].

Soil salinity is considered as a destructive environmental factor that has severe negative effects on seed germination, crop growth, and productivity [2]. Of the 230 million

 IntechOpen

hectares of farmland in use around the world, 20% is affected by salt followed by their increase annually as a result of various factors including (1) inappropriate crop irrigation practices, (2) excessive fertilization and plowing, and (3) natural phenomena such as salt intrusion into coastal zones resulting from rising sea levels [3, 4]. For instance, around 100 million hectares of farmland in China suffer salinity stress [5]. In arid and semi-arid regions, salinity is becoming a more and more dreadful threat due to limited rainfall, high evapotranspiration, and extreme temperatures coupled with poor water and soil management [6, 7]. Globally, salinity is an important stress that affects a large proportion of crop productivity and agricultural soil [8]. Since, demand for food production is growing by enhancing world population, making use of salty lands to compensate for food shortages is of great importance [9]. Thus, scientists are attempting to increase food production by 70% to avoid the risk of food shortages for 9.3 billion people by 2050 [10].

The fact that the continuously growing population and meeting its demand has attracted agricultural politicians to focus on crop productivity enhancement. In this route, environmental stresses could be recognized as major obstacles to crop production [11]. Among all abiotic constraints, salinity stress is one of the major limitations for crop production, which leads to ionic imbalances, osmotic stress, and secondary stresses for glycophytes [12]. Currently, climate change and environmental stresses negatively influence arable lands due to irrigation malpractices, which results in the accumulation of sodium chloride (NaCl) in soil [13]. Besides inaccurate cultivation practices, arable lands may confront some natural phenomena including salty water on the coasts and contamination resulting from parent rock and oceanic salts, which threaten the productivity of arable land under cultivation. Although Na^+ causes toxicity in soil under salinity conditions, Cl^- anion can play an important role in causing some plant species to be sensitive [14].

Salinity has a negative effect on growth and development processes in plants, followed by their seed/biomass yield [15]. Salt stress negatively influences plant growth by germination reduction, decrease of aerial organ growth and dry matter production, as well as disturbance of nutritional imbalances, the complex interaction of hormones, osmotic adjustment, and specific ion toxicity [16, 17]. Increasing salinity decreases the leaf area, followed by hindering leaf development and a decrease in light absorption and photosynthesis rate [18]. If the new leaves appear, they may be weak and show burning symptoms when they turn to mature. In total, there is a positive correlation between leaf number/leaf area and photosynthetic rate, which means the less leaf number/area, the less photosynthetic rate following the decline of dry weight matter [19, 20]. Salinity disturbs ions transport in different plant's organs, which causes a decline in vegetative and reproductive quality.

Potassium is an essential nutrient for plant growth, which plays the main role in creation of osmotic pressure and protein production of cells [21, 22]. Salinity (sodium chloride) interferes with the active absorption of this element, since it prevents the selective absorption of cells. Besides, the competition of sodium with potassium in cells brings about a decrease in inactive absorption of potassium due to more abundance of sodium than potassium (**Figure 1**) [23, 24].

Since crops play an important role in the economy of agricultural societies and salt stress has strong and diverse effects on overall crop growth, the objective of this chapter is to accurately detect and awareness of the biological processes involved in salinity stress to prevent severe damage in consideration of various aspects of morphological, physiological, and biochemical mechanisms. In addition, the introduction of salt-tolerant cultivars could be the other management strategy that leads to decreased damages resulting from salt stress.

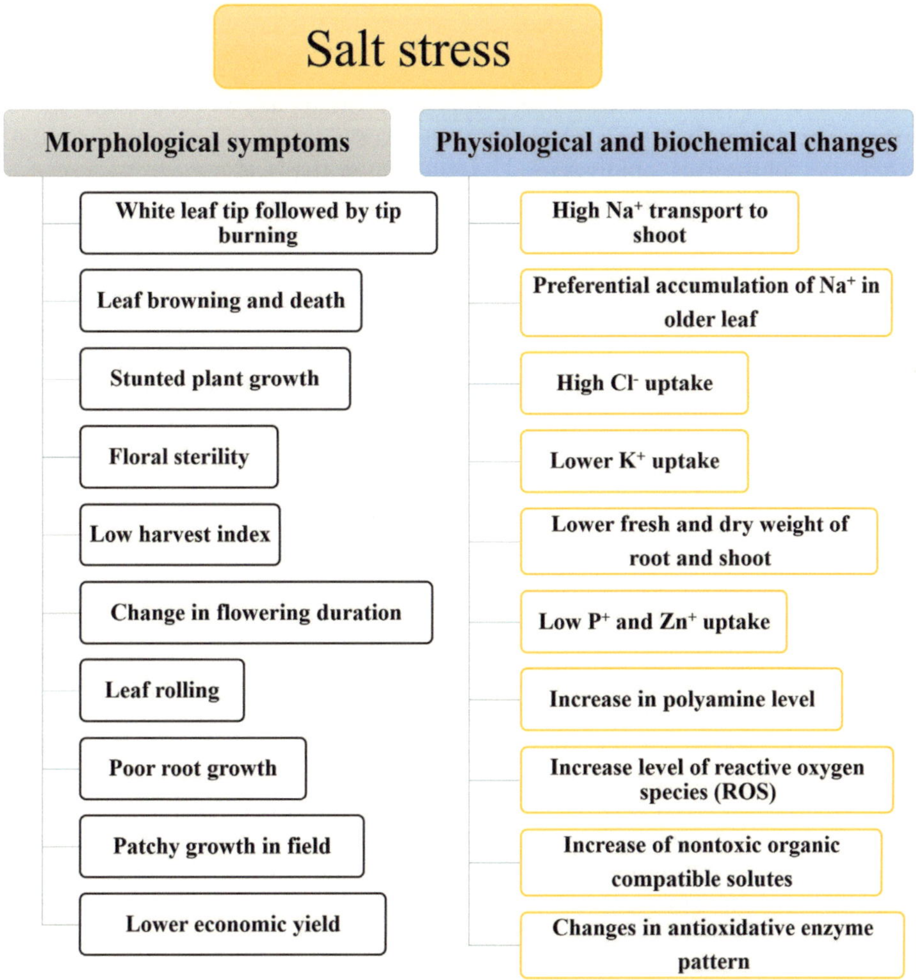

Figure 1.
A general sight of physiological, biochemical, and morphological effects of salt stress in plants.

2. Influence of salt stress

About 50% of agronomical lands are affected by different ranges of salt stress worldwide [25]. Among different climatic areas, it should be noted that arid and semi-arid zones are extremely sensitive to salt stress, which means salinity is considered the most significant obstacle to agronomical and horticultural production. Salt leads to decreased plant growth in two ways in both soil and water. In one way, the existence of salt in soil solution declines plants' capability in relation to water absorption, which is known as an osmotic effect or water shortage. In another way, extra salt entrance by transpiration results in damage to leaf cells [26].

Most morphological traits in plants such as leaf area, biomass, seed weight, root system, plant height, spikelet per panicle, and tillers per plant reduce under high salt concentration, leading to a reduction in grain yield and harvest index [27, 28]. Salt stress affects physiological and biochemical processes in plants at all developmental stages, from germination to senescence [29]. Salt stress has both osmotic and ionic or ion-toxicity effects on plants, ultimately causing oxidative stress and nutrient depletion in plant cells [30]. Continuous salt stress reduces plant cell turgor pressure, which in turn reduces cell growth; therefore, plants must osmotically adjust to maintain cell expansion and growth [30–32]. Osmotic stress (caused by the lower water potential of the external solution) is rapidly sensed by plants soon after exposure to saline conditions and leads to plant water and solute deficits [9, 33]. Osmotic stress also results in rapid stomatal closure, which reduces the plant's ability to assimilate CO_2 and inhibits photosynthesis [34]. Aging and death of plants are caused by the accumulation of sodium (Na^+) and chlorine (Cl) in plant cells due to ionic stress [35]. The most common explanation for Na^+ toxicity is that it has an inhibitory effect on enzyme activities, affecting metabolism negatively, including the Calvin cycle and other pathways [36, 37]. Excess sodium in the cytoplasm also interferes with the uptake and transport of potassium and other macro and micronutrients, including nitrogen, phosphorus, potassium, calcium, and zinc [38]. Various stresses, such as osmotic stress and salinity, damage the cells and vital molecules in plant cells due to the accumulation of reactive oxygen species [23]. Under salinity tension, plants should regulate biochemical and physiological pathways related to the regulation of osmotic and ionic homeostasis, nutrient balance, and oxidative stress [39].

The continuous transfer of salt into leaves during a long period will eventually lead to high sodium and chlorine accumulation, which causes devastating effects on old leaves, since the death rate of leaves is necessary for plant survival [40]. If the newly created leaves are more than dead leaves, the plant has enough photosynthesizing leaves to produce flowers and seeds, no matter how many of them have been decreased. However, if there are more dead leaves than produced leaves, then plants cannot survive to produce seeds [41].

The plant growth response to salinity stress includes two-phase reactions over time. The first stage of growth reduction, known as "initial growth decrease" or "early phase," is due to the osmotic effect of salt outside the root zone appearing quickly. Then, the second reduction stage, "late phase or ionic phase," occurs, which comes from internal damage and continues for a longer period of time. What creates a differentiation between salt-sensitive plants and salt-tolerant plants is the inability to prevent salt from reaching toxic levels in transpiration leaves over time [21].

In most crops, the critical growth stage is seed germination, which exhibits toxic effects on seed embryo water uptake inhibition under salinity. However, legumes, seedlings, and developmental stages are more sensitive than the germination stage. It is observed that salinity results in limitations such as a decrease in the water potential of tissue, photosynthesis, and different stages of growth. For example, the stomatal closure caused by less availability of water to cells brings about less photosynthesis and growth in soybeans under salinity stress [42]. In another study, it was found that the early stages of growth and development in lentils are sensitive to salt stress [43].

Ionic imbalance is the other indicator of salinity conditions that cause disturbance in competitive nutrient uptake, accumulation, and transport in plants. Indeed, the concentrations of cations including Ca^{2+}, K^+, and Mg^{2+} which are vital for photosynthetic activity, decrease by different levels of salt [44]. For instance, the survey of the Na^+/K^+ ratio in *Lotus creticus* showed Na^+ contents increase in aerial and root organs under salinity stress [45]. Moreover, the competitive uptake of Na^+ and K^+ ion flux caused an astonishing reduction of the Na^+/K^+ ratio in mungbean and chickpea, indicating a significant decrease in K^+ and yield [46]. In legumes, it was demonstrated that different species had various responses to saline conditions, since it was determined that soybean, mungbean, cowpea, and common bean indicated salt sensitivity as Na^+ toxicity, Cl^- toxicity, Na^+, and Cl^- toxicity, respectively [47]. Lopez-Aguilar et al. reported that the uptake and transport of Ca^{2+} and Mg^{2+} were significantly lower in cowpea and frijollilo plants under salt stress; however, these rates in Tepary bean plants showed no significant response to salinity treatments [48]. It was reported that the Na^+/K^+ ratio increased in the roots and shoots of Cheongcheong and IR28 cultivars of rice under salinity stress [49]. In addition, it was found that there was a significant correlation between grain dry matter and the K^+/Na^+ ratio under salt stress in wheat, which had an effect on the grain filling rate and its duration [21].

3. Different mechanisms of tolerance to salt stress

Plants show different salinity tolerance reactions, which are reflected in their different growth responses. To increment crop yield under salt stress, it is of great importance to recognize tolerance mechanisms consisting of ion homeostasis, compatible solute accumulation and osmotic protection, antioxidant regulation, and hormonal regulation.

The salt tolerance mechanisms in plants are divided into three groups: osmotic stress tolerance, Na^+ expulsion from leaf, and tissue tolerance [50]. With respect to osmotic stress tolerance, this type of stress decreases cell development on the tips of roots and young leaves. A decrease in response to osmotic stress leads to an increase in leaf growth and stomata conduction, while the increment of leaf area would be profitable for plants whose soil has enough water availability. The second group is related to repulsing Na^+ from leaves, which is assured that Na^+ ions with toxic concentration do not accumulate within leaves, since the deficiency of excretion of Na^+ toxic effects causes premature death of older leaves. The third mechanism is tissue tolerance, which appears to be tolerant of some tissues to Na^+ and Cl^- accumulation [51].

4. Salt tolerance strategies

Salt tolerance mechanisms include processes that allow plants to resist stress, take up nutrients from the soil, and complete their life cycle under high salt concentrations in the soil. Plants that are able to grow on higher levels of salt in the rhizosphere are termed as halophytes. To enhance salt tolerance, the following strategies will be briefly reviewed, aiming at crop salt tolerance under saline conditions.

In general, the physiological responses of plants to salinity stress are divided into two main phases. The first phase is ion-independent growth reduction. It occurs within a few minutes to a few days, resulting in closing stomata and cell expansion inhibition [52]. A second phase, which pertains to the build-up of cytotoxic ion levels, happens for a relatively long period, resulting in slowing down metabolic processes, reducing leaf water potential, premature senescence, and ultimately cell death [53, 54].

4.1 Ion homeostasis and its detention in tissue

Under salt stress, plants apply an ion homeostasis mechanism to maintain a high concentration of K^+ and a low concentration of Na^+ in the cytosol [55]. Ion homeostasis is enough for normal growth and a necessary process for growth during salt stress [56]. Regardless of the nature of halophytes and glycophytes, neither can tolerate high salt concentrations in their cytoplasm; thus, the focus of most projects is on the Na^+ ion transfer mechanism and its distribution in various parts of plants. After entering Na^+ into cytoplasm, it is transferred to vacuole through antiporter [57]. There are two types of H^+ pumps in vacuole membrane vacuolar H^+-ATPase (V-ATPase) and Vacuolar H^+-pyro-phosphatase [58]. The electrochemical proton gradient created by these two transferring enzymes in the vacuole provides the necessary power for ion transport. Under stress-free conditions, H^+ pump plays a significant role in preserving mineral homeostasis, secondary transferring energy, and facilitation of vesicle fusion. Under salt stress, the plant survival depends on the V-ATPase pump, which is dominant in the H^+ pump.

In another view, it is considered the vital role of salt overly sensitive gene network (SOS) to salinity for Na^+ ion regulation and salt tolerance [59]. The SOS signaling network consisted of three main proteins SOS1, SOS2, and SOS3. SOS1 gene codes anti-porter H^+/Na^+ of plasma membrane, which facilitates the transport of sodium from root tissue to stem. The SOS2 gene codes serine/threonine kinase which is activated by Ca^{2+} signals. The third protein involved in SOS signaling is SOS3, which is Ca^{2+} domain binding to N^- terminal of SOS2. By increasing Na^+ concentration, a severe increment of Ca^{2+} level in the intracellular area occurs, which in turn facilitates its binding to SOS3 protein. SOS3 activates SOS2 by releasing internal inhibitors. The combination of SOS2-SOS3 on plasma membrane is loaded where SOS1 phosphorylates, which causes an increase in sodium flow and a reduction in its toxicity. In legumes, the ion distribution, specifically the ratio of K^+/Na^+ in the cytosol, has been reported in various ways among their different species and cultivars.

For example, the response of various legumes to salt stress is determined by Cl^- toxicity (mungbean), Na^+ toxicity (soybean), and Cl^- and Na^+ toxicity (common bean and cowpea) [47]. In one study, it was reported that there were exceptional K^+/Na^+ transporters in the mash bean cultivar, which conserve lower levels of intracellular Na^+ [60]. However, it was observed that the uptake of K^+ resulted in the developing Na^+/K^+ ratio, which was regulated by a decrease of Na^+/Ca^{2+} ratio in pigeon peas under salinity [61]. According to Zhang et al., soybean salt tolerance was obtained by plasma membrane Na^+/H^+ exchanger GmSOS1, which maintains Na^+ homeostasis in soybean [62].

The mechanisms in relation to cytoplasmic Na^+ reduction include restriction of Na^+ uptake, increase of Na^+ efflux, and compartmentalization of Na^+ in the vacuole [63]. The rice plasma membrane Na^+/H^+ anti-porter (OsSOS1) excludes Na^+ from the shoot, promoting a lower cellular Na^+/K^+ ratio and increasing salt tolerance [64, 65]. The vacuolar Na^+/H^+ anti-porters OsNHX1, OsNHX2, OsNHX3, OsNHX4, OsNHX5, and OsARP/OsCTP play important roles in the vacuolar compartmentalization of Na^+ and K^+ that accumulate in the cytoplasm and thereby determine rice salt tolerance [66]. OsHKT1;1, OsHKT2;1, OsHKT2;3/OsHKT3, OsKAT1, OsKAT2, OsHAK5, and OsHAK21/qSE3 transport Na^+ or both Na^+ and K^+, helping to maintain Na^+/K^+ homeostasis in the cytoplasm and regulate rice response to salt stress [67].

Some genes that include transcription factors affect the salt tolerance of rice by regulating the expression of *CLC*, *NHX*, and *HKT* genes, activating the expression of the K^+/Na^+ transporter, and reducing sodium absorption into the cells [68].

4.2 Compatible solute accumulation and osmotic protection

The compatible solute (compatible osmolytes or osmoprotectants) are not toxic even at higher concentrations. Their effective role is to protect osmotic structure and balance within cells through continuous infiltration flow of water and water flow reduction. Related to their prominent feature, it can be referred to as being hydrophilic compounds with low molecular weight and no net charge at physiological pH. Under salt stress, there are various types of solute accumulation varying among different plant species including amino acids (proline), carbohydrates (sucrose and fructose), quaternary amines (glycine and betaine), ions (potassium), and organic acids (malate and nitrate) [69]. El-Sabagh et al. reported on the effect of exogenous osmoprotectants such as proline on soybeans under salinity stress in order to increase salinity resistance [70]. The accumulation of proline was observed in common bean (*Phaseolus vulgaris* L.) subjected to salt stress, followed by its increased accumulation in salt-stressed plants treated with effective microorganisms [71].

The other non-toxic osmolyte is beta-glycine, which raises osmorylation during stress conditions and plays an important role in osmotic adjustment [72], protein fixation [73], photosynthetic system protection [74], and reactive oxygen species (ROS) reduction [75]. Another research has demonstrated the existence of less protein and high protein and amino acids content in salt-tolerant than salt-sensitive beans [76]. In addition, it was shown that osmotic regulation in soybeans occurs by regulating the concentrations of trigonelline, proline, and potassium [70]. Also, the osmotic adjustment caused by accumulation of amino acids, reducing sugars, and ascorbic acid has been reported in pea plants subjected to salt stress [77]. Finally, osmoregulation can be controlled by various osmolytes, which are key factors to decrease osmotic stress in grain legumes.

Salt stress also causes osmotic stress and promotes the biosynthesis and accumulation of compatible osmolytes, which decrease cellular osmotic potential and stabilize cellular proteins and structures [78]. *OsP5CS1* and *OsP5CR* proline synthesis genes increase the accumulation of proline and improve the tolerance of rice to salt stress [79]. *OsGMST1*, which encodes a monosaccharide transporter, increases monosaccharide accumulation and improves salt tolerance in plants [80]. It is reported that sugars will eventually be exported to transporters (SWEETs) that can regulate sucrose transport/distribution and maintain sugar homeostasis in rice under drought and salinity stresses [81]. In addition, rice glycine betaine is the other osmolyte synthesized by choline monooxygenase *OsCMO,* and beating aldehyde dehydrogenase *OsBADH1* increases the tolerance of rice to salt stress by incrementing the accumulation of glycine betaine [82].

4.3 Regulation of antioxidants in order to tolerance to salt stress

Antioxidant metabolisms like enzymatic and non-enzymatic compounds play a critical role in preventing reactive oxygen species (ROS) resulting from salinity stress [83]. These lethal compounds contain hydrogen peroxide (H_2O_2), hydroxylradical (OH), superoxide-radical (O^{-2}), and singlet-oxygen (O^2), which damage cells, along with different proteins, nucleic acids, and membrane lipids, which lead to oxidative stress [84]. Salt tolerance has a positive correlation with antioxidant enzymes such as glutathione peroxidase (GPX), glutathione reductase (GR), superoxide dismutase (SOD), ascorbate peroxidase (APX), and catalase (CAT) as well as the accumulation of non-enzymatic antioxidant compounds. Under salt stress conditions, the activity of these enzymes increased in the tolerant variety [85]. In the previous research, APX and CAT activities were enhanced in bean salt-tolerant cultivars subjected to salt stress [86]. In conclusion, the enhanced activities of antioxidants in legumes perform as a protective process over oxidative stress.

Superoxide dismutase (SOD) can evacuate the surplus superoxide anions in the cells and hence act as the first line of defense. It is classified into three groups in higher plants. Based on the metal ions present on auxiliary sites, they are iron SOD (Fe-SOD), copper/zinc SOD (Cu/Zn-SOD), and manganese SOD (Mn-SOD). The SOD can dismutase the superoxide anion into the O_2 and H_2O_2 and abolish the superoxide toxicity [87].

Ascorbate peroxidase (APX) is one of the important enzymes that eradicate H_2O_2. The adverse effects of salt stress in different plant species are mitigated by exogenous application of ascorbate, and it endorses plant recovery from salt stress [88, 89].

Catalase (CAT) is another important enzyme that reduces H_2O_2 and is present in peroxisomes. It differs from APX in that it has a high affinity for hydrogen peroxide and requires a reducing substrate. Hydrogen peroxide from the light reaction is mainly removed by CAT. The effects of NaCl on H_2O_2 content and CAT activity were studied in various groups of plants, all of which show high tolerance to NaCl. In all cases, treatment of plants with high salinity levels increased H_2O_2 concentration, and as a result, CAT enzyme activity also increased [90].

A lower accumulation of O_2 and H_2O_2 was observed in overexpressing OsMn-SOD1 and OsCu/Zn-SOD lines under salt stress [91]. APX gene family in *Oryza sativa* increase APX activity, decrease H_2O_2 and malondialdehyde levels, reduce oxidative stress damage, and increase rice tolerance to salinity stress [92]. Glutathione responsive rice glyoxalase II in salt stress conditions with increased photosynthetic efficiency [93], the GR gene family in rice [94] and rice GDP-mannose pyrophosphorylase and dehydroascorbate reductase play an important role in increasing plant tolerance [95].

4.4 Polyamine role in salt tolerance

Polyamines are relatively polycationic low molecular weight compounds that are extremely widespread in all organisms from unicellular to multicellular including putrescine (Put), spermidine (Spd), and spermine (Spm). They are applicable at different development stages including cell division, nucleic acids stability, embryogenesis, dormancy termination, aging regulation, plant growth, and stress resistance [96]. In case of salinity, it was reported that the oxidative damage was decreased by the accumulation of putrescine in soybean [97], chickpea [98], and faba bean [99]. According to spermidine, it was exhibited that spermidine was applicable to alleviate the destructive effects of salinity condition and promote salinity tolerance in various legume species such as faba bean [100], soybean [101], and mungbean [102]. Thus, legumes make use of polyamines for their adaptation and defense mechanisms to cope with the lethal consequences of salinity.

4.5 Role of nitric oxide in salt tolerance

Nitric oxide (NO) is a volatile and small gas molecule that cause plant growth regulation, root growth development, transpiration, stomata closure, flowering, seed germination, and response to salt stress as a signaling molecule [103]. The NO effect in salt tolerance relies on the regulation of H^+-ATPase in cell membrane and Na^+/K^+. Under salinity stress, nitric acid is capable of ROS accumulation followed by prevention of thiobarbituric acid and other aldehyde production to decline disturbing effect of salinity stress on cell membrane [104]. In soybean, it has been reported that the use of NO increases the activity of ascorbate peroxidase enzyme, and also, in *Vigna angularis*, it increases the activity of oxidant enzymes and non-enzymatic compounds [105, 106].

4.6 Hormones role in salt tolerance

It is proved that crops utilize adaptive responses against salt stress (downstream of the initial salt signaling phase), which alters plant hormone levels and the salt-induced signaling cascade. Phytohormones, endogenous chemical signals, are one of the adaptive responses that coordinate plant growth and development under optimal conditions and environmental challenges [28]. Plant hormones are chemical compounds that exist in very low concentrations and play important key roles in controlling and regulating all aspects of growth and development. Hormones are classified into two main groups containing (1) growth promoters (gibberellins, cytokinins, and auxins) and (2) growth inhibitors (ABA and ethylene) that individually affect salt stress. Among all phytohormones, ABA has a significant role in saline conditions. It is considered as the central regulator of abiotic stress tolerance in plants by up-regulation of PYR, PYL, RCAR proteins complex, protein phosphatases 2C (PP2C), ABF transcription factors, and SnRK2 protein kinases, which can decrease the preventive saline effect on growth and photosynthesis. In fact, ABA is responsible for metabolic variations, stomatal closure, and stress--responsive gene regulation under salinity condition in crops, even at high concentrations of ABA, which are more helpful in osmotic adjustment using more stress protein production and salinity adaptation. For example, Shu et al. suggested that ABA could be a potential plant growth regulator in soybeans, which could improve soybean germination under salt stress [107]. In addition, it was reported that the stomata closure occurred in response to ABA in *Lupinus albus* L. [108]. Thus, ABA also plays a major role in mediating physiological responses to salt stress through alteration in osmotic adjustment and photosynthetic and plant developmental growth. The ABA biosynthesis genes *OsNCED3* and *OsNCED5*, encoding 9-cis-epoxycarotenoid dioxygenases, increase ABA levels and enhance salt stress tolerance [109]. The rice ABA receptor gene OsPYL/RCAR5 is a positive regulator of the ABA signal transduction pathway in abiotic stress tolerance [110]. *OsbZIP23* gene directly affects the gene responsible for ABA synthesis and the component gene of ABA signaling and can regulate ABA signaling and biosynthesis by regulating stress response pathways [111].

In one study, it was noticed that salinity tolerance incremented by promotion of ethylene biosynthesis and signal transduction in rice, while, their inhibition increased salinity sensitivity in Arabidopsis and soybean [112]. However, in tomato, the direct ethylene precursor 1-aminocyclopropane-1 carboxylic acid (ACC) causes Na^+ accumulation and oxidative stress in leaves and promotes leaf senescence under salinity stress [113, 114]. Meanwhile, protein salt intolerance 1 (*OsSIT1*) in rice increases sensitivity to salinity stress by increasing ethylene production in the plant [115]. DOF transcription factors and ethylene-responsive element binding protein (EREBP) transcription factor OsBIERF3 cause positive changes in plant root length by limiting ethylene synthesis [116].

Gibberellic acid (GA) is another important hormone that regulates plant growth and is associated with the regulation of growth under abiotic stress. Gibberellin 2-oxidase 5 (OsGA2ox5) and GA catabolic pathway genes reduce GA accumulation and increase plant salt tolerance with delayed growth [117].

In case of cytokinins (CKs), its function is to control plant adaptation to environmental stresses. In Arabidopsis, by the reduction of various levels of CKs, the salt tolerance increased [118], while it was exhibited that the exogenous CK application caused salt tolerance enhancement in *Solanum melongena*. Indeed, the enzyme results in CK degradation pathway called as cytokinin oxidase OsCKX2, which negatively regulates salt stress tolerance by controlling CK levels [119].

Jasmonic acid (JA) is the other hormone that reduces root length and activates antioxidant enzymes under salt stress conditions [120]. It was reported that JA accumulates endogenously in rice roots and exogenously increases salt stress tolerance in rice and wheat [121, 122].

Salicylic acid (SA) is also known as adaptive response in plants under salinity stress, which plays an important role in plant salt tolerance. Exogenous SA application in plants can control salinity conditions through (1) enhancement of antioxidant system, (2) increases in the synthesis of osmolytes, and (3) promotes photosynthesis and nitrogen fixation [123].

5. Genetic engineering for salinity stress tolerance

Salt tolerance is generally controlled by large numbers of genes and different physiological mechanisms. Recently, researchers have focused on analyzing genes controlling salt tolerance by introduction of hypothetical premise for augmentation of the stress signal system and upgrading plant tolerance to stress [124]. Genetic transformation technology allows genetic scientists to transfer genes in a precise and predictable way. Based on this, manipulation of the biosynthetic pathways involved in osmotic protection and physiological traits of plants could be efficient tools to (1) raise the level of tolerance to salinity, (2) upgrade the accumulation of molecules that remove ROS, (3) reduce the concentration of lipid peroxidation, (4) maintain the three-dimensional and functional structure of proteins, etc. [125]. To produce salt-resistant plants, plant engineering methods can be used to overexpress resistance genes. Gene expression studies using constitutive promoters provide limited biological information compared to using inducible promoters or cell type-specific promoters. In this direction, the selection of strong promoters inducing abiotic stress is very important in plant breeding programs. In addition, a wide range of stress-inducible promoters are available that vary in their ability for regulation of gene expression patterns to enhance salinity tolerance in crop plants. The selection of these promoters can significantly affect the results of a transgenic manipulation. For instance, salt tolerance is fundamentally enhanced by vacuolar Na^+/H^+ anti-porter gene AtNHX1 and plasma membrane Na^+/H^+ anti-porter gene SOS1 in transgenic *A. thaliana* [126]. A substantial number of genes, HKT and NHX, encoding K^+ transporters and channels have been distinguished and cloned in different plant species. The class 1 HKT transporters, distinguished in Arabidopsis, shield the plant from the antagonistic impacts of salinity by anticipating over-abundance aggregation of Na^+ in leaves [127]. Similar results were observed in an experiment with rice, where the class 1 HKT transporter removes excess sodium from the xylem, thus protecting photosynthetic leaf tissues against the toxic agent Na^+ [128]. Transgenic genes such as *Ta-UnP*, *TaZNF*, *TaSST*, *TaDUF1,* and *TaSP* can significantly increase salt tolerance in wheat [129]. Studies conducted by Sun et al. revealed that an overexpressing salt-responsive gene, *ZFP179*, encoding a Cys2/His2 zinc finger protein, enhanced salt tolerance in rice [130]. It has also been reported that overexpression of genes for signaling proteins such as kinases and phosphatases are involved in salt tolerance in various plants. One of these proteins is mitogen-activated protein kinases (MAPKs), which are conserved signaling proteins and play an essential role in improving plant tolerance to various stresses of crops. It was reported that different transgenic plants engineered with MAPK cascade are tolerant to salt stress [131].

6. Conclusions

Totally, various studies have been demonstrated that salt tolerance consists of different cellular, metabolical, molecular, and physiological responses in the whole plant. During evolution, plants have developed the ability to detoxify excessive levels of Na^+ and use multiple strategies to mitigate damages resulting from salinity stress. There are various ways to decline the effects of salt stress on plant growth, which require a thorough conception of which defensive responses to environmental stresses could be more appropriate. Thus, it is indispensable to apply integrated approaches such as molecular, biochemical, and physiological techniques in order to develop salt-tolerant cultivars. A comprehensive understanding of molecular mechanisms in crop salt tolerance has been gained through the development of molecular salt tolerance markers, complete sequencing of plant genomes, and the widespread use of microarray analysis and genome editing technology. In order to create salt-tolerant crops using genetic engineering methods, various tools and techniques could be distinguished, including an extensive number of salt-responsive genes, methods of gene transfer, and their promoter sequences. These powerful techniques can offer advantages and solutions to the complex and fascinating traits involved in salt tolerance plants.

Author details

Parastoo Majidian and Hamidreza Ghorbani*
Crop and Horticultural Science Research Department, Mazandaran Agricultural and Natural Resources Research and Education Center, AREEO, Sari, Iran

*Address all correspondence to: h.ghorbani@areeo.ac.ir

References

[1] Isayenkov SV. Physiological and molecular aspects of salt stress in plants. Cytology and Genetic. 2012;**46**:302-318. DOI: 10.3103/S0095452712050040

[2] Yang Y, Guo Y. Elucidating the molecular mechanisms mediating plant salt-stress responses. New Phytologist. 2018;**217**(2):523-539

[3] Shabala S. Learning from halophytes: Physiological basis and strategies to improve abiotic stress tolerance in crops. Annals of Botany. 2013;**112**(7):1209-1221

[4] Kamran M, Parveen A, Ahmar S, Malik Z, Hussain S, Chattha MS, et al. An overview of hazardous impacts of soil salinity in crops, tolerance mechanisms, and amelioration through selenium supplementation. International Journal of Molecular Sciences. 2019;**21**(1):148

[5] Naveed SA, Zhang F, Zhang J, Zheng TQ, Meng LJ, Pang YL, et al. Identification of QTN and candidate genes for salinity tolerance at the germination and seedling stages in rice by genome-wide association analyses. Scientific Reports. 2018;**8**(1):6505

[6] de Azevedo Neto AD, Prisco JT, Enéas-Filho J, de Abreu CEB, Gomes-Filho E. Effect of salt stress on antioxidative enzymes and lipid peroxidation in leaves and roots of salt-tolerant and salt-sensitive maize genotypes. Environmental and Experimental Botany. 2006;**56**(1):87-94

[7] Lodeyro AF, Carrillo N. Salt stress in higher plants: Mechanisms of toxicity and defensive responses. Stress Responses in Plants: Mechanisms of Toxicity and Tolerance. 2015:1-33

[8] Munns R. Comparative physiology of salt and water stress. Plant, Cell and Environment. 2002;**25**(2):239-250

[9] Qin H, Huang R. The phytohormonal regulation of Na^+/K^+ and reactive oxygen species homeostasis in rice salt response. Molecular Breeding. 2020;**40**(5):47

[10] Shabala S, Cuin TA. Potassium transport and plant salt tolerance. Physiologia Plantarum. 2008;**133**(4): 651-669

[11] Vinocur B, Altman A. Recent advances in engineering plant tolerance to abiotic stress: Achievements and limitations. Current Opinion in Biotechnology. 2005;**16**(2):123-132

[12] Ahmad R, Hussain S, Anjum MA, Khalid MF, Saqib M, Zakir I, et al. Oxidative stress and antioxidant defense mechanisms in plants under salt stress. In: Plant Abiotic Stress Tolerance. USA: Springer International Publishing; 2019. pp. 191-205

[13] Carillo P, Annunziata MG, Pontecorvo G, Fuggi A, Woodrow P. Salinity stress and salt tolerance. Abiotic Stress in Plants-mechanisms and Adaptations. 2011;**1**:21-38

[14] Islam F, Yasmeen T, Arif MS, Ali S, Ali B, Hameed S, et al. Plant growth promoting bacteria confer salt tolerance in *Vigna radiata* by up-regulating antioxidant defense and biological soil fertility. Plant Growth Regulation. 2016;**80**(1):23-36

[15] Ghassemi-Golezani K, Taifeh-Noori M, Oustan S, Moghaddam M. Response of soybean cultivars to salinity stress. Journal of Food Agricultural and Environment. 2009;7(2):401-404

[16] Nadeem M, Li J, Yahya M, Wang M, Ali A, Cheng A, et al. Grain legumes and fear of salt stress: Focus on mechanisms and management strategies. International Journal of Molecular Science. 2019;**20**(4):799. DOI: 10.3390/ijms20040799

[17] Tlahig S, Bellani L, Karmous I, Barbieri F, Loumerem M, Muccifora S. Response to salinity in legume species: An insight on the effects of salt stress during seed germination and seedling growth. Chemistry and Biodiversity. 2021;**18**(4):e2000917

[18] Manchanda G, Garg N. Salinity and its effects on the functional biology of legumes. Acta Physiologiae Plantarum. 2008;**30**:595-618

[19] Bybordi A. The influence of salt stress on seed germination, growth and yield of canola cultivars. Notulae Botanicae Horti Agrobotanici Cluj-Napoca. 2010;**38**(1):128-133

[20] Dadkhah A. Effect of salinity on growth and leaf photosynthesis of two sugar beet (*Beta vulgaris* L.) cultivars. Journal of Agricultural Science and Technology. 2011;**13**(7):1001-1012

[21] Poustini K, Siosemardeh A. Ion distribution in wheat cultivars in response to salinity stress. Field Crops Research. 2004;**85**(2-3):125-133

[22] Fan M, Bie Z, Krumbein A, Schwarz D. Salinity stress in tomatoes can be alleviated by grafting and potassium depending on the rootstock and K-concentration employed. Scientia Horticulturae. 2011;**130**(3):615-623

[23] Ahanger MA, Tomar NS, Tittal M, Argal S, Agarwal R. Plant growth under water/salt stress: ROS production; antioxidants and significance of added potassium under such conditions. Physiology and Molecular Biology of Plants. 2017;**23**(4):731-744

[24] Wu H, Li Z. The importance of Cl^- exclusion and vacuolar Cl– sequestration: Revisiting the role of Cl^- transport in plant salt tolerance. Frontiers in Plant Science. 2019;**10**:1418

[25] Jouyban Z. The effects of salt stress on plant growth. Technical Journal of Engineering and Applied Sciences. 2012;**2**(1):7-10

[26] Ondrasek G, Rengel Z, Veres S. Soil salinization and salt stress in crop production. Abiotic Stress in Plants-Mechanisms and Adaptations. 2011;**1**:171-190

[27] Razzaq A, Ali A, Safdar LB, Zafar MM, Rui Y, Shakeel A, et al. Salt stress induces physiochemical alterations in rice grain composition and quality. Journal of Food Science. 2020;**85**(1):14-20

[28] Van Zelm E, Zhang Y, Testerink C. Salt tolerance mechanisms of plants. Annual Review of Plant Biology. 2020;**71**:403-433

[29] Muranaka S, Shimizu K, Kato M. Ionic and osmotic effects of salinity on single-leaf photosynthesis in two wheat cultivars with different drought tolerance. Photosynthetica. 2002;**40**:201-207

[30] Zhu JK. Abiotic stress signaling and responses in plants. Cell. 2016;**167**(2):313-324

[31] Hakim MA, Juraimi AS, Hanafi MM, Ismail MR, Selamat A, Rafii MY, et al. Biochemical and anatomical changes and yield reduction in rice (*Oryza sativa* L.) under varied salinity regimes. BioMed Research International. 2014;**2014**:1-11

[32] Munns R, Guo J, Passioura JB, Cramer GR. Leaf water status controls day-time but not daily rates of leaf expansion in salt-treated barley. Functional Plant Biology. 2000;**27**(10):949-957

[33] Zhou J, Li Z, Xiao G, Zhai M, Pan X, Huang R, et al. CYP71D8L is a key regulator involved in growth and stress responses by mediating gibberellin homeostasis in rice. Journal of Experimental Botany. 2020;**71**(3):1160-1170

[34] Wegner LH, Stefano G, Shabala L, Rossi M, Mancuso S, Shabala S. Sequential depolarization of root cortical and stelar cells induced by an acute salt shock–implications for Na^+ and K^+ transport into xylem vessels. Plant Cell Environment. 2011;**34**(5):859-869

[35] Yang Y, Guo Y. Unraveling salt stress signaling in plants. Journal of Integrative Plant Biology. 2018;**60**(9):796-804

[36] Wu TM, Lin WR, Kao CH, Hong CY. Gene knockout of glutathione reductase 3 results in increased sensitivity to salt stress in rice. Plant Molecular Biology. 2015;**87**:555-564

[37] Cheeseman JM. The integration of activity in saline environments: Problems and perspectives. Functional Plant Biology. 2013;**40**(9):759-774

[38] Iqbal MN, Rasheed R, Ashraf MY, Ashraf MA, Hussain I. Exogenously applied zinc and copper mitigate salinity effect in maize (*Zea mays* L.) by improving key physiological and biochemical attributes. Environmental Science and Pollution Research. 2018;**25**:23883-23896

[39] Shahid MA, Sarkhosh A, Khan N, Balal RM, Ali S, Rossi L, et al. Insights into the physiological and biochemical impacts of salt stress on plant growth and development. Agronomy. 2020;**10**(7):938

[40] Abdelraheem A, Esmaeili N, O'Connell M, Zhang J. Progress and perspective on drought and salt stress tolerance in cotton. Industrial Crops and Products. 2019;**130**:118-129

[41] Ibrahim EA. Seed priming to alleviate salinity stress in germinating seeds. Journal of Plant Physiology. 2016;**1**(92):38-46

[42] Do TD, Vuong TD, Dunn D, Smothers S, Patil G, Yungbluth DC, et al. Mapping and confirmation of loci for salt tolerance in a novel soybean germplasm, Fiskeby III. Theoretical Applied Genetics. 2018;**131**(3):513-524

[43] Sidari M, Santonoceto C, Anastasi U, Preiti G, Muscolo A. Variations in four genotypes of lentil under NaCl-salinity stress. American Journal of Agriculture and Biological Science. 2008;**3**:410-416

[44] Yang CW, Xu HH, Wang LL, Liu J, Shi DC, Wang DL. Comparative effects of salt-stress and alkali-stress on the growth, photosynthesis, solute accumulation, and ion balance of barley plants. Photosynthetica. 2009;**47**(1):79-86

[45] Rejili M, Vadel AM, Guetet A, Mahdhi M, Lachiheb B, Ferchichi A, et al. Influence of temperature and salinity on the germination of *Lotus creticus* (L.) from the arid land of Tunisia. African Journal of Ecology. 2010;**48**(2):329-337

[46] Nandwal AS, Godara M, Kamboj DV, Kundu BS, Mann A, Kumar B, et al. Nodule functioning in trifoliate and Pentafoliate mungbean genotypes as influenced by salinity. Biologia Plantarum. 2000;**43**:459-462. DOI: 10.1023/A:1026704107525

[47] Le LTT, Kotula L, Siddique KH, Colmer TD. Na+ and/or Cl− toxicities determine salt sensitivity in soybean (*Glycine max* (L.) Merr.), mungbean (*Vigna radiata* (L.) R. Wilczek), cowpea (*Vigna unguiculata* (L.) Walp.), and common bean (*Phaseolus vulgaris* L.). International Journal of Molecular Sciences. 2021;**22**(4):1909

[48] López-Aguilar R, Orduño-Cruz A, Lucero-Arce A, Murillo-Amador B, Troyo-Diéguez E. Response to salinity of three grain legumes for potential cultivation in arid areas. Soil Science and Plant Nutrition. 2003;**49**(3):329-336

[49] Farooq M, Park JR, Jang YH, Kim EG, Kim KM. Rice cultivars under salt stress show differential expression of genes related to the regulation of Na^+/K^+ balance. Frontiers in Plant Science. 2021;**12**:680131

[50] Sairam RK, Tyagi A. Physiology and molecular biology of salinity stress tolerance in plants. Current Science. 2004;**86**:407-421

[51] Acosta-Motos JR, Ortuño MF, Bernal-Vicente A, Diaz-Vivancos P, Sanchez-Blanco MJ, Hernandez JA. Plant responses to salt stress: Adaptive mechanisms. Agronomy. 2017;**7**(1):18

[52] Rajendran K, Tester M, Roy SJ. Quantifying the three main components of salinity tolerance in cereals. Plant Cell Environment. 2009;**32**:237-249

[53] Khan MIR, Khan NA. Ethylene reverses photosynthetic inhibition by nickel and zinc in mustard through changes in PSII activity, photosynthetic nitrogen use efficiency, and antioxidant metabolism. Protoplasma. 2014;**251**:1007-1019

[54] Roy SJ, Negrão S, Tester M. Salt resistant crop plants. Current Opinion in Biotechnology. 2014;**26**:115-124

[55] Serrano R, Rodriguez-Navarro A. Ion homeostasis during salt stress in plants. Current Opinion in Cell Biology. 2001;**13**(4):399-404

[56] Manishankar P, Wang N, Köster P, Alatar AA, Kudla J. Calcium signaling during salt stress and in the regulation of ion homeostasis. Journal of Experimental Botany. 2018;**69**(17):4215-4226

[57] Basu S, Kumar A, Benazir I, Kumar G. Reassessing the role of ion homeostasis for improving salinity tolerance in crop plants. Physiologia Plantarum. 2021;**171**(4):502-519

[58] Silva P, Gerós H. Regulation by salt of vacuolar H^+-ATPase and H^+-pyrophosphatase activities and Na^+/H^+ exchange. Plant Signaling Behavior. 2009;**4**(8):718-726

[59] Liu M, Wang TZ, Zhang WH. Sodium extrusion associated with enhanced expression of SOS1 underlies different salt tolerance between *Medicago falcata* and *Medicago truncatula* seedlings. Environment Experimental Botany. 2015;**110**:46-55

[60] Mishra S, Panda SK, Sahoo L. Transgenic asiatic grain legumes for salt tolerance and functional genomics. Reveiws in Agricultural Sciences. 2014;**2**:21-36

[61] Subbarao GV, Johansen C, Jana MK, Rao JK. Effects of the sodium/calcium ratio in modifying salinity response of pigeonpea (*Cajanus cajan*). Journal of Plant Physiology. 1990;**136**(4):439-443

[62] Zhang M, Cao J, Zhang T, Xu T, Yang L, Li X, et al. A putative plasma membrane Na^+/H^+ antiporter GmSOS1 is critical for salt stress tolerance in Glycine max. Frontier in Plant Sciences. 2022;**16**(13):870695. DOI: 10.3389/fpls.2022.870695

[63] Tester M, Davenport R. Na^+ tolerance and Na^+ transport in higher plants. Annals of Botany. 2003;**91**(5):503-527

[64] El Mahi H, Pérez-Hormaeche J, De Luca A, Villalta I, Espartero J, Gámez-Arjona F, et al. A critical role of sodium flux via the plasma membrane Na^+/H^+ exchanger SOS1 in the salt tolerance of rice. Plant Physiology. 2019;**180**(2):1046-1065

[65] Martínez-Atienza J, Jiang X, Garciadeblas B, Mendoza I, Zhu JK, Pardo JM, et al. Conservation of the salt overly sensitive pathway in rice. Plant Physiology. 2007;**143**(2):1001-1012

[66] Fukuda A, Nakamura A, Hara N, Toki S, Tanaka Y. Molecular and functional analyses of rice NHX-type Na^+/H^+ antiporter genes. Planta. 2011;**233**:175-188

[67] Wang R, Jing W, Xiao L, Jin Y, Shen L, Zhang W. The rice high-affinity potassium transporter1; 1 is involved in salt tolerance and regulated by an MYB-type transcription factor. Plant Physiology. 2015;**168**(3):1076-1090

[68] Lee SK, Kim BG, Kwon TR, Jeong MJ, Park SR, Lee JW, et al. Overexpression of the mitogen-activated protein kinase gene *OsMAPK33* enhances sensitivity to salt stress in rice (*Oryza sativa* L.). Journal of Biosciences. 2011;**36**:139-151

[69] Rontein D, Basset G, Hanson AD. Metabolic engineering of osmoprotectant accumulation in plants. Metabolic Engineering. 2002;**4**(1):49-56

[70] El Sabagh A, Sorour S, Ragab A, Islam MS, Barutçular C, Ueda A, et al. Alleviation of adverse effects of salt stress on soybean (*Glycine max*. L.) by using osmoprotectants and organic nutrients. International Journal of Agricultural and Biosystems Engineering. 2015;**9**(9):1014-1018

[71] Nounjan N, Theerakulpisut P. Effects of exogenous proline and trehalose on physiological responses in rice seedlings during salt-stress and after recovery. Plant Soil and Environment. 2012;**58**(7):309-315

[72] Meloni DA, Oliva MA, Ruiz HA, Martinez CA. Contribution of proline and inorganic solutes to osmotic adjustment in cotton under salt stress. Journal of Plant Nutrition. 2001;**24**(3):599-612

[73] Mäkelä P, Kärkkäinen J, Somersalo SJBP. Effect of glycinebetaine on chloroplast ultrastructure, chlorophyll and protein content, and RuBPCO activities in tomato grown under drought or salinity. Biologia Plantarum. 2000;**43**(3):471-475

[74] Chou TS, Chao YY, Kao CH. Involvement of hydrogen peroxide in heat shock-and cadmium-induced expression of ascorbate peroxidase and glutathione reductase in leaves of rice seedlings. Journal of Plant Physiology. 2012;**169**(5):478-486

[75] Saxena M, Saxena J, Nema R, Singh D, Gupta A. Phytochemistry of medicinal plants. Journal of Pharmacognosy and Phytochemistry. 2013;**1**:168-182

[76] Khan MSA, Karim MA, Haque MM. Genotypic differences in growth and ions accumulation in soybean under NaCl salinity and water stress conditions. Bangladesh Agronomy Journal. 2014;**17**(1):47-58

[77] Zhu M, Zhou M, Shabala L, Shabala S. Linking osmotic adjustment and stomatal characteristics with salinity stress tolerance in contrasting barley accessions. Functional Plant Biology. 2014;**42**(3):252-263

[78] Zhao C, Zhang H, Song C, Zhu JK, Shabala S. Mechanisms of plant

responses and adaptation to soil salinity. The Innovation. 2020;**1**:1-41

[79] Sripinyowanich S, Klomsakul P, Boonburapong B, Bangyeekhun T, Asami T, Gu H, et al. Exogenous ABA induces salt tolerance in indica rice (*Oryza sativa* L.): The role of *OsP5CS1* and *OsP5CR* gene expression during salt stress. Environmental and Experimental Botany. 2013;**86**:94-105

[80] Cao H, Guo S, Xu Y, Jiang K, Jones AM, Chong K. Reduced expression of a gene encoding a Golgi localized monosaccharide transporter (*OsGMST1*) confers hypersensitivity to salt in rice (*Oryza sativa*). Journal of Experimental Botany. 2011;**62**(13):4595-4604

[81] Mathan J, Singh A, Ranjan A. Sucrose transport in response to drought and salt stress involves ABA-mediated induction of *OsSWEET13* and *OsSWEET15* in rice. Physiologia Plantarum. 2021;**171**(4):620-637

[82] Hasthanasombut S, Supaib-ulwatana K, Mii M, Nakamura I. Genetic manipulation of japonica rice using the OsBADH1 gene from Indica rice to improve salinity tolerance. Plant Cell Tissue and Organ Culture (PCTOC). 2011;**104**:79-89

[83] Irshad A, Rehman RNU, Abrar MM, Saeed Q, Sharif R, Hu T. Contribution of rhizobium–legume symbiosis in salt stress tolerance in *Medicago truncatula* evaluated through photosynthesis, antioxidant enzymes, and compatible solutes accumulation. Sustainability. 2021;**13**(6):3369

[84] Rubio MC, Bustos-Sanmamed P, Clemente MR, Becana M. Effects of salt stress on the expression of antioxidant genes and proteins in the model legume Lotus japonicus. New Phytologist. 2009;**181**(4):851-859

[85] Hernandez JA, Jimenezm A, Mullineaux P, Sevilla F. Tolerance of pea (*Pisum sativum* L.) to long term salt stress is associated with induction of antioxidant defenses. Plant Cell Environment. 2000;**23**:853-862

[86] Yasar F, Ellialtioglu S, Yildiz K. Effect of salt stress on antioxidant defense systems, lipid peroxidation, and chlorophyll content in green bean. Russian Journal of Plant Physiology. 2008;**55**:782-786. DOI: 10.1134/S1021443708060071

[87] Ma L, Zhang H, Sun L. NADPH oxidase AtrbohD and AtrbohF function in ROS-dependent regulation of Na+/K^+ homeostasis in Arabidopsis under salt stress. Journal of Experimental Botany. 2012;**63**:305-317

[88] Agarwal S, Shaheen R. Stimulation of antioxidant system and lipid peroxidation by abiotic stresses in leaves of *Momordica charantia*. Brazilian Journal of Plant Physiology. 2007;**19**:149-161

[89] Munir N. Aftab F enhancement of salt tolerance in sugarcane by ascorbic acid pre-treatment. African journal of Biotechnology. 2011;**10**:18362-18370

[90] Liu C, Mao B, Ou S, Wang W, Liu L, Wu Y, et al. OsbZIP71, a bZIP transcription factor, confers salinity and drought tolerance in rice. Plant Molecular Biology. 2014;**84**:19-36

[91] Guan Q, Liao X, He M, Li X, Wang Z, Ma H, et al. Tolerance analysis of chloroplast *OsCu/Zn-SOD* overexpressing rice under NaCl and NaHCO3 stress. PLoS One. 2017;**12**(10):e0186052

[92] Zhang Z, Zhang Q, Wu J, Zheng X, Zheng S, Sun X, et al. Gene knockout study reveals that cytosolic ascorbate peroxidase 2 (*OsAPX2*) plays a critical

role in growth and reproduction in rice under drought, salt and cold stresses. PLoS One. 2013;**8**(2):e57472

[93] Ghosh A, Pareek A, Sopory SK, Singla-Pareek SL. A glutathione responsive rice glyoxalase II, Os GLYII-2, functions in salinity adaptation by maintaining better photosynthesis efficiency and anti-oxidant pool. The Plant Journal. 2014;**80**(1):93-105

[94] Lima-Melo Y, Carvalho FE, Martins MO, Passaia G, Sousa RH, Neto MCL, et al. Mitochondrial GPX1 silencing triggers differential photosynthesis impairment in response to salinity in rice plants. Journal of Integrative Plant Biology. 2016;**58**(8):737-748

[95] Ushimaru T, Nakagawa T, Fujioka Y, Daicho K, Naito M, Yamauchi Y, et al. Transgenic Arabidopsis plants expressing the rice dehydroascorbate reductase gene are resistant to salt stress. Journal of Plant Physiology. 2006;**163**(11):1179-1184

[96] Todorova D, Sergiev I, Alexieva V, Karanov E, Smith A, Hall M. Polyamine content in *Arabidopsis thaliana* (L.) Heynh during recovery after low and high temperature treatments. Plant Growth Regulation. 2007;**51**:185-191

[97] Zhang GW, Xu SC, Hu QZ, Mao WH, Gong YM. Putrescine plays a positive role in salt-tolerance mechanisms by reducing oxidative damage in roots of vegetable soybean. Journal of Integrative Agriculture. 2014;**13**:349-357

[98] Sheokand S, Kumari A, Sawhney V. Effect of nitric oxide and putrescine on antioxidative responses under NaCl stress in chickpea plants. Physiology and Molecular Biology of Plants. 2008;**14**:355-362. DOI: 10.1007/s12298-008-0034-y

[99] Suleiman S. Effects of exogenous application of diamine (putrescine) on growth and mineral elements distribution in Faba bean plants under saline conditions. Tishreen University Journal of Research Science Studies. 2007;**30**:257-265

[100] Mahdi AHA. Improvement of salt tolerance in *Vicia faba* (L.) plants by exogenous application of polyamines. Egyptian Journal of Agronomy. 2016;**38**:1-21

[101] Radhakrishnan R, Lee IJ. Spermine promotes acclimation to osmotic stress by modifying antioxidant, abscisic acid, and jasmonic acid signals in soybean. Journal of Plant Growth Regulation. 2013;**32**:22-30

[102] Ghosh S, Mitra S, Paul A. Physiochemical studies of sodium chloride on mungbean (*Vigna radiata* L. Wilczek) and its possible recovery with spermine and gibberellic acid. The Scientific World Journal. 2015;**2015**:1-8. DOI: 10.1155/2015/858016

[103] Dadasoglu E, Ekinci M, Kul R, Shams M, Turan M, Yildirim E. Nitric oxide enhances salt tolerance through regulating antioxidant enzyme activity and nutrient uptake in pea. Legume Research. 2021;**44**(1):41-45

[104] Salahuddin M, Nawaz F, Shahbaz M, Naeem M, Zulfiqar B, Shabbir RN, et al. Effect of exogenous nitric oxide (NO) supply on germination and seedling growth of mungbean (cv. Nm-54) under salinity stress. Seed. 2017;**40**:1-7

[105] Ahanger MA, Aziz U, Alsahli AA, Alyemeni MN, Ahmad P. Influence of exogenous salicylic acid and nitric oxide on growth, photosynthesis, and ascorbate-glutathione cycle in salt stressed *Vigna angularis*. Biomolecules. 2019;**10**(1):42

[106] Sofo A, Scopa A, Nuzzaci M, Vitti A. Ascorbate peroxidase and catalase activities and their genetic regulation in plants subjected to drought and salinity stresses. International Journal of Molecular Sciences. 2015;**16**(6):13561-13578

[107] Shu K, Qi Y, Chen F, Meng Y, Luo X, Shuai H, et al. Salt stress represses soybean seed germination by negatively regulating GA biosynthesis while positively mediating ABA biosynthesis. Frontier in Plant Sciences. 2017;**10**(8):1372. DOI: 10.3389/fpls.2017.01372

[108] Wolf O, Jeschke WD. Long distance transport of abscisic acid in NaCl-treated intact plants of *Lupinus albus*. Journal of Experimental Botany. 1990;**41**:593-600

[109] Huang Y, Jiao Y, Xie N, Guo Y, Zhang F, Xiang Z, et al. *OsNCED5*, a 9-cis-epoxycarotenoid dioxygenase gene, regulates salt and water stress tolerance and leaf senescence in rice. Plant Science. 2019;**287**:110188

[110] Kim H, Lee K, Hwang H, Bhatnagar N, Kim DY, Yoon IS, et al. Overexpression of *PYL5* in rice enhances drought tolerance, inhibits growth, and modulates gene expression. Journal of Experimental Botany. 2014;**65**(2):453-464

[111] Zong W, Tang N, Yang J, Peng L, Ma S, Xu Y, et al. Feedback regulation of ABA signaling and biosynthesis by a bZIP transcription factor targets drought-resistance-related genes. Plant Physiology. 2016;**171**(4):2810-2825

[112] Ellouzi H, Hamed KB, Hernández I, Cela J, Müller M, Magné C, et al. Comparative study of the early osmotic, ionic, redox and hormonal signaling response in leaves and roots of two halophytes and a glycophyte to salinity. Planta. 2014;**240**:1299-1317

[113] Ghanem ME, Albacete A, Martínez-Andújar C, Acosta M, Romero-Aranda R, Dodd IC, et al. Hormonal changes during salinity-induced leaf senescence in tomato (*Solanum lycopersicum* L.). Journal of Experimental Botany. 2008;**59**(11):3039-3050

[114] Zhang M, Smith JAC, Harberd NP, Jiang C. The regulatory roles of ethylene and reactive oxygen species (ROS) in plant salt stress responses. Plant Molecular Biology. 2016;**91**:651-659

[115] Li CH, Wang G, Zhao JL, Zhang LQ, Ai LF, Han YF, et al. The receptor-like kinase SIT1 mediates salt sensitivity by activating MAPK3/6 and regulating ethylene homeostasis in rice. The Plant Cell. 2014;**26**(6):2538-2553

[116] Qin H, Wang J, Chen X, Wang F, Peng P, Zhou Y, et al. Rice *OsDOF15* contributes to ethylene-inhibited primary root elongation under salt stress. New Phytologist. 2019;**223**(2):798-813

[117] Shan C, Mei Z, Duan J, Chen H, Feng H, Cai W. *OsGA2ox5*, a gibberellin metabolism enzyme, is involved in plant growth, the root gravity response and salt stress. PLoS One. 2014;**9**(1):e87110

[118] Nishiyama R, Watanabe Y, Fujita Y, Le DT, Kojima M, Werner T, et al. Analysis of cytokinin mutants and regulation of cytokinin metabolic genes reveals important regulatory roles of cytokinins in drought, salt and abscisic acid responses, and abscisic acid biosynthesis. The Plant Cell. 2011;**23**(6):2169-2183

[119] Wu X, He J, Chen J, Yang S, Zha D. Alleviation of exogenous 6-benzyladenine on two genotypes of

eggplant (*Solanum melongena* mill.) growth under salt stress. Protoplasma. 2014;**251**:169-176

[120] Toda Y, Tanaka M, Ogawa D, Kurata K, Kurotani KI, Habu Y, et al. Rice salt sensitive forms a ternary complex with JAZ and class-C bHLH factors and regulates jasmonate-induced gene expression and root cell elongation. The Plant Cell. 2013;**25**(5):1709-1725

[121] Qiu Z, Guo J, Zhu A, Zhang L, Zhang M. Exogenous jasmonic acid can enhance tolerance of wheat seedlings to salt stress. Ecotoxicology and Environmental Safety. 2014;**104**:202-208

[122] Kang DJ, Seo YJ, Lee JD, Ishii R, Kim KU, Shin DH, et al. Jasmonic acid differentially affects growth, ion uptake and abscisic acid concentration in salt-tolerant and salt-sensitive rice cultivars. Journal of Agronomy and Crop Science. 2005;**191**(4):273-282

[123] Filgueiras CC, Martins AD, Pereira RV, Willett DS. The ecology of salicylic acid signaling: Primary, secondary and tertiary effects with applications in agriculture. International Journal of Molecular Sciences. 2019;**20**:1-19

[124] Liang W, Ma X, Wan P, Liu L. Plant salt-tolerance mechanism: A review. Biochemical and Biophysical Research Communications. 2018;**495**:286-291

[125] Abogadallah GM. Antioxidative defense under salt stress. Plant Signal Behavior. 2010;**5**:369-374

[126] He AL, Niu SQ, Zhao Q, Li YS, Gou JY, Gao HJ, et al. Induced salt tolerance of perennial ryegrass by a novel bacterium strain from the rhizosphere of a desert shrub Haloxylon ammodendron. International Journal of Molecular Sciences. 2018;**19**(2):69

[127] An BY, Luo Y, Li JR, Qiao WH, Zhang XS, Gao XQ. Expression of a vacuolar Na^+/H^+ antiporter gene of alfalfa enhances salinity tolerance in transgenic Arabidopsis. Acta Agronomical Sinica. 2008;**34**:557-564

[128] Schroeder JI, Delhaize E, Frommer WB. Using membrane transporters to improve crops for sustainable food production. Nature. 2013;**497**:60-66

[129] Liang W, Cui W, Ma X. Function of wheat *Ta-UnP* gene in enhancing salt tolerance in transgenic Arabidopsis and rice. Biochemical and Biophysical Research Communications. 2014;**450**:794-801

[130] Sun SJ, Guo SQ, Yang X, Bao YM, Tang HJ, Sun H. Functional analysis of a novel Cys2/His2 type zinc finger protein involved in salt tolerance in rice. Journal of Experimental Botany. 2010;**61**:2807-2818

[131] Zhang L, Xi D, Li S, Gao Z, Zhao S. Shi J a cotton group *CMAP* kinase gene, *GhMPK2*, positively regulates salt and drought tolerance in tobacco. Plant Molecular Biology. 2011;**77**:17-31

Chapter 3

Underlying Mechanisms of Action to Improve Plant Growth and Fruit Quality in Crops under Alkaline Stress

Fabián Pérez-Labrada, José Luis Espinoza-Acosta, Daniel Bárcenas-Santana, Elizabeth García-León and Mari Carmen López-Pérez

Abstract

The high content of carbonates (CO_3^{2-}), bicarbonate (HCO_3^-), and high pH (>7.5) causes environmental pressure and alkaline stress, impairs plant growth and development, and limits fruit quality by causing osmotic alterations and hindering nutrient absorption. Because of alkaline stress, plants are in an oxidative environment that alters their metabolic processes, impairing their growth, development, and fruit quality. In response to this situation, plants use several mechanisms to cope, including the alteration of osmolytes, induction of transcription factors, signal transduction, hormone synthesis, alteration of the antioxidant system, and differential gene expression. Current knowledge and understanding of the underlying mechanisms that promote alkalinity tolerance in crops may lead to new production strategies to improve crop quality under these conditions, while ensuring food security.

Keywords: antioxidants, calcium, pH, oxidative stress, potassium, transcriptomic

1. Introduction

Natural alkalinity (associated with arid and semi-arid regions) or alkalinity induced by agricultural activity is serious because it affects food productivity [1, 2]. Of the total global cultivated land, this stress affects 30–37% (438 million ha) [3]. High carbonate (CO_3^{2-}) and bicarbonate (HCO_3^-) contents [4] promote the soil buffering capacity, inducing a high pH [5]. Natural highly alkaline conditions (pH ≥ 9) are associated with parts of Western Asia, Eastern Europe, southern South America, and southern Australia, with increasing incidence in India and Egypt as a consequence of irrigation by the Ganges River and the Nile (**Figure 1**) [6]. This alkaline environment impairs plant and fruit growth and quality by negatively altering the root system, modifying the nutrient uptake system, and altering the relative water, chlorophyll (Chl), and soluble sugar contents [7]. In addition, there is usually a reduction in the

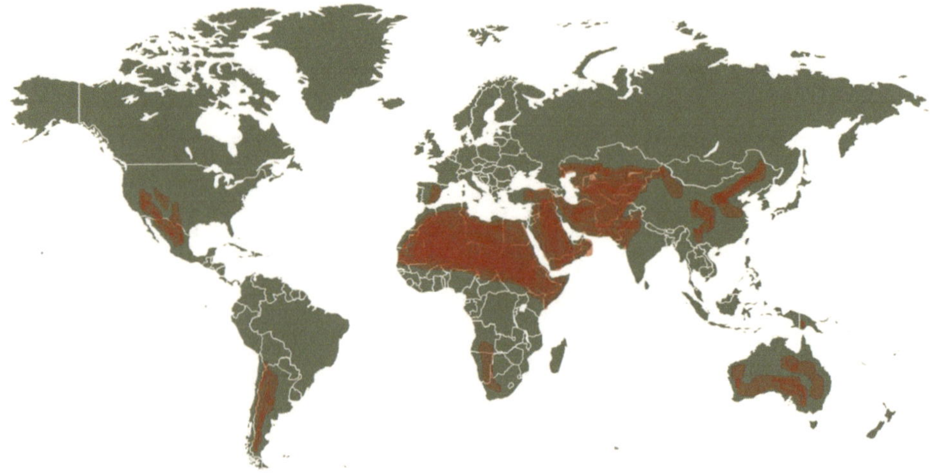

Figure 1.
Probable distribution of alkaline soils.

microbiological activity and bioavailability of mineral nutrients [5, 8]. Likewise, plants show damage to electron transport, and thus, photosynthetic activity [9]. Similarly, high pH induces the synthesis of reactive oxygen species (ROS) and malondialdehyde, which cause significant damage to the cell membrane and intracellular components [10].

Alkalinity has diversified impacts on plants, so the response and tolerance mechanisms are more complex [10]. Initially, the plant senses the stressful condition (elevated pH derived from carbonate content) via root apoplast, thanks to a sulfotyrosine of the *endogenous peptide root growth factor* (*RGF*) being deprotonated, destabilizing the *RGF-RGFR receptor-SERK* coreceptor complex. Considering that the target of *RGF* is a master regulator of root development (PLETHORA) [11], inhibition of this complex reduces root apical meristem growth [12, 13]. The same authors pointed out that this pH sensing system procures a growth or defense system, which significantly affects plant quality.

To combat this stress, plants develop mechanisms that allow them to cope with high pH, high carbonate content, ionic toxicity, and oxidizing environments. Initially, they promote an osmoregulatory system to reduce cellular Ψ [9] and maintain a high K^+/Na^+ ratio [14]. Likewise, it promotes the activity of enzymatic and non-enzymatic antioxidant systems to combat oxidative stress [15, 16] as well as the expression of hormones and a diversity of metabolites [17] derived from transcriptional modifications [18–21]. Alternatives in the management of crops developed in these environments include grafting [22], application of Ca^{2+}-based fertilizers (to replace Na^+ ions) [23], micronutrients [24], nanotechnology [25], biochar [26], microorganisms [27], growth regulators [28], and genetic engineering [29, 30].

In that sense, this study aimed to establish the main impacts of alkalinity on plants and to review the underlying mechanisms that could promote tolerance to this type of stress. It also highlights management alternatives that can potentiate crop quality in the face of environmental stress. This knowledge seeks to promote the development of innovative strategies to mitigate the negative impacts of alkalinity and safeguard the production of high-quality food.

2. Alkalinity environments

Alkaline stress in soils is characterized by a high pH (generally >7.5) caused by the dominance of $[HCO_3^-] + 2[CO_3^{2-}]$, salts, and carbonate alkalinity [8, 31]. These anions naturally originate from the dissolution of carbonate minerals from the parent material and by the combination of water and ($H_2O + CO_{2disolved} + CO_3^{2-} \leftrightarrow 2HCO_3^-$) which in turn can react with calcium (Ca^{2+}) in the soil: $Ca^{2+} + CO_3^{2-} \leftrightarrow CaCO_3$; $[Ca^{2+}][CO_3^{2-}]$ [8, 32]. Another pathway is high evaporation and low precipitation [33], which are common in arid and semi-arid areas. In soil solutions, when organic species are omitted, the alkalinity can be defined as $[HCO_3^-] + 2[CO_3^{2-}] + [H_2PO_4^-] + 2[HPO_4^-] + 3[PO_4^{3-}] + [B(OH)_4^-] + [OH^-] - [H^+]$ [8, 34]. Generally, soils can have between 0.4 and 3 mM HCO_3^-, a condition that increases and buffers the edaphic pH and alkalinizes the xylem medium [13]. However, high alkalinity (pH ≥ 9) can occur in soil solutions with $HCO_3^- + CO_3^{2-} > Ca^{2+} + Mg^{2+}$. Likewise, when sodium chloride (NaCl) salts react in the soil to release H^+, they tend to react with $CaCO_3$ to form sodium bicarbonate ($NaHCO_3$) or sodium carbonate (Na_2CO_3) salts. In addition to increasing exchangeable Na^+, it is important to mention that alkalinity is regulated by the concentrations of HCO_3^- and CO_3^{2-} and not by Na^+ [8, 35]. In this context, the increase in soil pH is associated with an increase in Na^+ concentration, an environment that reduces Ca^{2+} and magnesium (Mg^{2+}) cations, causing the dispersion of clays and loss of soil structure [8].

In water bodies, alkalinity is defined as $[HCO_3^-] + 2[CO_3^{2-}] + [OH^-] - [H^+]$ and is regulated by the dynamics of $CO_{2dissolved}$, carbonic acid (H_2CO_3) [$CO_2 + H_2O = H_2CO_3$], HCO_3^-, and CO_3^{2-}, as $H_2CO_3 \leftrightarrow H^+ + HCO_3^-$ and $HCO_3^- \leftrightarrow H^+ + CO_3^{2-}$, and CO_2 can react with $CaCO_3 + H_2O$ dissociating into $Ca^{2+} + 2HCO_3^-$ [36]. Therefore, water alkalinity is a determinant of agricultural production, as it can modify the pH of the substrate ($H_2O + CO_{2dissolved} + CO_3^{2-} \leftrightarrow 2HCO_3^-$), altering the growth, development, and quality of plants and fruits [37]. Although the negative effects of carbonates vary according to pH, the greatest damage by HCO_3^- occurs at pH 8–9 while CO_3^{2-} occurs at pH >9 [8]. When alkaline soils (with a sodic subsoil) present transient waterlogging or anoxic environments, the CO_2 concentration increases with increasing pH because of the increase in $HCO_3^- + CO_3^{2-} > Ca^{2+} + Mg^{2+}$ [8]. Waterlogging can induce toxicity in Mn, Fe, Al, and B.

Plants in alkaline environments are simultaneously exposed to salt action and elevated pH [9], causing osmotic stress, ionic toxicity, and elevated pH, which severely affect plant growth [1, 38]. High pH reduces cellular water content by altering the root system by reducing turgor pressure, which minimizes the extensibility processes of microfibrils and thus cell extension, water and nutrient uptake, and solute accumulation [39]. Similarly, the buffering capacity of HCO_3^- affects the solubility of essential nutrients, inducing deficits in iron (Fe^{2+}), manganese (Mn^{2+}), boron (B), zinc (Zn^{2+}), and phosphate (PO_4^{3-}) [8, 9, 40]. However, attenuation of alkaline effects in soil can be mitigated when pH < 9; soils with pH 8.5 or > 9 demand less H^+ to induce a reduction due to lower $CaCO_3$ content, which buffers the pH. In contrast, soils with pH ≈ 8.5 demand more H^+ because the concentration of $CaCO_3$ dissolves and buffers pH alteration [8, 41]. This is because $CaCO_3 \leftrightarrow Ca^{2+} + CO_3^{2-}$, and $CO_3^{2-} + H_2O \leftrightarrow HCO_3^- + OH^-$. Thus, one of the most feasible criteria for mitigating the negative effects of alkalinity is $HCO_3^- + CO_3^{2-} > Ca^{2+} + Mg^{2+}$, where one would ideally seek to reduce $HCO_3^- + CO_3^{2-}$ and maintain soil pH ≤ 8.5 [8].

3. Impact of alkalinity on agricultural crops and fruit quality

Alkaline stress has a significant impact on yield, plants, and fruit quality because it hinders the smooth uptake of nutrients [42]. The damage caused varies according to the type and phenological stage of the plant, time of exposure, and level of alkalinity of the medium; however, the main disorders are oxidative damage caused by alterations in ionic homeostasis and impairment of the photosynthetic process (**Figure 2a**) [43, 44].

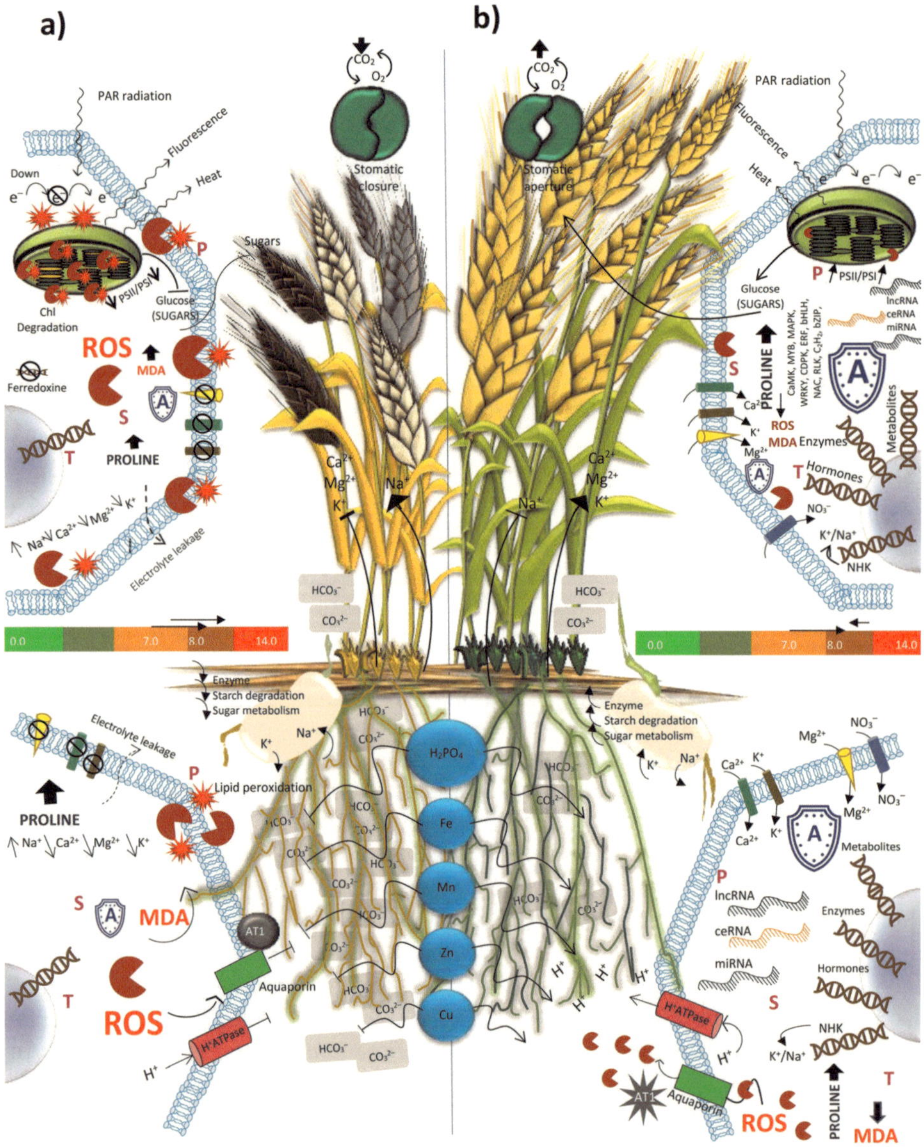

Figure 2.
Endogenous alterations (a) and response mechanisms (b) in plants under alkaline stress.

3.1 Process photosynthetic alterations

The photosynthetic capacity of plants under alkaline stress is negatively altered by reduced leaf area and Chl content [25, 45]. In *Medicago sativa* plants, Chl synthesis and photosystem activity (PSII/PSI) are altered, promoting non-photochemical quenching (NPQ) and cyclic electron transfer, and negative regulation of *PetA*, *PetB*, *petE*, and *petF* genes associated with the photosynthetic mechanism was observed [46]. A decrease in the maximum quantum yield of PSII photochemical parameters (F_v/F_m), PI, F_v, F_m, F_v/F_o, RC/ABS, and $(1\text{-}V_j)/V_j$, Chl fluorescence, and relative leaf water content in relation to $NaHCO_3$ levels during the vegetative and reproductive growth stages has been documented in strawberry plants. These conditions reduce fruit quality parameters, such as electrical conductivity, pH, total soluble solids, and color [9]. Likewise, a decrease in gas exchange rates (reduced transpiration, stomatal conductance, and CO_2 assimilation) can occur, reducing the maximum quantum yield of PSII primary photochemistry (φPo) and causing an increase in PSII energy dissipation and inhibiting PSII electron transport. This environment leads to overproduction and overaccumulation of ROS, mainly hydrogen peroxide (H_2O_2), hydroxyl ions ($OH^{\bullet}$), singlet oxygen ($^1O_2$) [47], and superoxide radicals (O_2^-). Increased concentrations of H_2O_2, $OH^{\bullet}$, $^1O_2$, and $O_2^{\bullet -}$ cause lipid peroxidation of the thylakoid and chloroplast membranes, degrading Chl and inhibiting the electron transport chain, leading to the collapse of the photosynthetic apparatus.

The growth, development, and quality of plants and fruits strongly depend on the synthesis of sugars (carbohydrate metabolism), such as hexose, glucose, and fructose, as they are the basis for the generation of metabolites involved in plant growth [48, 49]. The collapse of the photosynthetic process causes a reduction or no synthesis of sugars; therefore, alkalinity decreases the chemical characteristics of the fruit, which is cited in strawberry fruits under alkaline conditions, where the increase in $NaHCO_3$ concentration was associated with a decrease in the content of soluble sugars [9] coupled with intercellular ion imbalance [50]. Alkalinity stress decreases production, as well as total soluble solids and acidity, in Blanc Du Bois vines with their own roots, derived from the low nutritional status [51]. Similarly, reduction of the photosynthetic process by carbonate salts minimizes the synthesis of some pigments, such as anthocyanins, which may affect the quality of leaves, flowers, and fruits [9, 52]. In addition, the aforementioned soluble sugars participate in carbon stores and are critical for osmoregulation, osmotic adjustment, and ROS scavenging [14, 53].

3.2 Ionic disbalance

The affection in the ionic balance of the plant by alkalinity ($NaHCO_3$) is due to a differential modification in the absorption and storage of calcium (Ca^{2+}), potassium (K^+) and sodium (Na^+) or by the increase in pH that reduces the availability and absorption of phosphorus ($H_2PO_4^-$), magnesium (Mg^{2+}), K^+, iron (Fe), copper (Cu), zinc (Zn), and manganese (Mn) [25, 54–56]. For example, in oilseed rape (*Brassica napus* L.), inhibition of Ca^{2+} uptake in stems and leaves has been reported, which leads to ionic imbalance between K^+ and Na^+ [57].

Under 40 mM Na_2CO_3, a reduction in the growth of *B. napus* shoots was documented because of the high pH and pH-salt ion interaction, showing a high K^+ concentration and K^+/Na^+ ratio in leaves as well as an increase in Ca^{2+} and Mg^{2+} in roots [58]. In blueberry "Jewel" (*Vaccinium corymbosum*) plants, a reduction in H^+ flux from the root was documented, which may be associated with inhibition of the *VcHA* gene encoding H^+-ATPases derived from the alkaline condition, pH 8.0 [59]. In strawberry plants, increased Na^+ absorption was observed, accompanied by a decrease in K^+, Ca^{2+}, Mg^{2+}, and Fe, which negatively affected fruit color [47]. In *Ricinus communis* L. seeds exposed to 50, 100, and 150 mM $NaHCO_3$, the germination rate was reduced by an increase in Na^+ content and reduction in K^+ [60].

3.3 Oxidative stress

The burst in ROS production and concentration is a generalized response in plants subjected to alkaline stress, which causes lipid peroxidation, increases malondialdehyde (MDA) concentration, and electrolyte leakage [16, 28, 61], creating an oxidative environment [56] that can damage plant genetic material by arresting the cell cycle in the root [14]. This oxidative environment can reduce the germination process and root development in seedlings [62]. In this regard, inhibition of the germination power of *Hordeum vulgare* L. under alkaline stress has been cited, decreasing seedling vigor, which may be due to starch degradation, suppression of sugar metabolism, and reduction of enzymes involved in the Embden-Meyerhof-Parnas process, tricarboxylic acid cycle, pentose phosphate pathway, and the enzymes hexokinase and malic dehydrogenase [63].

In maize, increased lipid peroxidation, higher Na^+ content, and elevated H_2O_2 levels were documented, which is in agreement with the reduction of K^+ and Chl, decreased shoot and root fresh weight, root dry weight, plant length, number of leaves, and maximum leaf width, when grown under $NaHCO_3$:Na_2CO_3 (9:1) [64]. The oxidative environment prevailing in plants under alkaline stress can significantly reduce the level of phosphatidylcholine (the main phospholipid of the cell membrane) in addition to inducing alterations in the expression of non-specific phospholipase A and phospholipase D/phospholipase C encoding genes [65]. This phenomenon can lead to marked deterioration in plant and fruit quality.

Under alkaline conditions (Na_2CO_3 75 mM), *Trifolium alexandrinum* L. showed oxidative damage by H_2O_2, $O_2^{\bullet-}$, methylglyoxal, MDA, and oxidized glutathione, reducing shoot and root length, as well as Chl content, water, and nutrient uptake [54], reducing plant quality. In *Onobrychis viciaefolia*, there was an increase in MDA derived from the generation of ROS (mainly $O_2^{\bullet-}$ and H_2O_2), accompanied by an increase in proline content. This environment reduces fresh and dry biomass, Chl content, K^+ concentration, K^+/Na^+ ratio, and fructose and glucose contents [66]. This stress also affects fruit quality; it decreases nucleic acids, lipids, hydroxycinnamoyl derivatives, organic acids and derivatives, and vitamins in wolfberry (*Lycium barbarum*) under saline-alkaline conditions [67].

3.4 Enzyme activity and transcriptomic responses

Under alkaline stress, radical cells are severely damaged by the accumulation of ROS (mainly $O_2^{\bullet-}$ and H_2O_2), which inhibit plant growth. Although this condition increases the activation of antioxidant enzymes such as superoxide dismutase (SOD, EC 1.15.1.1), catalase (CAT, EC 1.11.1.6), peroxidase (POD, EC 1.11.1.7), ascorbate peroxidase (APX, EC 1.11.1.11), ascorbic acid (AsA), and glutathione (GSH), their synthesis is overcome by excess ROS. This phenomenon was associated with the positive expression of cell death-associated genes *OsKOD1*, *OsHsr203j*, *OsCP1*, and *OsNAC4* and negative regulation of the cell death suppressor gene (*OsBI1*) [68].

Suppression of enzyme activity implies a potential reduction in plant and fruit growth, development, and quality, although this response is cultivar- and alkalinity level-dependent. In mulberry seedlings, oxidative stress ($O_2^{\bullet-}$ and H_2O_2) induced by $NaHCO_3$ (100 mM) reduced CAT activity and CAT gene expression (*W9RJ43*), but improved POD and the AsA-GSH acid cycle, allowing the elimination of H_2O_2. The genes involved in the expression of the electron donor ferredoxin-thioredoxin reductase (*FTR*) and proteins such as thioredoxin M (*TrxM*), thioredoxin M4 (*TrxM4*), thioredoxin X (*TrxX*), *TrxF*, and *Trx CSDP* decreased their expression, that is, there was an inhibition of the thioredoxin-peroxiredoxin (*Trx-Prx*) elimination pathway, which did not decrease oxidative damage [69]. Higher POD, SOD, and CAT activities have been reported in maize plants [70].

Transcriptomic regulation related to the oxidation-reduction process and lipid metabolism, as well as the expression of genes encoding D-3-phosphoglycerate dehydrogenase 1, glutamyl-tRNA reductase 1, fatty acid hydroperoxide lyase, ethylene-insensitive protein 2, metal tolerance protein 11, and magnesium-chelatase subunit ChlI, are altered in sugar beets under this stress [10]. These transcriptional modifications can significantly reduce crop quality; however, some red cherry tomato varieties susceptible to alkaline soils (pH 7.9) exhibit acceptable fruit quality [71].

4. Plant mechanism against alkalinity stress

Plants sense alkaline environments through the root, where the sulfotyrosine of the endogenous peptide *RGF* is deprotonated and destabilizes the *RGF-RGFR receptor-SERK* receptor complex, causing alterations in the root apical meristem [12, 13]. Thus, *RGF* is a pH sensor that reacts to H^+ concentration in the medium, which could be the first signaling mechanism for plants under alkaline stress. Considering that this stress impacts plants in a multifaceted manner, plant response mechanisms are complex [10]. However, as shown in **Figure 2b**, the tolerance systems developed by plants can be generalized to osmoregulation, induction of the antioxidant defense system, and metabolites through transcriptomic modifications.

4.1 Osmoregulation

As previously mentioned, alkaline environments cause alterations in ionic homeostasis, and plants can extrude hydrogen ions (H^+) into the rhizosphere by means of proton pumps associated with the root plasma membrane (PM H^+-ATPases - HAs) acidifying the rhizosphere zone, allowing nutrient uptake [59] to achieve an adequate ionic balance. Although Na^+ can enter through non-selective K^+ channels, the latter (K^+) tends to be translocated to the cytoplasm to generate osmotic regulation [54], thereby reducing toxicity and generating an adequate flow of water and nutrients [72].

In canola roots under alkaline conditions (Na_2CO_3, 40 mM), the concentrations of K^+, Ca^{2+}, and Mg^{2+} are elevated, similar to the K^+/Na^+ ratio in the leaves [58]. In alkaline soils, Mn deficiency triggers the kinase pathway (*CPK21* and *CPK23*) and increases Ca^{2+} uptake mediated by the *NRAMP1* transporter [73], safeguarding the cation balance. A good ionic supply and K^+/Na^+ regulation can improve transpiration, Chl content, PSII quantum efficiency, relative water content, and water saturation [74], in addition to photosynthetic indicators [18].

Another osmoregulation alternative can be the increase in sugar concentration, in this regard it is mentioned that in fruits of Jujube "Junzao" irrigated with a moderate alkaline solution (90 mM of NaCl:$NaHCO_3$ (3:1)), an increase in sucrose, glucose, and fructose content was observed, in addition to an increase in sucrose phosphate synthase (SPS) and sucrose synthase (SS-I) activities [53] allowing a partial attenuation of stress and maintaining plant and fruit quality. In addition to the accumulation of soluble sugars [14], proline is associated with osmoregulatory mechanisms in plants under alkaline conditions [46].

4.2 Stimulation of antioxidant biocompounds

One of the mechanisms developed by plants to census and scavenge ROS under alkaline stress conditions is the promotion of the antioxidant system by SOD, POD, CAT, APX, glutathione reductase (GR, EC 1.6.4.2), dehydroascorbate reductase (DHAR, EC 1.8.5.1), and monodehydroascorbate reductase (MDHAR, EC 1.6.5.4) [16], as well as phenolic acids and flavonoids. Increased SOD, CAT, POD, and APX enzyme activities are in line with higher levels of SOD, CAT, POD, and APX gene expression [64]; however, this mechanism only operates when the ROS burst is below the enzyme response threshold and in line with osmotic regulation. For example, some cultivars of broom millet show tolerance to alkaline stress (80 and 40 mM) because of high antioxidant enzyme activity that reduces MDA concentration, as well as a high content of soluble osmolytes, which reduces electrolyte leakage [15]. Similarly, in soybeans grown in hydroponics ($NaHCO_3$ and Na_2CO_3; Na^+ from 25 to 50 mM), activation of the antioxidant enzymes SOD, POD, and APX was found [14]. In contrast, rapeseed line 2205 (*B. napus*) showed reduced plasma membrane damage due to osmoregulation and higher enzyme activity [75]. In the case of autotetraploid rice 93-11 T grown on saline-alkali soil, tolerance was found to be derived from soluble sugar accumulation and increased SOD and POD activity in leaves [17]. It appears that the level of polyploidization in rice may trigger alkaline tolerance by stimulating peroxide-associated genes, transcription factors [76], and possibly the antioxidant system.

Another type of compound that promotes ROS attenuation is phenolic acids, polyphenols, flavonoids, and alkaloids; in grapes, their content increases under alkaline conditions [16]. In addition to these compounds, some essential oils [25], amino acids, and organic acids (in *Lycium barbarum* L. berry) promote regulatory mechanisms [67]. In alkalinity-tolerant *Arabidopsis thaliana*, an increase in methionine-derived glucosinolates via jasmonate and salicylate has been reported [77]. Likewise, salicylic acid (SA) can increase alkalinity by promoting signaling mechanisms [17].

4.3 Transcriptomic regulation

Plants under alkaline stress can exhibit alterations in signaling and response processes through a complex transcriptomic network. In sugar beet (*Beta vulgaris* L.), an alkaline salt (Na_2CO_3) attenuating mechanism, differential expression of genes related to cutin, suberin, wax biosynthesis, sesquiterpenoid and triterpenoid biosynthesis, flavonoid biosynthesis and monoterpenoid biosynthesis, amino acid metabolism, and starch and sucrose metabolism have been documented [18]. Histologically, *B. napus* showed more integrated mesophyll cells as well as higher expression of genes associated with carbohydrate metabolism, photosynthetic processes, and ROS regulation [75].

Inhibition of ion uptake could be a mechanism used by plants to signal alkaline environments through the differential expression of genes related to hormone signal transduction, mitogen-activated protein kinase (MAPK) signaling, phenylpropanoid biosynthesis, and ATP-binding cassette (ABC) transport. Although phenylpropanoid and flavonoid biosynthesis pathways are exacerbated, the accumulation of metabolites varies according to the level of alkalinity in the environment [78]. Likewise, there may be positive regulation of specific root transporters (e.g., *NTR1*, *NRT2.1*, and *NRT2.5*) [79].

Surprisingly, alkalinity of alfalfa (*Medicago truncatula*) generates a redistribution of nutrients, a mechanism controlled by the differential expression of 985 genes associated with nutrient cycling, cellular amino acid metabolic, and catabolic processes of organic substances, and 1180 genes associated with the oxidoreductase complex, aerobic respiration, and ion transport. In addition to several transcription factors, *MYB*, *WRKY*, *bHLH*, and *NAC* are involved in stress resistance and senescence regulation [80]. In contrast, recombinant inbred lines (RILs) of rice showed tolerance to alkalinity due to a low Na^+/K^+ ratio derived from a reduction in Na^+ accumulation and increased K^+ accumulation in roots and shoots. Tolerance may also be due to the expression of expressed protein genes, glucan endo-1,3-beta-glucosidase precursor, F-box domain-containing proteins, double-stranded RNA-binding motif-containing proteins, aquaporins, receptor kinase-like proteins, semialdehyde, hydrogenase, and NAD-binding domain-containing protein genes [81].

Sodium-hydrogen exchangers (*NHX*) are fundamental regulators of cellular Na^+/K^+ and pH homeostasis. In soybean, the expression of the *GmNHX6* gene encodes a *sodium/hydrogen exchanger (NHX)* in the Golgi apparatus via a single nucleotide polymorphism in the NHX6 promoter region. In *A. thaliana*, the expression of *GmNHX6* enhances tolerance to alkaline salts derived from a high K^+ content and low Na^+/K^+ ratio [31, 82]. In contrast, the Gγ protein seems to regulate sensitivity to alkalinity, with the *alkaline tolerance 1 (AT1)* locus, which encodes the γ subunit of the atypical G protein. In this context, a model for the regulation of crop alkaline tolerance mediated by the *AT1* subunit of the G protein is proposed because under alkaline stress conditions, artificially or naturally removing *AT1*, as well as its homologs, modulates the phosphorylation of *PIP2*, which allows the outflow of H_2O_2, reducing its cellular concentration, improving tolerance, and promoting crop yield [61]. Several aquaporins can interact with *AT1*, including *SbPIP2;1* with *SbAT1* (sorghum) and *OsPIP2* with *1-OsGS3*, a homolog of *AT1* (rice). These findings indicate a vital role for aquaporins in ROS regulation. Thus, *SbAT1*, which encodes the γ subunit of the G protein, inhibits aquaporin phosphorylation, thus regulating H_2O_2 concentration and levels under alkaline stress [83]. This information applies to monocotyledonous crops; therefore, future research should be conducted on dicotyledonous plants. Similarly, when *TaAT1* was modified, higher wheat yields were observed in alkaline soils [84]. In sorghum plants tolerant to alkaline stress, *ATP1* (*alkaline tolerance 1*) encodes the G protein by negatively modulating the phosphorylation level of PIP2 (aquaporin with H_2O_2 transport activity), that is, when *ATP1* is eliminated, PIP2 is phosphorylated from the cell membrane, allowing greater transport of H_2O_2 from the cytosol to the apoplast, reducing oxidative stress [85].

Oilseed rape grown in alkalinized soil (Na_2CO_3; 8g kg^{-1}) shows strong genetic regulation associated with ionic K^+ transport as well as alterations in the amino acid metabolic pathway, leading to reduced Na^+, MDA, and relative conductance (REC), increased K^+ and Mg^{2+}, and the enzymes SOD, POD, and CAT. Interestingly, the positive regulation of phenylalanine ammonia lyase (PAL) transcription factors shows greater importance in mitigating alkaline stress [86]. Similarly, higher nitrogen content, soluble sugar, proline, and higher SOD and POD activities have been reported in plants under alkaline conditions [79].

In soybean (*Glycine max*, cultivar Talar), exposure to alkalinity (50 mmol L^{-1} $NaHCO_3$) affected the transcriptomic level associated with abiotic stress response processes, metabolic regulation, response to cellular stimuli, and signaling, highlighting 321 overexpressed genes [87]. *A. thaliana* is sensitive to alkalinity and induces genes associated with Fe deficiency in leaves [77]. Molecular regulation enhances the defense system and osmotic adjustment, by modifying genes associated with ubiquitin-mediated proteolysis, glycolysis/gluconeogenesis pathways, transcription factors (*C2H2*, *bZIP*, and *bHLH*), and protein kinases CaMK (calcium/calmodulin-dependent protein kinase), CDPK (calcium-dependent protein kinase), MAPK (mitogen-activated protein kinase), and RLK (receptor-like kinase) [88]. In wheat plants, approximately 20 genes are associated with salt stress control, abiotic stress tolerance, response to oxidative stress, and response to osmotic stress, including *TraesCS2B02G276500*, *TraesCS2B02G504900*, and *TraesCS2B02G276300* [55].

Hormone content may also be regulated in response to alkalinity; cotton plants differentially express 18,230 and 11,177 genes in the root and leaf tissue, respectively. These genes are associated with plant hormone signal transduction, amino acid biosynthesis, and secondary metabolites, the latter of which is related to ROS scavenging [89]. In broomcorn millet (*Panicum miliaceum* L.), there is negative regulation of gibberellic acid (GA) synthesis genes and positive regulation of GA inactivation and abscisic acid (ABA) synthesis genes; therefore, resistant genotypes have higher GA/ABA ratios, in addition to having a more robust antioxidant system [90]. Similarly, polyploid rice plants under high salt–alkali stress conditions (50 mM Na_2CO_3, pH = 11.39) showed positive hormone regulation, particularly of genes associated with auxin (*AUX*), SA signal transduction (roots), lignin biosynthesis (leaves or roots), and wax biosynthesis (leaves) [17].

Some plant 3-ketoacyl-CoA synthase gene families (*KCS; GBKCS3*, *GBKCS8*, *GBKCS20*, *GBKCS34*) may be involved in hormonal signaling, defense, and stress responses that allow *Gossypium barbadense* to improve its fiber quality [91]. In lentil plants grown in alkaline soil, an increased number of genes related to ABA signaling (dehydrin 1, 9-cis-epoxycarotenoid dioxygenase, ABA-responsive protein 18, and BEL1-like homeodomain protein 1) as well as increased synthesis of secondary metabolites were found [20], which could promote a higher mitotic rate, allowing the organization of the internal structures (root and shoot) of the plant, avoiding the disintegration of the internal structure and the distortion of the root and shoot architecture. This environment allows for an increase in K^+ accumulation, osmolytes, antioxidant enzymes, relative water content, membrane stability index, and abscisic acid. In canola, tolerance to alkaline salts is mediated by modifications in Ca^{2+} signaling, ABA, and ROS pathways, which together promote gene regulation, organic acid synthesis (osmoregulators), and the transcription factors *bHLH*, *WRKY*, *ERF*, *MYB*, and *NAC* [92]. In the cereal *Chenopodium quinoa* Willd, the differential expression of genes associated with ROS homeostasis, hormone signaling (auxin and ethylene), cell wall synthesis (hemicellulose), and transcription factors has been documented [93].

Metabolite accumulation is a plant response to alkaline events; however, these transcriptomic alterations can reduce plant production and fruit quality [28]. In this regard, canola plants stressed with Na_2CO_3 showed accumulation of L-proline, L-glutamate, L-histidine, L-phenylalanine, L-citrulline, L-tyrosine, L-saccharopine, L-tryptophan, linoleic acid, dihomo gamma linolenic acid, alpha linolenic acid, Eric acid, oleic acid, neuronic acid, sucrase, alpha, alpha trehalose, and polyol (ribitol), associated with a decrease in UDP-D-galactose, D-mannose, D-fructose, and D-glucose 6-phosphate [21]. In broom millet, differential accumulation of volatile and non-volatile metabolites has been reported, namely phenylpropanoids, flavonoids, flavones, flavanols, valine, leucine, isoleucine, arginine, proline, tryptophan, and ascorbate [78]. Under alkaline conditions (50, 100, and 150 mM $NaHCO_3$), castor seeds promoted between 31 and 113 differential metabolites and 138–694 genes associated with amino acid, carbohydrate, lipid, and nucleotide metabolism. In this sense, castor seeds in alkaline conditions improve amino acid metabolism and the tricarboxylic acid cycle, allowing the regulation of ROS levels and promotion of osmotic balance [60].

Alkalinity induced higher expression of *GmABF4*, *GmAAO3*, *GmABI5*, *GmNCED5*, and *GmNCED9* genes in soybean [62]; however, ribonucleic acid editing in rice chloroplast and mitochondria mRNA increased under alkaline stress; four types of RNA editing were notable—A-G (A-I), C-T (C-U), G-A, and T-C. Pentatricopeptide repeats (*PPR*) and organelle zinc-fingers (*OZI*) are involved in various biological functions and molecular processes. This phenomenon may contribute to amino acid changes in genes, which could lead to tolerance to this type of stress [94].

In *Glycine max* (L.) Merri, the gene coding for DNAJ chaperone proteins (coding for *GmDNAJC7* gene) showed overexpression allowing stress tolerance; higher Chl content and relative water content [82]. In soybean, the gene associated with chaperone proteins *GmDnaJA6* seems to coordinate resistance to alkaline stress; likewise, *DnaJ* interacting with *Hsp70* and *GmHsp70* generates greater tolerance to alkaline stress (100 mmol L^{-1} $NaHCO_3$) [95]. For *Gossypium hirsutum* under alkaline conditions, differential transcriptional regulation is associated with carbohydrate metabolism and the oxidation-reduction process (stage SS48), while in roots, there is an alteration of the oxidation-reduction process (stages SS3, SS12, and SS48). Specifically, the *GhSOS3/GhCBL10-SOS2* network is notoriously related to the response to this type of stress [96]. Finally, in apples, overexpression of the *MdATG18a* gene under alkaline conditions provides tolerance to alkaline stress by increasing biomass, photosynthetic rate, and antioxidant capacity. Similarly, an increase in glutamate promotes the derivation of γ-aminobutyric acid. Similarly, upregulation of *MdATG18a* increases autophagosomes, allowing the recovery of small molecules from substances damaged by alkaline stress [2].

Recently, it has been reported that long non-coding RNAs (lncRNAs) can act as competitive endogenous RNAs (ceRNAs) that participate in the response to alkaline stress. Rice plants showed alterations in differentially expressed lncRNAs, miRNAs, and mRNA (75, 34, and 1725, respectively). Likewise, the triplet NONOSAT000455-osa_miR5809b-LOC_Os11g01210 was found to be a central node in the response to saline-alkali stress. These lncRNAs act as competitive endogenous RNAs (ceRNAs) and constitute a novel set of genetic controls under saline-alkaline stress conditions [97]. It is important to note that the response to this type of stress may exhibit differences according to the level of polyploidization and expression of transcription factors (phytohormone pathways, salt resistance, signal transduction, and physiological and biochemical metabolism) [98].

5. Alkalinity management strategies

Several alternatives have been studied to reduce alkali stress damage in crops. The use of grafting ameliorates this stress by enhancing the desired characteristics of the rootstock and improving the Ca, Mg, P, S, Cu, Fe, and Mn contents [22]. The use of rootstocks can increase the total soluble solids and nutritional status of grapevine fruits grown in alkaline environments [51]. In contrast, the use of light-emitting diodes in monochromatic blue (460 nm), monochromatic red (660 nm), and dichromatic blue/red (1:3) light at 200 μmol m^{-2} s^{-1} enhances electron transport per reaction center and the probability of transferring electrons from the exciton into the electron transport system, promoting gas exchange in alkalinity-stressed strawberry plants [47, 99].

The efficient use of fertilizers (macro- and micronutrients) can ameliorate the negative effects of alkalinity by stimulating plant and soil enzymatic activity, improving nutrient content, microbiological activity, total Chl, protein [26, 42, 100], antioxidant activity [70], ion homeostasis (K^+/Na^+) [57], photosynthetic activity [24, 101, 102], and gene expression [103]. Potassium-ferrite-nanocoated diammonium phosphate [104], organic polymer composite materials [105], carbon nanodots [106], zinc nanocomplexes [25], and nano-Ca^{2+} [107] can ameliorate alkalinity damage. Similarly, biochar combined with daily fertigation in alkaline soil can improve cucumber yield [101]. Similarly, microorganisms are another window of action to mitigate alkalinity damage, as they increase antioxidant enzymes and modify plant gene expression [108], regulate enzyme activity and soil microbial load [27, 109], decrease pH [110], and regulate soil ionic balance [111, 112], improving nutrient availability [113]. All of these management techniques improve plant productivity and promote better fruit quality by increasing the total soluble solids, acid ratio, sugar, and phenol content [42, 114].

Some compounds, such as spermine [115], taurine [54], auxin (IAA, indole-3-acetic acid)/TIBA (2,3,5-triiodobenzoic acid, which inhibits the transport of auxins) [116], hormone-like growth regulator (AUT) [117], yeast extract and gibberellic acid (GA) [118], ABA [119], strigolactone [62], and brassinosteroid [28] improve plant development under alkaline conditions by inducing alterations at the transcriptomic level, promoting Chl content, photosynthetic rate, enzymatic activity, and reduction of prooxidant compounds. Finally, transgenic plants can show stress tolerance [29, 30, 56, 61, 83, 84, 120, 121], which is a further strategy to promote quality foods under alkaline conditions. The use of vinas and molases favors yield, total soluble solids, and total acidity in strawberries grown in calcareous soils [122]. In addition, several genotypes (VRO-112, VRO-110, Kashi Kranti, VROB-178, AE-70, and VRO-108) can tolerate alkaline conditions (pH 8–9.5), thereby increasing yield [123].

6. Conclusions and outlook

Alkalinity stress as an environmental stress, which is associated with high pH and high CO_3^{2-} and HCO_3^- content, causes considerable decreases in crop quality. To attenuate stress, plants censor the edaphic environment and promote osmotic regulation (soluble sugars and proline), ionic homeostasis (K^+), activation of secondary messengers, differential gene expression (associated with the antioxidant system), and accumulation of secondary metabolites. Such mechanisms may promote tolerance to alkalinity by reducing the burst and accumulation of ROS, consistent with the high efficiency of the photosynthetic process. Plant type, phenological stage, and the level and time of exposure to an alkaline environment are fundamental to the response to this stress.

The climatic alterations of the last decades, the excessive use of fertilizers, and poor management of water use can increase alkalinity stress, so it is important to understand the underlying mechanisms that plants develop to cope with this stress. Future research should be conducted to mitigate the negative effects of alkaline salts and produce high-quality crops in a sustained manner.

Acknowledgements

The authors are highly thankful to the researchers for their findings in this area, which served directly or indirectly in the preparation of this manuscript.

Funding

The authors received no direct funding for this study.

Conflict of interest

The authors declare no conflicts of interest.

Data availability

All the data generated are included in this article reference's part.

Author details

Fabián Pérez-Labrada[1], José Luis Espinoza-Acosta[2], Daniel Bárcenas-Santana[2], Elizabeth García-León[3] and Mari Carmen López-Pérez[2]*

1 Universidad Autónoma Agraria Antonio Narro, Saltillo, Coahuila, Mexico

2 Universidad Estatal de Sonora, Navojoa, Sonora, Mexico

3 Instituto Nacional de Investigación Forestal, Agrícola y Pecuaria, Los Mochis, Sinaloa, Mexico

*Address all correspondence to: mari.lopez@ues.mx; mcloper@outlook.com

References

[1] Liu X, Liang W, Xing Li Y, Jun Li M, Quan Ma B, Hai Liu C, et al. Transcriptome analysis reveals the effects of alkali stress on root system architecture and endogenous hormones in apple rootstocks. Journal of Integrative Agriculture. 2019;**18**(10): 2264-2271. DOI: 10.1016/S2095-3119(19) 62706-1

[2] Li Y, Liu C, Sun X, Liu B, Zhang X, Liang W, et al. Overexpression of MdATG18a enhances alkaline tolerance and GABA shunt in apple through increased autophagy under alkaline conditions. Tree Physiology. 2020; **40**(11):1509-1519. DOI: 10.1093/ treephys/tpaa075

[3] Wei WF, Wang C, Sun Y, Wang N, Wei LX, Yuan DY, et al. Overexpression of vacuolar proton pump ATPase (V-H^+-ATPase) subunits B, C and H confers tolerance to salt and saline-alkali stresses in transgenic alfalfa (*Medicago sativa* L.). Journal of Integrative Agriculture. 2016; **15**(10):2279-2289. Available from: https://linkinghub.elsevier.com/retrieve/ pii/S209531191661399-0

[4] Singh K, Trivedi P, Singh G, Singh B, Patra DD. Effect of different leaf litters on carbon, nitrogen and microbial activities of sodic soils. Land Degradation and Development. 2016; **27**(4):1215-1226. Available from: https:// onlinelibrary.wiley.com/doi/full/ 10.1002/ldr.2313

[5] Liu J, Tang L, Gao H, Zhang M, Guo C. Enhancement of alfalfa yield and quality by plant growth-promoting rhizobacteria under saline-alkali conditions. Journal of the Science of Food and Agriculture. 2019;**99**(1): 281-289. Available from: https:// onlinelibrary.wiley.com/doi/full/ 10.1002/jsfa.9185

[6] Jobbágy EG, Tóth T, Nosetto MD, Earman S. On the fundamental causes of high environmental alkalinity ($pH \geq 9$): An assessment of its drivers and global distribution. Land Degradation and Development. 2017;**28**(7):1973-1981. Available from: https://onlinelibrary. wiley.com/doi/full/10.1002/ldr.2718

[7] Damodaran T, Mishra VK, Jha SK, Pankaj U, Gupta G, Gopal R. Identification of rhizosphere bacterial diversity with promising salt tolerance, PGP traits and their exploitation for seed germination enhancement in sodic soil. Agricultural Research. 2019;**8**(1):36-43. Available from: https://link.springer. com/article/10.1007/s40003-018-0343-5

[8] Rengasamy P, De Lacerda CF, Gheyi HR. Salinity, sodicity and alkalinity. In: Subsoil Constraints for Crop Production. Berlin, Heidelberg, Germany: Springer International Publishing; 2022. pp. 83-108. Available from: https://link.springer.com/chapter/ 10.1007/978-3-031-00317-2_4

[9] Malekzadeh Shamsabad MR, Roosta HR, Esmaeilizadeh M. Responses of seven strawberry cultivars to alkalinity stress under soilless culture system. Journal of Plant Nutrition. 2021; **44**(2):166-180. Available from: https:// www.tandfonline.com/doi/abs/10.1080/ 01904167.2020.1822401

[10] Zou C, Liu D, Wu P, Wang Y, Gai Z, Liu L, et al. Transcriptome analysis of sugar beet (*Beta vulgaris* L.) in response to alkaline stress. Plant Molecular Biology. 2020;**102**(6):645-657. Available from: https://link.springer.com/article/ 10.1007/s11103-020-00971-7

[11] Galinha C, Hofhuis H, Luijten M, Willemsen V, Blilou I, Heidstra R, et al. PLETHORA proteins as dose-dependent

master regulators of Arabidopsis root development. Nature. 2007;**449**(7165): 1053-1057. Available from: https://www.nature.com/articles/nature06206

[12] Liu L, Song W, Huang S, Jiang K, Moriwaki Y, Wang Y, et al. Extracellular pH sensing by plant cell-surface peptide-receptor complexes. Cell. 2022;**185**(18): 3341-3355.e13. Available from: http://www.cell.com/article/S0092867422009151/fulltext

[13] Vélez-Bermúdez IC, Schmidt W. Plant strategies to mine iron from alkaline substrates. Plant and Soil. 2023; **483**(1–2):1-25. Available from: https://link.springer.com/article/10.1007/s11104-022-05746-1

[14] Wang G, Shen W, Zhang Z, Guo S, Hu J, Feng R, et al. The effect of neutral salt and alkaline stress with the same Na^+ concentration on root growth of soybean (*Glycine max* (L.) Merr.) seedlings. Agronomy. 2022;**12**(11):2708. Available from: https://www.mdpi.com/2073-4395/12/11/2708/htm

[15] Ma Q, Wu C, Liang S, Yuan Y, Liu C, Liu J, et al. The alkali tolerance of broomcorn millet (*Panicum miliaceum* L.) at the germination and seedling stage: The case of 296 broomcorn millet genotypes. Frontiers in Plant Science. 2021;**12**:711429. DOI: 10.3389/fpls.2021.711429

[16] Lu X, Chen G, Ma L, Zhang C, Yan H, Bao J, et al. Integrated transcriptome and metabolome analysis reveals antioxidant machinery in grapevine exposed to salt and alkali stress. Physiologia Plantarum. 2023; **175**(3):e13950. Available from: https://onlinelibrary.wiley.com/doi/full/10.1111/ppl.13950

[17] Zhang C, Meng W, Wang Y, Zhou Y, Wang S, Qi F, et al. Comparative analysis of physiological, hormonal and transcriptomic responses reveal mechanisms of saline-alkali tolerance in autotetraploid rice (*Oryza sativa* L.). International Journal of Molecular Sciences. 2022;**23**(24):16146. Available from: https://www.mdpi.com/1422-0067/23/24/16146

[18] Geng G, Li R, Stevanato P, Lv C, Lu Z, Yu L, et al. Physiological and transcriptome analysis of sugar beet reveals different mechanisms of response to neutral salt and alkaline salt stresses. Frontiers in Plant Science. 2020;**11**:571864. DOI: 10.3389/fpls.2020.571864

[19] Ma Q, Wang H, Wu E, Yuan Y, Feng Y, Zhao L, et al. Comprehensive physiological, transcriptomic, and metabolomic analysis of the response of *Panicum miliaceum* L. roots to alkaline stress. Land Degradation and Development. 2023;**34**(10):2912-2930. Available from: https://onlinelibrary.wiley.com/doi/full/10.1002/ldr.4656

[20] Singh D, Singh CK, Taunk J, Gaikwad K, Singh V, Sanwal SK, et al. Linking genome wide RNA sequencing with physio-biochemical and cytological responses to catalogue key genes and metabolic pathways for alkalinity stress tolerance in lentil (*Lens culinaris* Medikus). BMC Plant Biology. 2022; **22**(1):1-17. Available from: https://bmcplantbiol.biomedcentral.com/articles/10.1186/s12870-022-03489-w

[21] Wang W, Pang J, Zhang F, Sun L, Yang L, Zhao Y, et al. Integrated transcriptomics and metabolomics analysis to characterize alkali stress responses in canola (*Brassica napus* L.). Plant Physiology and Biochemistry. 2021;**166**:605-620. DOI: 10.1016/j.plaphy.2021.06.021

[22] Ulas F, Aydın A, Ulas A, Yetisir H. The efficacy of grafting on alkali stressed

watermelon cultivars under hydroponic conditions. Gesunde Pflanzen. 2021; **73**(3):345-357. Available from: https://link.springer.com/article/10.1007/s10343-021-00559-1

[23] Machado RMA, Serralheiro RP. Soil salinity: Effect on vegetable crop growth. Management practices to prevent and mitigate soil salinization. Horticulturae. 2017;**3**(2):30. Available from: https://www.mdpi.com/2311-7524/3/2/30/htm

[24] Nandita K, Kundu M, Rakshit R, Nahakpam S. Effects of foliar application of micronutrients on growth, yield and quality of sweet orange (*Citrus sinensis* (l.) Osbeck). Bangladesh Journal of Botany. 2022;**51**(1):57-63. Available from: https://www.banglajol.info/index.php/BJB/article/view/58821

[25] Ghodrati-Tazangi MJ, Samani RB, Tavallali V, Alizadeh A, Honarvar M. Responses of *Hyssopus officinalis* to bicarbonate stress and foliar application of green synthesized zinc nano-complex formed on *Medicago sativa* extract. Scientia Horticulturae. 2023;**319**:112197. DOI: 10.1016/j.scienta.2023.1121197

[26] He K, Xu Y, He G, Zhao X, Wang C, Li S, et al. Combined application of acidic biochar and fertilizer synergistically enhances *miscanthus* productivity in coastal saline-alkaline soil. Science of the Total Environment. 2023;**893**:164811. DOI: 10.1016/j.scitotenv.2023.164811

[27] Mishra SK, Misra S, Dixit VK, Kar S, Chauhan PS. Ochrobactrum sp. NBRISH6 inoculation enhances *Zea mays* productivity, mitigating soil alkalinity and plant immune response. Current Microbiology. 2023;**80**(10):1-13. Available from: https://link.springer.com/article/10.1007/s00284-023-03441-7

[28] Ma Q, Wu E, Wang H, Feng Y, Zhao L, Feng B. Comparative transcriptomic and metabolomic study reveal that exogenous 24-epiandrosterone mitigate alkaline stress in broomcorn millet (*Panicum miliaceum* L.) via regulating photosynthesis and antioxidant capacity. GCB Bioenergy. 2023;**15**(4):494-507. Available from: https://onlinelibrary.wiley.com/doi/full/10.1111/gcbb.13032

[29] Liu X, Xie X, Zheng C, Wei L, Li X, Jin Y, et al. RNAi-mediated suppression of the abscisic acid catabolism gene OsABA8ox1 increases abscisic acid content and tolerance to saline–alkaline stress in rice (*Oryza sativa* L.). Crop Journal. 2022;**10**(2):354-367. DOI: 10.1016/j.cj.2021.06.011

[30] Wang F, Xu H, Zhang L, Shi Y, Song Y, Wang X, et al. The lipoxygenase OsLOX10 affects seed longevity and resistance to saline-alkaline stress during rice seedlings. Plant Molecular Biology. 2023;**111**(4–5):415-428. Available from: https://link.springer.com/article/10.1007/s11103-023-01334-8

[31] Jin T, An J, Xu H, Chen J, Pan L, Zhao R, et al. A soybean sodium/hydrogen exchanger GmNHX6 confers plant alkaline salt tolerance by regulating Na^+/K^+ homeostasis. Frontiers in Plant Science. 2022;**13**:938635. DOI: 10.3389/fpls.2022.938635

[32] Zhao K, Wu Y. Effects of Zn deficiency and bicarbonate on the growth and photosynthetic characteristics of four plant species. PLoS One. 2017;**12**(1):e0169812. Available from: https://journals.plos.org/plosone/article?id=10.1371/journal.pone.0169812

[33] Tsai HH, Schmidt W. The enigma of environmental pH sensing in plants. Nature Plants. 2021;**7**(2):106-115.

Available from: https://www.nature.com/articles/s41477-020-00831-8

[34] Sposito G. The chemical composition of soils. In: Sposito G, editor. The Chemistry of Soils. New York, United States of America: Oxford University Press Inc.; 2008. pp. 2-24. Available from: https://books.google.com/books/about/The_Chemistry_of_Soils.html?hl=es&id=XCJnDAAAQBAJ

[35] Tavakkoli E, Rengasamy P, Smith E, Mcdonald GK. The effect of cation-anion interactions on soil pH and solubility of organic carbon. European Journal of Soil Science. 2015;**66**(6):1054-1062. Available from: https://onlinelibrary.wiley.com/doi/full/10.1111/ejss.12294

[36] Boyd CE, Tucker CS, Somridhivej B. Alkalinity and hardness: Critical but elusive concepts in aquaculture. Journal of the World Aquaculture Society. 2016; **47**(1):6-41. Available from: https://www.researchgate.net/publication/292185663

[37] Roosta HR. Interaction between water alkalinity and nutrient solution pH on the vegetative growth, chlorophyll fluorescence and leaf magnesium, iron, manganese, and zinc concentrations in lettuce. Journal of Plant Nutrition. 2011;**34**(5):717-731. Available from: https://www.tandfonline.com/doi/abs/10.1080/01904167.2011.540687

[38] Jiao X, Li Y, Zhang X, Liu C, Liang W, Li C, et al. Exogenous dopamine application promotes alkali tolerance of apple seedlings. Plants. 2019;**8**(12):580. Available from: https://www.mdpi.com/2223-7747/8/12/580/htm

[39] Yang CW, Wang P, Li CY, Shi DC, Wang DL. Comparison of effects of salt and alkali stresses on the growth and photosynthesis of wheat. Photosynthetica. 2008;**46**(1):107-114. Available from: https://link.springer.com/article/10.1007/s11099-008-0018-8

[40] Roosta HR, Manzari Tavakkoli M, Hamidpour M. Comparison of different soilless media for growing gerbera under alkalinity stress condition. Journal of Plant Nutrition. 2016;**39**(8):1063-1073. Available from: https://www.tandfonline.com/doi/abs/10.1080/01904167.2015.1109112

[41] Brautigan DJ, Rengasamy P, Chittleborough DJ. Amelioration of alkaline phytotoxicity by lowering soil pH. Crop and Pasture Science. 2014; **65**(12):1278-1287. Available from: https://www.publish.csiro.au/cp/CP13435

[42] Nandita K, Kundu M, Nahakpam S, Rakshit R. Micronutrients spray to combat granulation and improve fruit quality of sweet orange [*Citrus sinensis* (L.) Osbeck] cv. Mosambi under nontraditional citrus growing track. Journal of Plant Nutrition. 2023;**46**(17): 4207-4223. Available from: https://www.tandfonline.com/doi/abs/10.1080/01904167.2023.2224384

[43] Dawood MFA, Sohag AAM, Tahjib-Ul-Arif M, Abdel Latef AAH. Hydrogen sulfide priming can enhance the tolerance of artichoke seedlings to individual and combined saline-alkaline and aniline stresses. Plant Physiology and Biochemistry. 2021;**159**:347-362. DOI: 10.1016/j.plaphy.2020.12.034

[44] Hitti Y, MacPherson S, Lefsrud M. Separate effects of sodium on germination in saline–sodic and alkaline forms at different concentrations. Plants. 2023;**12**(6):1234. Available from: https://www.mdpi.com/2223-7747/12/6/1234/htm

[45] Liu D, Ma Y, Rui M, Lv X, Chen R, Chen X, et al. Is high ph the key factor of

alkali stress on plant growth and physiology? A case study with wheat (*Triticum aestivum* L.) seedlings. Agronomy. 2022;**12**(8):1820. Available from: https://www.mdpi.com/2073-4395/12/8/1820/htm

[46] Wang Y, Wang J, Guo D, Zhang H, Che Y, Li Y, et al. Physiological and comparative transcriptome analysis of leaf response and physiological adaption to saline alkali stress across pH values in alfalfa (*Medicago sativa*). Plant Physiology and Biochemistry. 2021;**167**: 140-152. DOI: 10.1016/j.plaphy.2021.07.040

[47] Malekzadeh Shamsabad MR, Esmaeilizadeh M, Roosta HR, Dąbrowski P, Telesiński A, Kalaji HM. Supplemental light application can improve the growth and development of strawberry plants under salinity and alkalinity stress conditions. Scientific Reports. 2022;**12**(1):1-13. Available from: https://www.nature.com/articles/s41598-022-12925-8

[48] Gill PK. Effect of various abiotic stresses on the growth , soluble sugars and water relations of Sorghum seedlings grown in light and darkness. Bulgarian Journal of Plant Physiology. 2014; **27**(1–2):72-84

[49] Deng C, Zhang G, Pan X. Photosynthetic responses in reed (*Phragmites australis* (CAV.) TRIN. ex Steud.) seedlings induced by different salinity-alkalinity and nitrogen levels. Journal of Agricultural Science and Technology. 2011;**13**(5):687-699. Available from: http://jast.modares.ac.ir/article-23-1590-en.html

[50] Yang CW, Xu HH, Wang LL, Liu J, Shi DC, Wang DL. Comparative effects of salt-stress and alkali-stress on the growth, photosynthesis, solute accumulation, and ion balance of barley plants. Photosynthetica. 2009;**47**(1): 79-86. Available from: https://link.springer.com/article/10.1007/s11099-009-0013-8

[51] Scheiner JJ, Labay A, Kamas J. Rootstocks improve blanc Du Bois vine performance and fruit quality on alkaline soil. Catalyst: Discovery into Practice. 2020;**4**(2):63-73. Available from: https://www.ajevonline.org/content/4/Suppl_2/63

[52] Roura S, Davidovich L, Del Valle C. Quality loss in minimally processed swiss chard related to amount of damaged area. LWT. 2000;**33**(1):53-59. DOI: 10.1006/fstl.1999.0615

[53] Wang Y, Feng Y, Yan M, Yu J, Zhou X, Bao J, et al. Effect of saline–alkali stress on sugar metabolism of jujube fruit. Horticulturae. 2022;**8**(6): 474. Available from: https://www.mdpi.com/2311-7524/8/6/474/htm

[54] Rasheed R, Ashraf MA, Ahmad SJN, Parveen N, Hussain I, Bashir R. Taurine regulates ROS metabolism, osmotic adjustment, and nutrient uptake to lessen the effects of alkaline stress on *Trifolium alexandrinum* L. plants. South African Journal of Botany. 2022;**148**: 482-498. DOI: 10.1016/j.sajb.2022.05.023

[55] Mourad AMI, Farghly KA, Börner A, Moursi YS. Candidate genes controlling alkaline-saline tolerance in two different growing stages of wheat life cycle. Plant and Soil. 2023;**493**:283-307. Available from: https://link.springer.com/article/10.1007/s11104-023-06232-y

[56] Navarro-León E, Grazioso A, Atero-Calvo S, Rios JJ, Esposito S, Blasco B. Evaluation of the alkalinity stress tolerance of three *Brassica rapa* CAX1 TILLING mutants. Plant Physiology and Biochemistry. 2023;**198**:107712. DOI: 10.1016/j.plaphy.2023.107712

[57] Cao X, Sun L, Wang W, Zhang F. Exogenous calcium application mediates K^+ and Na^+ homeostasis of different salt-tolerant rapeseed varieties under $NaHCO_3$ stress. Plant Growth Regulation. 2023;**102**:367-378. Available from: https://link.springer.com/article/10.1007/s10725-023-01066-1

[58] Wang W, Zhang F, Sun L, Yang L, Yang Y, Wang Y, et al. Alkaline salt inhibits seed germination and seedling growth of canola more than neutral salt. Frontiers in Plant Science. 2022;**13**: 814755. DOI: 10.3389/fpls.2022.814755

[59] Chen L, Zhao R, Yu J, Gu J, Li Y, Chen W, et al. Functional analysis of plasma membrane H^+-ATPases in response to alkaline stress in blueberry. Scientia Horticulturae. 2022; **306**:111453. DOI: 10.1016/j.scienta.2022.111453

[60] Han P, Li S, Yao K, Geng H, Liu J, Wang Y, et al. Integrated metabolomic and transcriptomic strategies to reveal adaptive mechanisms in castor plant during germination stage under alkali stress. Environmental and Experimental Botany. 2022;**203**:105031. DOI: 10.1016/j.envexpbot.2022.105031

[61] Zhang L, Qian Q, Song S. Combating alkaline stress with alkaline tolerance 1, which encodes a conserved Gγ protein in multiple crops. Plant Communications. 2023;**4**(3):100603. DOI: 10.1016/j.xplc.2023.100603

[62] Zaib-un-Nisa MX, Anwar S, Chen C, Jin X, Yu L, et al. Strigolactone enhances alkaline tolerance in soybean seeds germination by altering expression profiles of ABA biosynthetic and signaling genes. Journal of Plant Biology. 2023;**66**(4):373-381. Available from: https://link.springer.com/article/10.1007/s12374-022-09357-2

[63] Li J, Jin Y, Liu Z, Ba T, Wang W, Xu S. Morphology and metabolism of storage substances contribution to alkali stress responses in two contrasting barley cultivars during germination stage. Journal of Seed Science. 2023;**45**: e202345030. Available from: https://www.scielo.br/j/jss/a/nQKBBGLq8Lq5zbFrRzTsZdP/?lang=en

[64] Fatima A, Hussain S, Hussain S, Ali B, Ashraf U, Zulfiqar U, et al. Differential morphophysiological, biochemical, and molecular responses of maize hybrids to salinity and alkalinity stresses. Agronomy. 2021;**11**(6):1150. Available from: https://www.mdpi.com/2073-4395/11/6/1150/htm

[65] Xu X, Zhang J, Yan B, Wei Y, Ge S, Li J, et al. The adjustment of membrane lipid metabolism pathways in maize roots under saline–alkaline stress. Frontiers in Plant Science. 2021;**12**: 635327. DOI: 10.3389/fpls.2021.635327

[66] Wu GQ, Li H, Zhu YH, Li SJ. Comparative physiological response of sainfoin (*Onobrychis viciaefolia*) seedlings to alkaline and saline-alkaline stress. The Journal of Animal and Plant Sciences. 2020;**31**(4):2021. DOI: 10.36899/JAPS.2021.4.0299

[67] Liang X, Wang Y, Li Y, An W, He X, Chen Y, et al. Widely-targeted metabolic profiling in *Lycium barbarum* fruits under salt-alkaline stress uncovers mechanism of salinity tolerance. Molecules. 2022;**27**(5):1564. Available from: https://www.mdpi.com/1420-3049/27/5/1564/htm

[68] Zhang H, Liu XL, Zhang RX, Yuan HY, Wang MM, Yang HY, et al. Root damage under alkaline stress is associated with reactive oxygen species accumulation in rice (*Oryza sativa* L.). Frontiers in Plant Science. 2017;**8**: 288138. DOI: 10.3389/fpls.2017.01580

[69] Huihui Z, Xin L, Yupeng G, Mabo L, Yue W, Meijun A, et al. Physiological and proteomic responses of reactive oxygen species metabolism and antioxidant machinery in mulberry (*Morus alba* L.) seedling leaves to NaCl and $NaHCO_3$ stress. Ecotoxicology and Environmental Safety. 2020;**193**:110259. DOI: 10.1016/j.ecoenv.2020.110259

[70] Sriramachandrasekharan MV, Priya NG, Manivannan R, Prakash M. Ameliorative role of silicon on osmoprotectants, antioxidant enzymes and growth of maize grown under alkaline stress. Silicon. 2022;**14**(12): 6577-6585. Available from: https://link.springer.com/article/10.1007/s12633-021-01434-4

[71] Todevska D, Kovacevik B, Kostadinvik-Velickovska S, Markova-Ruzdik N, Mihajlov L. Evaluation of the quality traits of red cherry tomato varieties grown in alkaline soil. Zemdirbyste. 2023;**110**(2):157-164. DOI: 10.13080/z-a.2023.110.019

[72] Andrioli RJ. Adaptive mechanisms of tall wheatgrass to salinity and alkalinity stress. Grass and Forage Science. 2023; **78**(1):23-36. Available from: https://onlinelibrary.wiley.com/doi/full/10.1111/gfs.12585

[73] Fu D, Zhang Z, Wallrad L, Wang Z, Höller S, Ju CF, et al. Ca^{2+}-dependent phosphorylation of NRAMP1 by CPK21 and CPK23 facilitates manganese uptake and homeostasis in Arabidopsis. Proceedings of the National Academy of Sciences of the United States of America. 2022;**119**(40):e2204574119. Available from: https://www.pnas.org/doi/abs/10.1073/pnas.2204574119

[74] Müller B, Sterner VAC, Papp L, May Z, Orlóci L, Gyuricza C, et al. Alkaline salt tolerance of the biomass plant *Arundo donax*. Agronomy. 2022; **12**(7):1589. Available from: https://www.mdpi.com/2073-4395/12/7/1589/htm

[75] Xu Y, Tao S, Zhu Y, Zhang Q, Li P, Wang H, et al. Identification of alkaline salt tolerance genes in *Brassica napus* L. by transcriptome analysis. Genes. 2022; **13**(8):1493. Available from: https://www.mdpi.com/2073-4425/13/8/1493/htm

[76] Wang N, Fan X, Lin Y, Li Z, Wang Y, Zhou Y, et al. Alkaline stress induces different physiological, hormonal and gene expression responses in diploid and autotetraploid Rice. International Journal of Molecular Sciences. 2022;**23**(10):5561. Available from: https://www.mdpi.com/1422-0067/23/10/5561/htm

[77] Pérez-Martín L, Busoms S, Tolrà R, Poschenrieder C. Transcriptomics reveals fast changes in salicylate and jasmonate signaling pathways in shoots of carbonate-tolerant *Arabidopsis thaliana* under bicarbonate exposure. International Journal of Molecular Sciences. 2021;**22**(3):1-24. Available from: https://www.mdpi.com/1422-0067/22/3/1226/htm

[78] Ma Q, Wang H, Wu E, Zhang H, Feng Y, Feng B. Widely targeted metabolomic analysis revealed the effects of alkaline stress on nonvolatile and volatile metabolites in broomcorn millet grains. Food Research International. 2023;**171**:113066. DOI: 10.1016/j.foodres.2023.113066

[79] Geng G, Wang G, Stevanato P, Lv C, Wang Q, Yu L, et al. Physiological and proteomic analysis of different molecular mechanisms of sugar beet response to acidic and alkaline pH environment. Frontiers in Plant Science. 2021;**12**:682799. DOI: 10.3389/fpls.2021.682799

[80] Dong S, Pang W, Liu Z, Li H, Zhang K, Cong L, et al. Transcriptome analysis of leaf senescence regulation under alkaline stress in *Medicago truncatula*. Frontiers in Plant Science. 2022;**13**:881456. DOI: 10.3389/fpls.2022.881456

[81] Singh L, Coronejo S, Pruthi R, Chapagain S, Subudhi PK. Integration of QTL mapping and whole genome sequencing identifies candidate genes for alkalinity tolerance in Rice (*Oryza sativa*). International Journal of Molecular Sciences. 2022;**23**(19):11791. Available from: https://www.mdpi.com/1422-0067/23/19/11791/htm

[82] Jin T, Shan Z, Zhou S, Yang Q, Gai J, Li Y. GmDNAJC7 from soybean is involved in plant tolerance to alkaline-salt, salt, and drought stresses. Agronomy. 2022;**12**(6):1419. Available from: https://www.mdpi.com/2073-4395/12/6/1419/htm

[83] Zhang H, Yu F, Xie P, Sun S, Qiao X, Tang S, et al. A Gγ protein regulates alkaline sensitivity in crops. Science. 2023;**379**(6638):eade8416. Available from: https://www.science.org/doi/10.1126/science.ade8416

[84] Sun W, Zhang H, Yang S, Liu L, Xie P, Li J, et al. Genetic modification of Gγ subunit AT1 enhances salt-alkali tolerance in main graminaceous crops. National Science Review. 2023;**10**(6):nwad075. DOI: 10.1093/nsr/nwad075

[85] Wang P, Ma JF. Knockout of a gene encoding a Gγ protein boosts alkaline tolerance in cereal crops. aBIOTECH. 2023;**4**(2):180-183. Available from: https://link.springer.com/article/10.1007/s42994-023-00106-8

[86] Li Z, An M, Hong D, Chang D, Wang K, Fan H. Transcriptomic and metabolomic analyses reveal the differential regulatory mechanisms of compound material on the responses of *Brassica campestris* to saline and alkaline stresses. Frontiers in Plant Science. 2022;**13**:820540. DOI: 10.3389/fpls.2022.820540

[87] Seraj RGM, Mohammadi M, Shahbazi M, Tohidfar M. Co-expression network analysis of soybean transcriptome identify hub genes under saline-alkaline and water deficit stress. Iranian Journal of Biotechnology. 2022;**20**(4):110-121. Available from: https://www.ijbiotech.com/article_157961.html

[88] Zhang P, Duo T, Wang F, Zhang X, Yang Z, Hu G. De novo transcriptome in roots of switchgrass (*Panicum virgatum* L.) reveals gene expression dynamic and act network under alkaline salt stress. BMC Genomics. 2021;**22**(1):82. Available from: https://bmcgenomics.biomedcentral.com/articles/10.1186/s12864-021-07368-w

[89] Fan Y, Lu X, Chen X, Wang J, Wang D, Wang S, et al. Cotton transcriptome analysis reveals novel biological pathways that eliminate reactive oxygen species (ROS) under sodium bicarbonate ($NaHCO_3$) alkaline stress. Genomics. 2021;**113**(3):1157-1169. DOI: 10.1016/j.ygeno.2021.02.022

[90] Ma Q, Yuan Y, Wu E, Wang H, Dang K, Feng Y, et al. Endogenous bioactive gibberellin/abscisic acids and enzyme activity synergistically promote the phytoremediation of alkaline soil by broomcorn millet (*Panicum miliaceum* L.). Journal of Environmental Management. 2022;**305**:114362. DOI: 10.1016/j.jenvman.2021.114362

[91] Rui C, Chen X, Xu N, Wang J, Zhang H, Li S, et al. Identification and structure analysis of KCS family genes suggest their reponding to regulate fiber development in long-staple cotton under

salt-alkaline stress. Frontiers in Genetics. 2022;**13**:812449. DOI: 10.3389/fgene.2022.812449

[92] Wang W, Pang J, Zhang F, Sun L, Yang L, Fu T, et al. Transcriptome analysis reveals key molecular pathways in response to alkaline salt stress in canola (*Brassica napus* L.) roots. Journal of Plant Growth Regulation. 2023;**42**(5): 3111-3127. Available from: https://link.springer.com/article/10.1007/s00344-022-10774-3

[93] Han H, Qu Y, Wang Y, Zhang Z, Geng Y, Li Y, et al. Transcriptome and small rna sequencing reveals the basis of response to salinity, alkalinity and hypertonia in quinoa (*Chenopodium quinoa* Willd.). International Journal of Molecular Sciences. 2023;**24**(14):11789. Available from: https://www.mdpi.com/1422-0067/24/14/11789/htm

[94] Rehman O, Uzair M, Chao H, Khan MR, Chen M. Decoding RNA editing sites through transcriptome analysis in rice under alkaline stress. Frontiers in Plant Science. 2022;**13**: 892729. DOI: 10.3389/fpls.2022.892729

[95] Zhang B, Liu Z, Zhou R, Cheng P, Li H, Wang Z, et al. Genome-wide analysis of soybean DnaJA-family genes and functional characterization of GmDnaJA6 responses to saline and alkaline stress. Crop Journal. 2023;**11**(4): 1230-1241. DOI: 10.1016/j.cj.2023.06.005

[96] Xu Y, Magwanga RO, Yang X, Jin D, Cai X, Hou Y, et al. Genetic regulatory networks for salt-alkali stress in *Gossypium hirsutum* with differing morphological characteristics. BMC Genomics. 2020;**21**(1):1-19. Available from: https://bmcgenomics.biomedcentral.com/articles/10.1186/s12864-019-6375-9

[97] Rehman OU, Uzair M, Farooq MS, Saleem B, Attacha S, Attia KA, et al. Comprehensive insights into the regulatory mechanisms of lncRNA in alkaline-salt stress tolerance in rice. Molecular Biology Reports. 2023;**50**(9): 7381-7392. Available from: https://link.springer.com/article/10.1007/s11033-023-08648-2

[98] Wang N, Wang Y, Wang C, Leng Z, Qi F, Wang S, et al. Evaluating the differential response of transcription factors in diploid versus autotetraploid Rice leaves subjected to diverse saline–alkali stresses. Genes. 2023;**14**(6):1151. Available from: https://www.mdpi.com/2073-4425/14/6/1151/htm

[99] Shamsabad MRM, Esmaeilizadeh M, Roosta HR, Dehghani MR, Dąbrowski P, Kalaji HM. The effect of supplementary light on the photosynthetic apparatus of strawberry plants under salinity and alkalinity stress. Scientific Reports. 2022; **12**(1):13257. Available from: https://www.nature.com/articles/s41598-022-17377-8

[100] Xudong G, Fengju Z, Teng W, Xiaowei X, Xiaohui J, Xing X. Effects of nitrogen and phosphorus addition on growth and leaf nitrogen metabolism of alfalfa in alkaline soil in Yinchuan plain of Hetao Basin. PeerJ. 2022;**10**:e13261. Available from: https://peerj.com/articles/13261

[101] Zhang P, Yang F, Zhang H, Liu L, Liu X, Chen J, et al. Beneficial effects of biochar-based organic fertilizer on nitrogen assimilation, antioxidant capacities, and photosynthesis of sugar beet (*Beta vulgaris* L.) under saline-alkaline stress. Agronomy. 2020;**10**(10): 1562. Available from: https://www.mdpi.com/2073-4395/10/10/1562/htm

[102] Rehman B, Hussain S, Zulfiqar A. Zinc sulfate biofortification enhances

physio-biochemical attributes and oxidative stress tolerance in rice varieties grown in zinc deficient alkaline soil. South African Journal of Botany. 2023; **162**:271-281. DOI: 10.1016/j.sajb.2023.09.023

[103] Danaeifar A, Khaleghi E, Zivdar S, Mehdikhanlou K. Physiological, biochemical, and gene expression of Sour Orange (*Citrus aurantium* L.) to Iron(II)-Arginine Chelate under salinity, alkalinity, and salt–alkali combined stresses. Scientia Horticulturae. 2023; **319**:112146. DOI: 10.1016/j.scienta.2023.112146

[104] Saleem I, Maqsood MA, Aziz T, Bhatti IA, Jabbar A. Potassium ferrite nano coated dap fertilizer improves wheat (*Triticum aestivum* L.) growth and yield under alkaline calcareous soil conditions. Pakistan Journal of Agricultural Sciences. 2021;**58**(2): 485-492. DOI: 10.21162/PAKJAS/21.939

[105] Wang X, An M, Wang K, Fan H, Shi J, Chen K. Effects of organic polymer compound material on K^+ and Na^+ distribution and physiological characteristics of cotton under saline and alkaline stresses. Frontiers in Plant Science. 2021;**12**:636536. DOI: 10.3389/fpls.2021.636536

[106] Chen Q, Cao X, Li Y, Sun Q, Dai L, Li J, et al. Functional carbon nanodots improve soil quality and tomato tolerance in saline-alkali soils. Science of the Total Environment. 2022;**830**:154817. DOI: 10.1016/j.scitotenv.2022.154817

[107] El-temsah ME, Abd-Elkrem YM, El-Gabry YA, Abdelkader MA, Morsi NAA, Taha NM, et al. Response of diverse peanut cultivars to nano and conventional calcium forms under alkaline sandy soil. Plants. 2023;**12**(14): 2598. Available from: https://www.mdpi.com/2223-7747/12/14/2598/htm

[108] Li F, Wu Z, Zuo S, Fan L, Wei Z, Ma L, et al. Application of enzymatic hydrolysate of *Ulva clathrata* as biostimulant improved physiological and metabolic adaptation to salt-alkaline stress in wheat. Journal of Applied Phycology. 2022;**34**(3):1779-1789. Available from: https://link.springer.com/article/10.1007/s10811-022-02684-4

[109] Dixit VK, Misra S, Mishra SK, Tewari SK, Joshi N, Chauhan PS. Characterization of plant growth-promoting alkalotolerant *Alcaligenes* and *bacillus* strains for mitigating the alkaline stress in *Zea mays*. Antonie van Leeuwenhoek, International Journal of General and Molecular Microbiology. 2020;**113**(7):889-905. Available from: https://link.springer.com/article/10.1007/s10482-020-01399-1

[110] Youssef SM, El-Serafy RS, Ghanem KZ, Elhakem A, Abdel Aal AA. Foliar spray or soil drench: Microalgae application impacts on soil microbiology, morpho-physiological and biochemical responses, oil and fatty acid profiles of chia plants under alkaline stress. Biology. 2022;**11**(12):1844. Available from: https://www.mdpi.com/2079-7737/11/12/1844/htm

[111] Fu J, Xiao Y, Wang YF, Liu ZH, Zhang YF, Yang KJ. *Trichoderma asperellum* alters fungal community composition in saline–alkaline soil maize rhizospheres. Soil Science Society of America Journal. 2021;**85**(4):1091-1104. Available from: https://onlinelibrary.wiley.com/doi/full/10.1002/saj2.20245

[112] Cabral-Miramontes JP, Olmedo-Monfil V, Lara-Banda M, Zúñiga-Romo ER, Aréchiga-Carvajal ET. Promotion of plant growth in arid zones by selected *Trichoderma* spp. strains with adaptation plasticity to alkaline pH. Biology. 2022; **11**(8):1206. Available from: https://

www.mdpi.com/2079-7737/11/8/1206/htm

[113] Farghaly FA, Nafady NA, Abdel-Wahab DA. The efficiency of arbuscular mycorrhiza in increasing tolerance of *Triticum aestivum* L. to alkaline stress. BMC Plant Biology. 2022;**22**(1):1-17. Available from: https://bmcplantbiol.biomedcentral.com/articles/10.1186/s12870-022-03790-8

[114] İpek M, Eşitken A. Effects of rhizobacteria on plant growth and fruit quality of blackberry in alkaline soil. Selcuk Journal of Agricultural and Food Sciences. 2022;**36**(3):387-392. DOI: 10.15316/SJAFS.2022.050

[115] Xu J, Yang J, Xu Z, Zhao D, Hu X. Exogenous spermine-induced expression of SlSPMS gene improves salinity–alkalinity stress tolerance by regulating the antioxidant enzyme system and ion homeostasis in tomato. Plant Physiology and Biochemistry. 2020;**157**:79-92. DOI: 10.1016/j.plaphy.2020.09.033

[116] Ma C, Yuan S, Xie B, Li Q, Wang Q, Shao M. IAA plays an important role in alkaline stress tolerance by modulating root development and ROS detoxifying systems in rice plants. International Journal of Molecular Sciences. 2022;**23**(23):14817. Available from: https://www.mdpi.com/1422-0067/23/23/14817/htm

[117] Chang D, Liu H, An M, Hong D, Fan H, Wang K, et al. Integrated transcriptomic and metabolomic analysis of the mechanism of foliar application of hormone-type growth regulator in the improvement of grape (*Vitis vinifera* L.) coloration in saline-alkaline soil. Plants. 2022;**11**(16):2115. Available from: https://www.mdpi.com/2223-7747/11/16/2115/htm

[118] Youssef SM, Abdella EMM, Al-Elwany OA, Alshallash KS, Alharbi K, Ibrahim MTS, et al. Integrative application of foliar yeast extract and gibberellic acid improves morpho-physiological responses and nutrient uptake of *Solidago virgaurea* plant in alkaline soil. Life. 2022;**12**(9):1405. Available from: https://www.mdpi.com/2075-1729/12/9/1405/htm

[119] Xu Z, Wang J, Zhen W, Sun T, Hu X. Abscisic acid alleviates harmful effect of saline–alkaline stress on tomato seedlings. Plant Physiology and Biochemistry. 2022;**175**:58-67. DOI: 10.1016/j.plaphy.2022.01.018

[120] Qu D, Show PL, Miao X. Transcription factor chbzip1 from alkaliphilic microalgae *Chlorella* sp. BLD enhancing alkaline tolerance in transgenic *Arabidopsis thaliana*. International Journal of Molecular Sciences. 2021;**22**(5):1-16. Available from: https://www.mdpi.com/1422-0067/22/5/2387/htm

[121] Wei L, Zhang R, Zhang M, Xia G, Liu S. Functional analysis of long non-coding RNAs involved in alkaline stress responses in wheat. Journal of Experimental Botany. 2022;**73**(16):5698-5714. DOI: 10.1093/jxb/erac211

[122] Cankurt K, İpek M. The effects of some organic compounds on yield and fruit quality in albion strawberry (*Fragaria x ananassa* Duch) Cultivar. Selcuk Journal of Agricultural and Food Sciences. 2023;**37**(1):19-124. DOI: 10.15316/SJAFS.2023.003

[123] Sanwal SK, Mann A, Kesh H, Kaur G, Kumar R, Rai AK. Genotype environment interaction analysis for fruit yield in okra (*Abelmoschus esculentus* L.) under alkaline environments. Indian Journal of Genetics and Plant Breeding. 2021;**81**(1):101-110. DOI: 10.31742/IJGPB.81.1.11

Chapter 4

High-Temperature Effect on Plant Development and Tuber Induction and Filling in Potato (*Solanum tuberosum* L.)

María del Socorro Sánchez-Correa, María del Rocío Reyero-Saavedra, Gladys Edith Jiménez-Nopala, Manuel Mandujano Piña and Juan Gerardo Ortiz-Montiel

Abstract

Heat stress significantly impacts the physiology of potato plants (*Solanum tuberosum* L.). Regardless of the phenological stage at which thermal stress occurs, the most notable effect consistently emerges in tuber production. Heat stress can disrupt photosynthesis, transpiration, and overall plant metabolism. It is known that various genes encoding enzymes and glucose and sucrose transporters respond to temperature changes, influencing the partitioning of carbohydrates toward storage organs. This disturbance leads to disruptions in starch formation in tubers, affecting their development, and subsequently, the yield and quality of tubers. Understanding these interactions is crucial for developing heat-tolerant potato varieties in the face of climate change resulting from global warming.

Keywords: potato, heat stress, tuberization, tuber physiology, thermal tolerance, glycoalkaloids

1. Introduction

The Intergovernmental Panel on Climate Change predicts a gradual increase in global temperatures, which will alter the geographical distribution of climate zones. This change will impact the distribution and abundance of many plant species, modify crop phenology, and reduce agricultural yields [1]. It is estimated that climate change will result in a reduction in global potato production by 18–32% [2].

Potatoes, ranking fourth in global food crops after rice, wheat, and corn, are crucial for worldwide food production. This crop yields more energy per unit area than any other and boasts the second-highest protein content after soy, providing an excellent balance of essential amino acids, particularly lysine [3].

Optimal potato development occurs in temperatures between 15 and 20°C [4], with yields ranging from 12 to 60 tons of tubers per hectare. However, observations show a decline in yield when the ambient temperature exceeds 24°C. This reduction is attributed to a combination of independent morphological, physiological, and developmental issues.

The generation of potato plants from seeds is vital for plant breeding. Although there is growing interest in true potato seed (TPS) production in some tropical and subtropical regions [5], most countries cultivate potatoes from seed tubers.

Potato cultivation undergoes various stages characterized by physiological and morphological changes. The sprouting of seed tubers (**Figure 1**), considered the interruption of dormancy, occurs when seed tubers have initial sprouts with identifiable nodes, recognizable by visible scale leaves. The description and interruption of the dormancy state in potatoes have been extensively documented [6]. Seed tubers can be planted with or without sprouts or cut into two or three pieces, which are planted independently. The conditions of humidity, light, and temperature in which they are stored for sprouting are crucial, as these sprouts determine many of the plant's developmental characteristics.

The number of sprouts that a seed tuber produces depends on both genotype and its previous development. According to the concept of the "physiological age" of the tuber: a "young" tuber exhibits apical dominance and produces a single sprout, whereas a tuber with more time loses apical dominance and forms several sprouts [7].

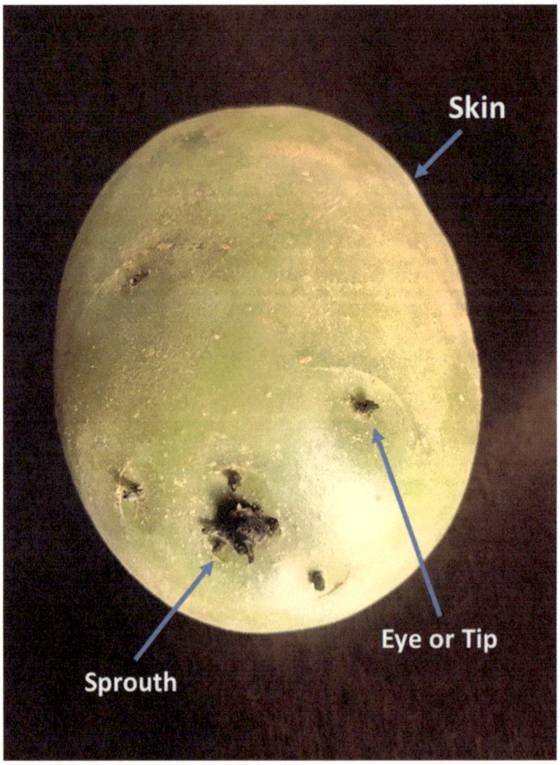

Figure 1.
Tuber anatomy. Sprout started and, in the future, a whole plant will grow.

Plant emergence is defined as the appearance of a sprout from the soil, and its development and growth are described based on the appearance of individual leaves up to the stem. It is preferred to define the stages of shoot growth and expansion based on the number of nodes, as they are easily identified as points of leaf insertion into the stem and can be counted along the shoot [6].

Shoot growth is determined by the number of nodes along the stem from the soil surface, as no morphological differences defining the moment of maturity of individual leaves are observed [8].

Flowering begins with the appearance of the floral bud and extends until the first flower is fully open, marking the end of vegetative development [9].

Tuberization is a stage that occurs shortly before flowering. In plants derived from seed tubers, the initiation of stolons generally occurs soon after plant emergence when it has a developed shoot. In most cultivars, stolons initiate in the axillary buds of the basal nodes of the stem and then elongate. The extension of stolons depends on genotype and environment. Stolons generated near the soil surface tend to be longer and initiate tubers later than stolons formed more deeply in the soil and closer to the original tuber. Moreover, the transition from stolon to tuber involves significant changes in both cell division and growth events, influenced by hormonal levels, with specific transcription factors and small RNAs playing crucial roles in tuberization. Stolon elongation ceases at the initiation of the tuber, and thickening occurs at its apical part, defined as the point at which the stolon thickens to double its initial diameter [10]. Tubercle growth is not considered a distinct developmental stage but is defined as when tubers on 80% of stolons are over 10 mm in diameter and progress.

Tuber maturation begins with the suberization of the periderm to form distinct skin, and development concludes when tubers detach from the stolons. For the optimal development of crops of this species, recommended temperatures are 13°C during the sprouting period, 12–14°C for shoot growth of foliar development, 18°C during the flowering process, and 16–20°C during tuber formation [11]. On the other hand, 20°C is recommended as an optimal temperature for photosynthesis in some European potato varieties [12]. From an ecological perspective, considering the potato's center of origin, its cultivation thrives at relatively cool temperatures. In the Andean region, likely the native habitat of the potato, the average temperature of the warmest month is 16°C [13]. In the United States, the highest yields are obtained in regions with cooler summer temperatures, while in Scotland and northern Europe, they occur in July when the average temperature is around 15°C [14].

This chapter will explore the effect of temperatures exceeding 24°C on the metabolism and development at each phenological stage of this crop and the implications for tuber formation.

2. Sprouting of the seed tuber and sprout growth

The seed is a crucial factor in every agricultural production process, and for potatoes, the tuber plays a pivotal role in generating a new plant, earning it the label "seed tuber". The health and physiological condition of these tubers significantly influence crop yield [15]. Seed tubers can be planted intact or divided; to a lesser extent, cuttings and microtubers obtained through in vitro culture are also utilized (**Figure 2**). In all these scenarios, the genetic makeup remains unchanged, as clonal reproduction is involved. The utilization of sexual seeds is primarily reserved for plant breeding purposes.

Figure 2.
Microtubers structure. Microtubers obtained by in vitro culture.

Seed tubers are temporarily stored at low temperatures and ambient humidity to suppress tuber metabolic activity and inhibit the development of microorganisms [16]. To break dormancy, the relative humidity of the environment is usually increased, as sprouts tend to emerge at low temperatures [17]. It is essential for seed tubers to reach their optimal sprouting state when planting for better yields. Sprout formation depends on the potato variety, determining the rest or dormancy period and physiological age, which, in turn, relies on storage conditions.

As a vegetative reproductive organ, part of its physiological development in the plant and post-harvest, the tuber goes through a state of rest or dormancy, a period of relative inactivity before sprout formation. The latency period is characteristic of each variety, and its duration is crucial for planting [18]. This period ends with the initiation of the first sprout's growth. The dormancy period is generally considered from the end of harvest until the emergence of sprouts [19].

It has been observed that during the dormancy period of tuber storage, the physiological maturation process accelerates at high temperatures, consequently reducing the dormancy period. This phenomenon is not limited to high temperatures; low temperatures can also shorten this process. The end of dormancy is characterized by the initiation of vegetative bud growth, often with the apical bud emerging first in a state of dominance, followed by the formation of sprouts. However, apical growth is not desirable as it may result in a single-sprout plant. Therefore, it is common practice to remove the tip of the sprout, inducing a state of multiple-sprout growth, stored at low temperatures and diffuse light [20]. Nevertheless, seed tubers in multiple sprouting may age rapidly, developing etiolated and weak sprouts that form unproductive plants [21]. Physiologically young seed tubers, unlike old ones, realize their full potential yield, and potato cultivation extends over a longer period, resulting in higher overall yields [22].

Environmental conditions can influence the duration of the dormancy period in seed tubers during their development. The most crucial condition is high temperature during tuber development; alterations in moisture content or nutrient depletion in the soil can modify seed tuber development and physiology, affecting dormancy [17].

Drought, salinity, and above all, heat, are environmental factors that can severely impact the growth and yield of potato crops [23]. The development of seed tubers requires a temperature between 6°C and 18°C for optimal stem elongation [24].

3. Emergency and plant development

The vegetative part of the potato comprises a generally short and shallow root system, stems capable of growing submerged in the soil, known as stolons, where tubers develop, and stems emerging from the soil bearing compound leaves, representing the photosynthetic portion of the plant.

The development of the potato plant is complex, whether it originates from True Potato Seed (TPS) or a seed tuber. When TPS germinates to form a potato plant, the root system develops from the radicle, growing into an axonomorphic root and later creating a system of adventitious roots through branching [25]. From seed germination, a single stem develops after cotyledon emergence, which then branches at different heights from the axils of the leaves of the main stem, forming a sympodium. Additionally, it branches at basal nodes, forming a variable number of stolons where tubers develop. In this case, a single inflorescence develops at the apex of the main stem [26].

When the plant originates from a seed tuber, the vegetative part consists of a variable number of stems, depending on the number of sprouts in the original tuber. These stems can develop either above or below the ground. Leaves grow on the aerial stems, and an inflorescence can form at the apex of each. In the submerged stems or stolons, the tubers develop [27]. Sprouting must be complete and adequate to generate productive plants with optimal growth. Altered sprouting can result in a single shoot with apical dominance, leading to abnormal stem development.

One of the crucial phenomena for vegetative development and tuber production is photosynthesis. The most suitable temperature for photosynthesis in potato plants is between 16°C and 20°C [28]. The measurement of chlorophyll a fluorescence, as a parameter of photosynthetic efficiency, has allowed deductions that plants cultivated under a low-temperature regime (18/16°C, day/night) experience less stress during the initiation stage of stolon growth compared to higher temperatures [29].

The optimal temperature for canopy-level photosynthesis is determined to be around 24°C [30]. However, temperatures above 25°C promote the formation of taller plants with more lateral branches and smaller leaves [31]. Although this delays or even inhibits the formation of stolons and the induction and growth of the tuber [32]. On the other hand, the combination of high temperatures with long days inhibits tuber growth, resulting in a significant reduction in yield [33]. This negatively affects metabolic processes at different stages of plant development, stem growth, and morphology, particularly the tuberization process [30, 32].

High temperatures adversely affect photosynthesis, transpiration rate, photochemical efficiency, leaf structure, stem growth, tuber growth, and overall yield. These negative impacts are further exacerbated by concurrent water stress [34].

During plant development under thermal stress, leaves increase stomatal conductance to reduce leaf temperature. However, if water is scarce, leaves decrease stomatal conductance to prevent transpiration loss, leading to reduced evaporative cooling and increased tissue temperature [35].

Well-irrigated potato plants exposed to moderate heat (30/20°C Day/Night) exhibit a 3.5-fold increase in stomatal conductance compared to low temperatures, accompanied by an improved net carbon assimilation rate [34].

High temperatures affect tuber growth more drastically than shoot growth because more dry matter is allocated to shoots than roots, stolons, and tubers [33, 36].

4. Flowering

Among all developmental stages of potatoes, flowering is the least studied, particularly regarding the potential impact of temperature increases. However, a recent study by Zhu and colleagues [37] revealed that the heat shock transcription factor SpHsfA4c, isolated from wild potatoes (*Solanum pinnatisectum* Dun), is predominantly expressed in the leaves and flowers of these plants. The study observed that the expression of the SpHsfA4c gene was induced by thermal stress, and plants with this gene demonstrated increased heat resistance, suggesting its potential for developing heat-tolerant potato plants.

In several plant species, a "florigen" exists—a molecule that induces flowering. This protein was first described in *Arabidopsis thaliana* and named FT (FLOWERING LOCUS T), considered a universal florigen [38]. In potatoes, there are FT-like proteins, such as StSP3D, acting as a florigen, and another crucial one, StSP6A, described as a "tuberigen" [39, 40].

5. Tuberization

Optimal tuberization concerning weight, number, and tuber quality per plant is promoted by long nights (less than 12 hours of daylight), low temperatures, and a moderate nitrogen supply. Conversely, tuberization is diminished by short nights, high temperatures, and nitrogen-rich fertilization [41, 42].

While numerous studies have explored factors influencing tuberization, with a particular emphasis on photoperiod, only a few have delved into understanding how high temperatures impact this process [43]. Tuberization is affected when temperatures rise, with observations indicating that temperatures exceeding 25°C strongly alter assimilate distribution, causing a delay in tuber formation and reduced tuber growth [4]. These responses vary among different cultivars [44], but it has been confirmed in various potato cultivars that temperature affects tuberization. However, the exact regulatory mechanisms remain unclear [43]. Therefore, comprehending how temperature increases affect the genetic regulation of tuberization [45] is crucial.

The FT-like protein StSP6A (SELF-PRUNING 6A) is the "tuberigen", crucial for tuber formation [39, 40]. StSP6A is expressed in the leaves and translocates to the tips of stolons when tuberization begins [40]. Despite reports indicating StSP6A's involvement in photoperiod-induced tuberization, it has been observed that silencing this gene in short-day cultivars delays tuberization. Conversely, overexpression in long-day cultivars induces tuberization, even in those not dependent on photoperiod, emphasizing that its regulation is primarily attributed to temperature [39]. Its expression decreases as this abiotic factor increases [34, 46, 47].

The transcript of StSP6A is suppressed by heat [34, 46]. Studies subjecting potato plants to different temperatures (22°C, 30°C, and 35°C) have shown that heat impacts tuberization [43]. The number and size of tubers were measured at 5 and 9 weeks, revealing a delay in tuberization (**Figure 1**).

The expression of StSP6A in leaves gradually increases in plants at 22°C but diminishes in those subjected to high temperatures [43]. However, when measuring gene expression, it remains relatively consistent across different temperature treatments. This suggests that the expression of StSP6A alone is insufficient to explain the tuberization delay. Thus, tuberization may result from the transcriptional and post-transcriptional regulation of StSP6A. This was corroborated in experiments with transgenic potatoes overexpressing StSP6A, where, under high temperatures, the number of tubers per plant recovered, but the overall yield did not [43].

The regulation of StSP6A occurs through a microRNA called SUPPRESSING EXPRESSION OF SP6A (SES), which targets StSP6A and is induced by high temperatures [43, 46]. SES regulates StSP6A at high temperatures by degrading it (targeting it for degradation). Additionally, it is known that StSP6A's expression occurs in leaves, but the protein moves to the stolon, indicating a potential post-translational regulation as well [48].

Other studies have proposed additional regulators of StSP6A, such as the homolog of TOC1 (TIMING OF CAB EXPRESSION 1) described in *A. thaliana*. TOC1 responds to temperature as a transcriptional regulator suppressing StSP6A expression [34, 47]. Silencing StTOC1 increases the expression of StSP6A and enhances tuberization, suggesting that StTOC1 likely directly binds to the promoter of StSP6A [47]. Other regulators affect StSP6A, including CYCLING DOF FACTOR (StCDF1), StSP5G (SELF-PRUNING 5G), StCOL1 (CONSTANS-LIKE 1), and GIGANTEA, whose expression increases in high temperatures. However, their role in tuberization has been primarily described in the induction of tuberization by photoperiod [34, 40, 48–50].

This intricate network of regulators underscores the complexity of the molecular mechanisms governing tuberization in response to temperature, providing valuable insights for future studies aiming to enhance potato crop resilience under changing climatic conditions.

In the transcriptomic analysis conducted by Park et al., [43], other potential genes involved in the thermo-regulation of tuberization are suggested. Genes such as StCEN1 (CENTRORADIALIS) and StGA2ox1-1 show decreased expression, while StCOL1 increases its expression with temperature, along with StERF5-3 (ETHYLENE RESPONSE FACTOR 5-3). This list includes genes related to plant hormones like gibberellins, which negatively affect tuberization [43], and ethylene, which also delays tuberization [51].

Another factor influencing tuberization is sugar transport. Mokrani and colleagues [52] reported a higher accumulation of glucose, sucrose, and starch in leaves at low temperatures. In contrast, plants grown at higher temperatures exhibited lower carbohydrate levels in leaves, leading to early tuberization, likely due to accelerated tuber initiation metabolism. The authors note that gene expression levels in leaves during temperature decrease favor the conversion of sucrose into starch during tuber maturation. In many cultivars, temperatures above 23°C resulted in increased photosynthesis, causing a noticeable shift in metabolites between leaves and tubers, generally resulting in decreased tuber yield [34].

The tuber-filling stage, dependent on carbohydrate metabolism, is highly sensitive to heat. The carbohydrate content in the stem (glucose, fructose, and sucrose) is altered under heat stress. High acid invertase activity has been observed in the stems of plants under heat stress, reducing assimilate partitioning to the tubers and increasing transport to the leaves [4, 36]. While this increase in invertase activity promotes stem growth, it reduces the assimilates that should contribute to tuber filling, leading to increased starch synthesis in the tuber. This phenomenon was observed by reducing the activity of enzymes involved in starch metabolism at soil temperatures of 30°C [52].

Tuberization induction begins with the thickening of stolon tips, as assimilate transport shifts from apoplastic to symplastic in the subapical region of the stolon.

This is accompanied by an increase in sucrose concentration in the symplast, inducing the expression of sucrose synthase [53, 54]. This shift decreases soluble invertase activity in the stolon [54].

Among the enzymes involved in tuber growth through the assimilation of translocated photosynthates to the stolon tip is AGPase (Adenine Diphosphoglucose Pyrophosphorylase), which is activated at low temperatures [55].

During photosynthesis, a series of primary reactions occur, involving sunlight, water, chloroplasts, pigment molecules associated with proteins forming antenna complexes, and two reaction centers. Enzymes are also involved in the formation of ATP and NADPH2. Additionally, Ribulose 1,5-bisphosphate Carboxylase-Oxygenase can catalyze both carboxylation and oxygenation of Ribulose 1,5-bisphosphate in the Calvin-Benson cycle. Carboxylation of Ribulose 1,5-bisphosphate results in two molecules of 3-phosphoglycerate, characterizing the cycle as C3 type. Oxygenation produces one molecule of 3-phosphoglycerate and another of 2-phosphoglycerate, which participates in the photorespiration process [35].

The potato plant typically undergoes the C3 photosynthetic process. It has been confirmed that high temperatures lead to increased photorespiration [56], consequently causing a considerable loss of carbohydrates in the photosynthetic leaves or storage organs [57]. This energy expenditure also results in a reduction in sucrose export to the tuber [36].

5.1 Tuber initiation

For tuber development, it is not only necessary to have stolons but also to form a sufficient leaf area to sustain the carbon fixation rate. The growth of tubers relies primarily on net photosynthesis rather than on previously accumulated photosynthetic products in the foliage [57]. The initiation stage is crucial, marked by the increase and subsequent suppression of StSP6A expression, facilitating tuberization [43].

Temperature plays a pivotal role in both the initiation and growth of tubers. Cool nighttime temperatures, coupled with a short photoperiod, favor tuber initiation [58]. The inhibitory impact of high temperatures on tuber initiation involves a common hormonal control that disrupts the balance between endogenous gibberellins and inhibitors such as abscisic acid [59].

Elevated temperatures can lead to excessive elongation of the stolon and its branching, resulting in chain tuberization and stolon growth in a process known as heat sprouting [60].

High temperatures promote the partitioning of dry matter in the stems while reducing it in roots, stolons, and tubers, thereby decreasing the number of formed tubers. Consequently, the negative feedback effect of high temperature, after stolon development begins, includes a reduction in tuber initiation, diminishing the demand for photosynthate translocation to them and adversely affecting the photosynthetic system [60, 61].

5.2 Tuber growth

Sucrose serves as the primary sugar translocated in many plants, with its synthesis catalyzed by sucrose phosphate synthase and sucrose-6-phosphate. Degradation, on the other hand, is facilitated by sucrose synthase or invertase. The control of photosynthate partitioning between sucrose and starch in leaves lies with sucrose phosphate synthase, while sucrose synthase is considered the principal enzyme for sucrose cleavage in various tissues.

Growing tubers exhibit heightened sucrose synthase activity and ADP glucose pyrophosphorylase, both of which decrease as tubers develop. The conversion of sucrose to starch and subsequent starch deposition in tubers is regulated by starch synthase and ADP-glucose pyrophosphorylase [62].

Under heat stress, carbohydrate partitioning in potato plants shifts from shoots to tubers, resulting in an overall reduction in plant yield [36]. The impact of heat stress on enzymes that control sucrose metabolism in potatoes is not yet fully understood. Studies by [43] indicate that the photosynthetic performance of StSP6A-overexpressing potato plants remains unchanged, suggesting that photosynthetic capacity does not account for the lack of tuber growth. However, at later stages (9 weeks), sucrose, glucose, and fructose content increases in leaves at high temperatures, regardless of StSP6A overexpression, indicating a correlation with sugar transport rather than accumulation [43, 48].

Inhibiting sugar transport may occur at the level of transporters, exemplified by the interaction between StSWEET11 transporter and StSP6A. This interaction hinders tuber filling, promoting sucrose transport through the symplastic pathway. StSP6A RNAi plants exhibit a more abundant sugar flow in the stem, indicating increased flow. Understanding how sugar transport to the tuber occurs at high temperatures requires further investigation [43].

Activities of enzymes like starch synthase and ADP-glucose pyrophosphorylase decrease at temperatures above 25°C, resulting in less conversion of sucrose into starch in initiated tubers [62, 63].

Sucrose transport from leaves to photosynthate-demanding tissues is facilitated by sucrose transporters susceptible to high temperatures [64]. However, the regulation of sucrose transporter genes and their impact on sucrose transport and loading in tubers at high temperatures remain areas of limited understanding. Elevated temperatures lead to decreased total tuber yield, with low nighttime temperatures increasing the number of tubers per plant [65].

In some cultivars, larger tubers are obtained at higher temperatures when fewer tubers are formed per plant. While elevated daytime and nighttime temperatures reduce tuber yield, high nighttime temperatures seem to have a more detrimental effect. Potato plants grown at 31/29°C compared to 19/17°C Day/Night exhibit reduced sucrose synthase activity, ADP-glucose pyrophosphorylase, and tuber growth rate [52].

Krauss and Marschner [66] found decreased starch synthase activity, lower starch content, and slower tuber growth rates in three potato cultivars when exposed to 30°C compared to constant temperatures of 20°C.

5.3 Tuber maturation

During the tuber maturation stage, elevated temperatures (25/22°C Day/Night) enabled plants to produce increased amounts of glucose and fructose in photosynthate-demanding organs. Conversely, plants grown under low-temperature conditions exhibited high sucrose phosphate synthase activity. The temperature rise during tuber maturation also impacts the biosynthesis of steroidal glycoalkaloids (SGA), compounds that are toxic to human health and remain unaffected by cooking [67]. Therefore, the temperature increase not only influences potato crop productivity but also poses risks to human health [68].

6. Conclusions

Heat stress significantly disrupts tuber production in potato plants, marking the most affected stage among all phenological stages. Heat stress induces changes in crucial physiological processes, including photosynthesis, transpiration, and overall plant metabolism, directly affecting crop performance. Genes encoding enzymes and glucose and sucrose transporters respond to temperature fluctuations, influencing the allocation of carbohydrates to storage organs and causing disruptions in starch formation in tubers. Given the vulnerability of potato plants to thermal stress, unraveling these interactions becomes imperative for the development of heat-tolerant varieties, particularly in the context of climate change and global warming.

Author details

María del Socorro Sánchez-Correa[1], María del Rocío Reyero-Saavedra[2], Gladys Edith Jiménez-Nopala[2], Manuel Mandujano Piña[3] and Juan Gerardo Ortiz-Montiel[4*]

1 Faculty of Higher Studies-Iztacala (FES-I), Scientific Investigation I Laboratory, National Autonomous University of Mexico (UNAM), México City, Mexico

2 Center for Genome Sciences (CCG), Functional Genomics of Eukaryotes, National Autonomous University of Mexico (UNAM), Cuernavaca, State of Morelos, Mexico

3 Faculty of Higher Studies-Iztacala (FES-I), Laboratory of Plant Ecophysiology, UMF, National Autonomous University of Mexico (UNAM), México City, Mexico

4 Faculty of Higher Studies-Iztacala (FES-I), Laboratory of Plant Tissues Culture, UMF, National Autonomous University of Mexico (UNAM), México City, Mexico

*Address all correspondence to: sscaronte@gmail.com

References

[1] IPCC. Climate change 2023: Synthesis report. Contribution of Working Groups I, II and III to the Sixth Assessment Report of the Intergovernmental Panel on Climate Change. In: Lee, Romero J editors. IPCC Switzerland, 2023;35-115. DOI: 10.59327/IPCC/AR6-9789291691647

[2] Hijmans RJ. The effect of climate change on global potato production. American Journal of Potato Research. 2003;**80**:271-279. DOI: 10.1007/BF02855363

[3] Bajaj S. Biotechnology of nutritional improvement of potato. In: Bajaj YPS, editor. Biotechnology in Agriculture and Forestry 3: Potato. Berlin: Springer-Verlag; 1987. pp. 143-145. DOI: 10.1007/978-3-642-72773-3_10

[4] Van Dam J, Kooman PL, Struik PC. Effects of temperature and photoperiod on early growth and final number of tubers in potato (*Solanum tuberosum* L.). Potato Research. 1996;**39**:51-62. DOI: 10.1007/BF02358206

[5] Ababaker SKH, Esho KB. Producing potato from true seed review. Agricultural Science. 2022;**6**(3):51-63. DOI: 10.31080/ASAG.2022.06.1111

[6] Tizio R. Physiology of dormancy in potato tubers and its relationships with the hormonal mechanism of tuberization. Agricultural Science. 1982;**3**:91-105. DOI: 10.31047/1668.298x.v3.n1.905

[7] Ewing EE, Struik PC. Tuber formation in potato: Induction, initiation, and growth. Horticultural Reviews. 1992;**14**:89-198. DOI: 10.1002/9780470650523.ch3

[8] Gong H, Dusengemungu L, Igiraneza C, Placide P. Molecular regulation of potato tuber dormancy and sprouting: A mini-review. Plant Biotechnology Reports. 2021;**15**(4):1-18. DOI: 10.1007/S11816-021-00689-Y

[9] Kumar R. The canon of potato science: 42. Flowering Potato Research. 2007;**50**:383-385. DOI: 10.1007/s11540-008-9066-5

[10] Kondhare KR, Natarajan B, Banerjee AK. Molecular signals that govern tuber development in potato. International Journal of Developmental Biology. 2019;**64**(1):133-140. DOI: 10.1387/IJDB.190132AB

[11] Sonnewald U. Control of potato tuber sprouting. Trends in Plant Science. 2001;**6**(8):333-335. DOI: 10.1016/s1360-1385(01)02020-9

[12] Burton WG. Challenges for stress physiology in potato. American Potato Journal. 1981;**58**(1):3-14. DOI: 10.1007/BF02855376

[13] Sanabria J, Calanca P, Alarcón C, Canchari G. Potential impacts of early twenty-first century changes in temperature and precipitation on rainfed annual crops in the Central Andes of Peru. Regional Environmental Change. 2014;**14**(4):1533-1548. DOI: 10.1007/S10113-014-0595-Y

[14] Guedes ML, Haynes KG, Vinyard BT, et al. Heat tolerance in diploid wild potato species *in vitro*. American Journal of Potato Research. 2019;**96**:294-302. DOI: 10.1007/s12230-019-09716-9

[15] Kwambai TK, Griffin D, Nyongesa M, et al. Dormancy and physiological age of seed tubers from a diverse set of potato cultivars grown at different altitudes and in different seasons in Kenya.

Potato Research. 2023;**66**:1091-1115. DOI: 10.1007/s11540-023-09617-9

[16] Zhang J, Yao J, Mao L, et al. Low temperature reduces potato wound formation by inhibiting phenylpropanoid metabolism and fatty acid biosynthesis. Frontiers in Plant Science. 2023;**13**:1109953. DOI: 10.3389/fpls.2022.1109953

[17] Wszelaczyńska E, Pobereżny J, Dudek S, et al. The effects of fertilizers, irrigation, and storage on the properties of potato tubers and their constituent starches. Starch-Starke. 2015;**67**:478-492. DOI: 10.1002/STAR.201400196

[18] Prossy N, Aijuka S, Arinda O, Mateeka B, Barekye A. Early generation seed starter materials and approaches to seed production: Challenge for improving the potato seed system in Uganda. Crop Science. 2023;**27**:1-9. DOI: 10.1002/csc2.20969

[19] Haider MW, Nafees M, Iqbal R, et al. Postharvest starch, and sugars adjustment in potato tubers of wide-ranging dormancy genotypes subjected to various sprout forcing techniques. Scientific Reports. 2023;**13**(1):1-22. DOI: 10.1038/s41598-023-37711-y

[20] Qiannan H, Tang C, Zhou X, et al. Potatoes dormancy release and sprouting commencement: A review on current and prospects. Food Frontiers. 2023;**4**(3):1001-1018. DOI: 10.1002/fft2.228

[21] Kara K, Imanparast F, Kara T. The effect of pre-sprouting and planting time on different sized potato tuber yields (*Solanum tuberosum* L.). Journal of Agricultural Science. 2022;**32**(4):754-765. DOI: 10.29133/yyutbd.1136457

[22] Paul V, Buckseth T, Singh RK, et al. Alternative methods of seed potato (*Solanum tuberosum*) production: Indian perspective—A review. Current Horticulture. 2022;**10**:3-11. DOI: 10.5958/2455-7560.2022.00019.X

[23] Demirel U. Environmental requirements of potato and abiotic stress factors. In: Çalişkan ME, Bakhsh A, Jabran K, editors. Potato Production Worldwide London. London: Academic Press; 2023. pp. 71-86. DOI: 10.1016/B978-0-12-822925-5.00011-6

[24] Kumar M, Rahul L, Tiwari K, et al. Mechanistic concept of physiological, biochemical, and molecular responses of the potato crop to heat and drought stress. Plants. 2022;**11**(21):2857-2857. DOI: 10.3390/plants11212857

[25] Joshi M, Ginzberg I. Adventitious root formation in crops – potato as an example. Physiologia Plantarum. 2021;**172**(1):124-133. DOI: 10.1111/PPL.13305

[26] Almekinders CJM, Struik PC. Shoot development and flowering in potato (*Solanum tuberosum* L). Potato Research. 1996;**39**(4):581-607. DOI: 10.1007/BF02358477

[27] Struik PC. Plant Development in Potato. In: Çalişkan ME, Bakhsh A, Jabran K, editors. Potato Production Worldwide. London: Academic Press; 2023. pp. 57-70. DOI: 10.1016/B978-0-12-822925-5.00025-6

[28] Winkler E. Kartoffelbau in Tirol. II. Photosynthesevermögen und Respiration von verschiedenen Kartoffelsorten. Potato Research. 1971;**14**:1-18. DOI: 10.1007/BF02355914

[29] Havaux M. Rapid photosynthetic adaptation to heat stress triggered in potato leaves by moderately elevated temperatures. Plant, Cell & Environment. 1993;**16**(4):461-467. DOI: 10.1111/j.1365-3040.1993.tb00893.x

[30] Fleisher DH, Timlin D. Modeling Expansion of Individual Leaves in the Potato Canopy. Agricultural Meteorology. 2006;**139**:84-93. DOI: 10.1016/j.agrformet.2006.06.002

[31] Desta B, Tena N, Amare G. Response of potato to heat stress and strategic methods to cope with it. Indian Journal of Agricultural Science. 2023:1-9. DOI: 10.22159/ijags.2023.v11i1.46889

[32] Struik PC. Responses of the potato plant to the temperature. In: Vreugdenhil D, Bradshaw J, Gebhardt C, Govers F, Taylor M, MacKerron D, Ross H, editors. Potato Biology and Biotechnology. Amsterdam, The Netherlands: Elsevier; 2007. pp. 367-393. DOI: 10.1016/B978-044451018-1/50060-9

[33] Ewing EE. Heat stress and the tuberization stimulus. American Potato Journal. 1981;**58**:31-49. DOI: 10.1007/BF02855378

[34] Hancock RD, Morris WL, Ducreux L, et al. Physiological, biochemical, and molecular responses of the potato (*Solanum tuberosum* L.) plant to moderately elevated temperature. Plant, Cell & Environment. 2014;**37**(2):439-450. DOI: 10.1111/pce.12168

[35] Taiz L, Moller IM, Murphy A, Zeiger E. Plant Physiology and Development. 7th ed. Sunderland: Sinauer Associates; 2022. pp. 432-435

[36] Wolf S, Olesinski AA, Rudich J, Marani A. Effect of high temperature on photosynthesis in potatoes. Annals of Botany. 1990;**65**(2):179-185. DOI: 10.1093/oxfordjournals.aob.a087922

[37] Zhu W, Li PP, Xue C, Chen M, Wang Z, Yang Q. Potato plants overexpressing SpHsfA4c exhibit enhanced tolerance to high-temperature stress. Russian Journal of Plant Physiology. 2021;**68**(6):1-10. DOI: 10.1134/S1021443721060248

[38] Putterill J, Varkonyi-Gasic E. FT and florigen long-distance flowering control in plants. Current Opinion in Plant Biology. 2016;**33**:77-82. DOI: 10.1016/j.pbi.2016.06.008

[39] Navarro C, Abelenda JA, Cuéllar C, Tamaki S, Shimamoto K, Prat S. Control of flowering and storage organ formation in potato by FLOWERING LOCUS T. Nature. 2011;**478**:119-122. DOI: 10.1038/nature10431

[40] Morris WL, Hancock RD, Ducreux LJ, et al. Day length dependent restructuring of the leaf transcriptome and metabolome in potato genotypes with contrasting tuberization phenotypes. Plant, Cell & Environment. 2014;**37**(6):1351-1363. DOI: 10.1111/pce.12238

[41] Zhang LX, Nie B, et al. Nitrogen management can inhibit or induce the sprouting of potato tubers: Consequences of regulation tuberization. Post Biology Technology. 2022;**183**:111722. DOI: 10.1016/j.postharvbio.2021.111722

[42] Rodríguez-Falcón M, Bou J, Prat S. Seasonal control of tuberization in potato: Conserved elements with the flowering response. Annual Review of Plant Biology. 2006;**57**:151-180. DOI: 10.1146/annurev.arplant.57.032905.105224

[43] Park JS, Park SJ, Kwon SY, et al. Temporally distinct regulatory pathways coordinate thermo-responsive storage organ formation in potato. Cell Reports. 2022;**38**(13):110579. DOI: 10.1016/j.celrep.2022.110579

[44] Rykaczewska K. The effect of high temperature occurring in subsequent stages of plant development on potato

yield and tuber physiological defects. American Journal of Potato Research. 2015;**92**:339-349. DOI: 10.1007/s12230-015-9436-x

[45] George TS, Taylor MA, Dodd IC, White PJ. Climate change and consequences for potato production: A review of tolerance to emerging abiotic stress. Potato Research. 2017;**60**(3):239-268. DOI: 10.1007/S11540-018-9366-3

[46] Lehretz GG,Sonnewald S, Hornyik C, Corral JM, Sonnewald U. Post-transcriptional regulation of FLOWERING LOCUS T modulates heat-dependent source-sink development in potato. Current Biology. 2019;**29**:1-11. DOI: 10.1016/j.cub.2019.04.027

[47] Morris WL, Ducreux LJM, Morris JJ, et al. Identification of TIMING OF CAB EXPRESSION 1 as a temperature-sensitive negative regulator of tuberization in potato. Journal of Experimental Botany. 2019;**70**:5703-5714. DOI: 10.1093/jxb/erz336

[48] Abelenda JA, Cruz-Oro E, Franco-Zorrilla JM, Prat S. Potato StCONSTANS-like 1 suppresses storage organ formation by directly activating the FT-like StSP5G repressor. Current Biology. 2016;**26**:872-881. DOI: 10.1016/j.cub.2016.01.066

[49] Singh A, Siddappa S, Bhardwaj V, Singh B, Kumar D, Singh BP. Expression profiling of potato cultivars with contrasting tuberization at elevated temperature using microarray analysis. Plant Physiology and Biochemistry. 2015;**97**:108-116. DOI: 10.1016/j.plaphy.2015.09.014

[50] Gonzalez-Schain ND, Diaz-Mendoza M, Zurczak M, Suarez-Lopez P. Potato CONSTANS is involved in photoperiodic tuberization in a graft-transmissible manner. The Plant Journal. 2012;**70**:678-690. DOI: 10.1111/j.1365-313X.2012.04909.x

[51] Mingo-Castel AM, Smith OE, Kumamoto J. Studies on the carbon dioxide promotion and ethylene inhibition of tuberization in potato ex plants cultured *in vitro*. Plant Physiology. 1976;**57**:480-485. DOI: 10.1104/pp.57.4.480

[52] Mokrani K, Kühn C, Tarchoun N. Temperature effects on tuber production and carbohydrates partitioning in different cultivars during consecutive stages of potato (*Solanum tuberosum* L.). Growth. Potato Research. 2023;**66**:75-299. DOI: 10.1007/s11540-022-09543-2

[53] Salanoubat M, Belliard G. The steady-state level of potato sucrose synthase mRN a is dependent on wounding, anaerobiosis and sucrose concentration. Gene. 1989;**84**(1):181-185. DOI: 10.1016/0378-1119(89)90153-4

[54] Viola R, Roberts A, Haupt S, et al. Tuberization in potato involves a switch from apoplastic to symplastic phloem unloading. The Plant Cell. 2001;**13**(2):385-398. DOI: 10.1105/tpc.13.2.385

[55] Jaiswal S, Paul K, Raman KV, et al. Differential expression pattern of UGPase gene homologs (StUGPase1 and StUGPase2) in potato (*Solanum tuberosum* L.) during tuberization process and post-harvest storage conditions. Russ. Journal of Plant Physiology. 2023;**70**(71):38-46. DOI: 10.1134/S1021443722602749

[56] Dahal K, Li XQ, Tai H, Creelman A, Bizimungu B. Improving potato stress tolerance and tuber yield under a climate change scenario-a current overview. Frontiers in Plant Science. 2019;**10**:1-16. DOI: 10.3389/fpls.2019.00563

[57] Farrar JF, Williams ML. The effects of increased atmospheric carbon

dioxide and temperature on carbon partitioning, source-sink relations, and respiration. Plant, Cell & Environment. 1991;**14**(8):819-830. DOI: 10.1111/j.1365-3040.1991.tb01445.x

[58] Menzel CM. Tuberization in potato at high temperatures: Responses to gibberellin and growth inhibitors. Annals of Botany. 1980;**46**(3):259-265. DOI: 10.1093/oxfordjournals.aob.a085916

[59] Brown PH. The canon of potato science: 37. Stolonization, tuber induction and tuberization. Potato Research. 2007;**50**(3-4):363-365. DOI: 10.1007/s11540-008-9070-9

[60] Momčilović I. Effects of heat stress on potato productivity and nutritive quality. Hrana Ishra. 2019;**60**(2):43-48. DOI: 10.5937/hraIsh1902043M

[61] Zhang G, Tang R, Niu S, et al. Heat-stress-induced sprouting, and differential gene expression in growing potato tubers: Comparative transcriptomics with that induced by postharvest sprouting. Horticulture Research. 2021;**8**(1):226-226. DOI: 10.1038/S41438-021-00680-2

[62] Geigenberger P, Stitt M, Fernie AR. Metabolic control analysis and regulation of the conversion of sucrose to starch in growing potato tubers. Plant, Cell & Environment. 2004;**27**(6):655-673. DOI: 10.1111/j.1365-3040.2004.01183.x

[63] Denyer K, Hylton CM, Smith AM. The effect of high temperature on starch synthesis and the activity of starch synthase. Functional Plant Biology. 1994;**21**(6):783-789. DOI: 10.1071/PP9940783

[64] Kühn C. A comparison of the sucrose transporter systems of different plant species. Plant Biology. 2003;**5**(3):215-232. DOI: 10.1055/s-2003-40798

[65] Levy D, Veilleux RE. Adaptation of potato to high temperatures and salinity—A review. American Journal of Potato Research. 2007;**84**:487-506. DOI: 10.1007/BF02987885

[66] Krauss A, Marschner H. Influence of nitrogen nutrition, day length and temperature on contents of gibberellic and abscisic acid and on tuberization in potato plants. Potato Research. 1982;**25**:13-21. DOI: 10.1007/BF02357269

[67] Musita CN, Okoth MW, Abong' GO, Omayio DG. Glycoalkaloids in commercial potato varieties traded in Nairobi, Kenya. F1000 Research. 2020;**9**(423):1-12. DOI: 10.12688/f1000research.23783.2

[68] Fogelman E, Oren-Shamir M, Hirshberg J, et al. Nutritional value of potato (*Solanum tuberosum* L.) in hot climates: Anthocyanins, carotenoids, and steroidal glycoalkaloids. Planta. 2019;**249**(4):1143-1155. DOI: 10.1007/s00425-018-03078-y

Chapter 5

Heavy Metal Contamination in Soil: Implications for Crop Resilience and Abiotic Stress Management

Hany Almotairy

Abstract

This chapter rigorously examines soil toxic metal/metalloid contamination and its profound implications on crop resilience, focusing on abiotic stress conditions. It begins by elucidating the natural and anthropogenic origins of soil contamination, illustrating how plants absorb these toxicants, and elaborating on their physio-molecular responses. The chapter accentuates the detrimental manifestations of impaired photosynthesis, nutrient uptake, and oxidative stress management, underscoring the urgent need for effective mitigation strategies. Phytoremediation and genetic engineering advancements are explored as promising strategies to optimize plant resilience in contaminated environments. Novel methodologies, including phytochelatins and the strategic application of genetic engineering, demonstrate potential in improving plant growth and resilience, showcasing significant advancements toward sustainable agricultural practices. Moreover, the interaction between plants and soil microbes is dissected, revealing a symbiotic relationship that influences the bioavailability of toxic metals/metalloids and optimizes plant health under stress conditions. This insight into microbial assistance opens new avenues for research and application in crop management and soil remediation. This chapter contributes essential knowledge toward bolstering crop resilience against toxic metal/metalloid contamination by presenting cutting-edge research findings and sophisticated mitigation techniques. It emphasizes the critical role of innovative research in overcoming the challenges posed by soil contamination, paving the way for achieving sustainable agricultural productivity and food security in the face of environmental stressors.

Keywords: soil contamination, toxic metals/metalloids, crop resilience, abiotic stress, phytoremediation, genetic engineering, sustainable agricultural practices

1. Introduction to soil toxic metals/metalloids

Soil toxic metal/metalloid contamination is a paramount global issue that presents considerable challenges to food security, sustainable agriculture, and human health. Principal factors contributing to the contamination of soils with toxic metals/metalloids encompass mining operations, industrial processes, sewage sludge application, and agricultural practices involving fertilizers and pesticides [1].

Lead (Pb), cadmium (Cd), arsenic (As), mercury (Hg), and chromium (Cr) are among the prevalent toxic metals/metalloids frequently encountered in soils that have undergone contamination [2].

Toxic metal/metalloid contamination in soils is globally widespread, with hotspots often found in regions with high industrial and mining activities. In the context of China, it is observed that the levels of soil toxic metal/metalloid pollution surpass the established standard rate of 16.1%. Notably, the Yunnan-Guizhou area, the Pearl River Delta, and southeast Fujian exhibit a significant prevalence of Pb pollution [3]. In the United States, an estimated 600,000 hectares of land are affected by toxic metal/metalloid contamination, while in Europe, countries such as Germany, the United Kingdom, Italy, and Spain, collectively account for approximately 400,000 contaminated sites [3]. A study conducted in southern Tibet, China, found that soil Cu, Zn, and Cd contents near national and nonnational highways were significantly higher than the corresponding background values in the region [3].

Similarly, a study in an industrial agglomeration area in China revealed high-resolution risk mapping of toxic metals/metalloids in soil, identifying high-risk hotspots [4]. These global hotspots of toxic metal/metalloid contamination in soils pose significant risks to agricultural productivity, food safety, and public health. Therefore, developing effective strategies for mitigating toxic metal/metalloid stress in soils and ensuring a safe and sustainable agricultural future is crucial.

Understanding the intricate ways toxic metals/metalloids exacerbate abiotic stress in crops is crucial as it directly impacts agricultural resilience, necessitating innovative approaches for effective management and remediation of contaminated soils.

The term "heavy metals" has been used in environmental and pollution studies for decades, yet its ambiguous nature and the absence of a standardized definition have ignited debates. It frequently amalgamates metals, metalloids, and nonmetals, absent a clear delineation, and is used as a group name for metals associated with contamination and potential toxicity [5]. The controversy over the definition of "heavy metals" complicates scientific discourse and influences public perception and policymaking by obscuring the distinctions between truly hazardous elements and those that are essential or benign at specific concentrations. Consequently, toxic metals/metalloids, rather than merely metal/metalloids, are traditionally delineated as metallic elements with a greater density than water. They are naturally occurring elements with a considerable atomic weight and at least five times the density of water. These elements are often associated with contamination, potential toxicity on living organisms and plants, and their negative impact on human health [6].

Metals can be classified as essential to a living organism and nonessential metals. Essential toxic metals/metalloids are required by living organisms, often functioning as protein cofactors in various biological processes. They are nontoxic when present in trace amounts in an organism but can become toxic if concentrations exceed a certain threshold. Examples of essential metals include copper (Cu), Fe, and zinc [7]. On the other hand, nonessential toxic metals/metalloids have no biological function and are toxic to the organism, even in trace amounts. Nonessential metal ions are chemically like their essential metal counterparts. They can interact similarly with the proteins and pathways that regulate necessary metal ions, although generally disrupting, rather than enabling, these [7]. For instance, essential metals, such as Fe, Cobalt (Co), Cu, Mn, Zn, and molybdenum (Mo), play crucial roles in plant development by participating in metabolic reactions and acting as micronutrients. However, when they exceed their threshold concentrations, their actions are considered toxic to plant development [8].

The toxicity thresholds for toxic metals/metalloids in plants encompass the critical concentration levels beyond which these elements transition from benign or beneficial to detrimental for plant growth and development. This delineation is vital as it underscores the narrow margin within which the biochemical balance of metal ions is maintained in plant systems. Elements, such as As, Ca, Pb, and Cr, among others, exemplify the broad spectrum of metals that, upon surpassing their respective threshold concentrations, initiate adverse effects on plant physiology and development [8].

Central to the detrimental impact of exceeding these toxicity thresholds is generating reactive oxygen species (ROS), a suite of highly reactive molecules derived from molecular oxygen. While integral to normal cellular processes, ROS can inflict significant damage when their production overwhelms the plant's antioxidative defense mechanisms. Some toxic metals/metalloids, such as Hg and As, can directly alter protein/enzyme activities by targeting their -SH group, further impeding cellular metabolism [9].

A prominent manifestation of toxic metal/metalloid-induced stress is the inhibition of photosynthesis, a process foundational to plant growth and energy production. The mechanism underlying this inhibition is twofold: firstly, toxic metals/metalloids can optimize the activity of chlorophyllase, an enzyme responsible for the degradation of chlorophyll molecules, and secondly, they can replace the central magnesium (Mg) ion in the chlorophyll molecule's porphyrin ring. Such replacements affect the molecule's structural integrity and functional capability, culminating in reduced photosynthetic capacity, and, by extension, diminished plant growth and yield [9].

The threshold concentrations of toxic metals/metalloids in plants vary depending on the specific metal. For instance, the World Health Organization (WHO) has set the permissible limits/MRL of their concentrations ($mg{\cdot}Kg^{-1}$) in plants: Cu (10), Pb (2), Zn (0.6), and Cd (0.02). Metals with high permissible limits are assumed to be safe. In plants, the permissible limits of Cu are the highest, followed by Pb and Zn, while the permissible limits of Cd are the lowest. These limited values mean that the accumulation of Cd in the plants, even at a lower concentration, is more toxic than the others [10].

Comparative analysis with existing literature reveals a consensus on the mechanism of ROS generation as a pivotal factor in metal/metalloid-induced toxicity. Studies across various plant species and metal/metalloid stressors consistently report oxidative stress, characterized by an overproduction of ROS, as an expected outcome of exposure to toxic metal concentrations beyond the threshold levels. This oxidative stress parallels findings in related studies, which underscore the universal challenge of managing metal/metalloid toxicity across diverse plant systems. However, variations in species-specific responses, metal uptake efficiencies, and antioxidative defense capacities highlight the complexity of metal/metalloid-plant interactions and underscore the significance of tailored research and mitigation strategies for managing metal/metalloid stress in plants.

In the broader context of toxic metal/metalloid research, the intricate balance between metals' beneficial and harmful effects underscores the complexity of plant responses to abiotic stress. Future research directions could include elucidating the molecular and genetic mechanisms underpinning these responses and focusing on identifying essential genes and pathways involved in metal uptake, sequestration, and detoxification. Furthermore, exploring the potential for genetic engineering and biotechnological interventions to optimize plant resilience against metal/metalloid toxicity presents a promising avenue for sustainable agricultural practices in contaminated environments.

The escalation of soil toxic metal/metalloid contamination is a critical environmental concern attributable to a complex interplay between natural phenomena and human activities (**Figure 1**). This bifurcation of sources, each with its unique contribution to soil metal levels, underscores the complex dynamics of environmental contamination. An in-depth examination of these sources reveals the inherent variability in contamination mechanisms and the exacerbating influence of human activities on the dispersal and concentration of these toxicants beyond natural background levels.

Intrinsic to the Earth's biogeochemical cycles, natural sources such as volcanic emissions, ocean salt sprays, eolian dispersion of soil particles, forest fires, and rock weathering constitute primary pathways through which toxic metals/metalloids infiltrate the soil matrix [11]. While fundamental to ecological balances, these phenomena introduce metals such as As, Hg, and Cd into terrestrial environments, albeit at concentrations typically within natural baseline levels. For instance, volcanic eruptions and rock weathering are pivotal in cycling essential and nonessential metals through the environment, subtly influencing soil composition over geological timescales [11, 12].

Contrastingly, anthropogenic activities amplify the presence and impact of toxic metals/metalloids in soil, often elevating concentrations well above natural background levels. Industrial processes, mining operations, and agricultural practices emerge as dominant anthropogenic vectors, introducing hazardous elements, such as As, Cd, Cr, Cu, Hg, and Pb, into the soil. These elements, deriving from sewage discharge, paints, alloys, electronic waste, and mining effluents, not only elevate soil metal concentrations but also pose considerable environmental and health hazards due to their persistence and bioaccumulative nature [11].

A comparative analysis underscores the disproportionate impact of human activities on soil contamination levels. For example, a notable study by Ma et al. [13] conducted in Tianjin, China, illuminates the extensive contribution of industrial and traffic-related factors to elevated soil concentrations such as Cd, Pb, and zinc (Zn). Such findings, indicative of widespread anthropogenic influence, resonate with observations in Saudi Arabia and Mayabeque, Cuba, where mining and agricultural practices significantly exacerbate soil metal concentrations. These examples, drawn

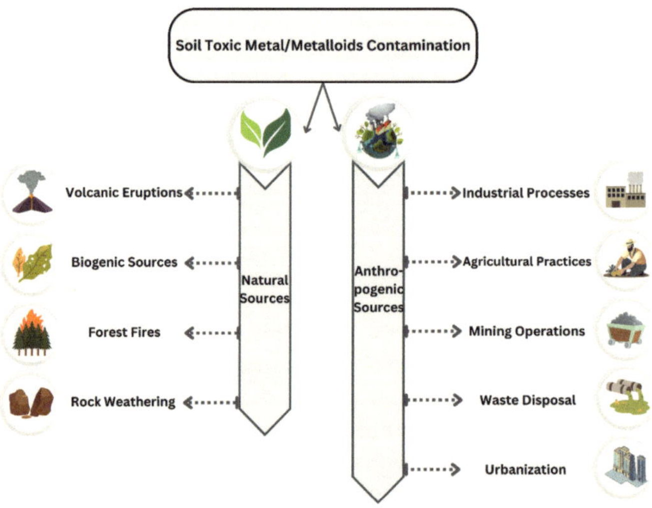

Figure 1.
Sources of soil toxic metal/metalloid contamination. This diagram depicts the main natural and anthropogenic sources contributing to soil contamination.

from diverse geographical contexts, elucidate the global scope of anthropogenic soil contamination, highlighting regional variations in sources and impacts [13].

Mining operations also play a critical role in soil contamination with toxic metals/metalloids. Research by Al-Swadi et al. [14] examined the influence of industrial and mining activities on soil and dust in urban and suburban locales in Saudi Arabia. The findings revealed that toxic metals/metalloids, such as Cd, Cu, Zn, and Pb, in mining areas were largely anthropogenic in origin. Contamination levels were markedly higher near mining sites, highlighting the substantial impact of mining activities on the escalation of soil toxic metal/metalloid concentrations [14].

Agricultural practices, especially fertilizers and pesticides, are another significant source of soil toxic metal/metalloid contamination. An investigation by Sosa et al. [15] into the soil composition in Mayabeque, Cuba, revealed the presence of toxic metals/metalloids such as Cu, Pb, Zn, and Hg, which are significantly influenced by agricultural activities. The study indicated that agricultural runoff, laden with these metals, considerably contributes to their accumulation in soil [15].

The advent of industrialization and the relentless pace of urbanization have precipitated a marked escalation in soil toxic metal/metalloid contamination, engendering significant environmental and public health challenges. This phenomenon, characterized by the rapid expansion of industrial operations and urban sprawls, has catalyzed the pervasive dissemination of pollutants, notably toxic metals/metalloids such as Cu, Zn, As, and Hg, into terrestrial ecosystems. The activities central to this dissemination—ranging from the combustion of fossil fuels to the smelting and diverse manufacturing processes—distort the biosphere's equilibrium, leading to the pronounced accumulation of toxic metals/metalloids in soil matrixes [8].

Urbanization, characterized by extensive anthropogenic activity, contributes to soil toxic metal/metalloid pollution. The conversion of rural land for urban development and population growth leads to environmental degradation and diverse pollution [16]. Discharging wastewater, which often contains toxic metals/metalloids, is a significant source of soil contamination in urban areas [16].

Urbanization further amplifies this contamination through substantial anthropogenic activities. The transformation of rural landscapes into urban and industrial zones contributes to environmental degradation and facilitates the introduction of varied pollutants into the soil [16]. A significant vector of this contamination is the discharge of wastewater enriched with toxic metals/metalloids from urban and industrial sources. This effluent, often a byproduct of myriad industrial processes, including plating, electroplating, battery manufacturing, pesticide production, and more, embodies a composite of harmful substances that, when released untreated into the environment, jeopardizes ecological integrity and human health [16, 17].

The ramifications of these activities extend beyond immediate environmental degradation, heralding long-term ecological and health consequences. Wastewater discharge, in particular, serves as a conduit for toxic metals/metalloids to infiltrate terrestrial and aquatic ecosystems, disrupting the ecological balance and posing bioaccumulation risks to flora and fauna. This bioaccumulation, in turn, facilitates the entry of toxic metals/metalloids into the human food chain, manifesting in a spectrum of adverse health outcomes ranging from acute poisoning to chronic conditions such as neurological disorders, cardiovascular diseases, and cancer.

Current research in environmental pollution and public health directly correlates soil contamination with toxic metals/metalloids and various health issues. Studies highlight the insidious nature of these contaminants, underscoring the necessity for vigilant monitoring and remediation strategies. The role of industrialization and urbanization in exacerbating soil contamination necessitates reevaluating waste management practices and industrial processes, advocating for sustainable development paradigms prioritizing environmental health and safety.

In light of these considerations, the impact of industrialization and urbanization on soil toxic metal/metalloid contamination embodies a multifaceted challenge. It necessitates a concerted effort among policymakers, industry stakeholders, and the scientific community to devise and implement comprehensive strategies to mitigate contamination levels and safeguard public health and ecological systems for future generations.

Other anthropogenic activities contribute to soil toxic metal/metalloid contamination besides industrialization and urbanization. These include farming practices, mining, and the use of domestic effluents [8].

The inadvertent release of toxic metals/metalloids through accidental spills or leaks during industrial operations or transportation constitutes a significant source of environmental contamination [11]. Such incidents, often precipitated by deficiencies in infrastructure integrity or lapses in waste management protocols, underscore the vulnerability of natural ecosystems to anthropogenic disturbances. Due to aging infrastructure or oversight in management practices, releasing toxicants in these contexts compromises soil health and poses broader ecological risks [18].

A critical concern within agriculture is using inorganic fertilizers derived from raw materials, such as phosphate rock [19]. While essential for enhancing crop yield and soil fertility, these fertilizers can also serve as conduits for introducing toxic metals/metalloids into agricultural soils. Phosphate rock, for instance, contains trace amounts of cadmium (Cd) and other potentially harmful elements that, when processed into fertilizers and applied to fields, can accumulate in the soil and subsequently be overtaken by crops [19]. This pathway threatens the integrity of soil ecosystems and raises significant concerns for food safety and human health as the bioaccumulation of toxic metals/metalloids in edible plant parts can lead to their entry into the food chain.

The ramifications of such contamination extend beyond immediate health risks to encompass broader challenges for agricultural sustainability. The persistent nature of toxic metals/metalloids in soils can impair soil biodiversity and fertility over time, undermining efforts to achieve sustainable agricultural practices. Furthermore, the accumulation of toxic metals/metalloids in agricultural produce can compromise food safety, necessitating rigorous monitoring and management measures to protect consumer health.

Recent studies have sought to quantify the impacts of inorganic fertilizer use on soil metal concentrations and the resultant implications for crop safety. For example, research has highlighted the potential for significant increases in soil Cd levels following prolonged phosphate fertilizer application, with consequent risks to crop quality and food safety [19]. These findings underscore the need for a balanced approach to fertilizer use that considers both nutritional efficacy and environmental safety to ensure the long-term viability of agricultural systems and the health of the consuming public.

In light of these considerations, addressing the contributions of various anthropogenic sources to soil toxic metal/metalloid contamination requires a multifaceted strategy. This strategy should encompass enhanced regulatory oversight of industrial and agricultural practices, investment in infrastructure maintenance and waste management technologies, and the promotion of sustainable agricultural inputs.

Only through such a comprehensive approach can the dual objectives of environmental protection and food safety be effectively realized.

2. Mechanisms of toxic metals/metalloids uptake in plants

2.1 Root absorption and translocation

Plant roots assimilate toxic metals/metalloids *via* two primary pathways: passive diffusion (apoplastic pathway) and active transport (symplastic pathway) [20]. Passive absorption entails transporting metals along a concentration gradient without energy input. This process is primarily influenced by the concentration of toxic metals/metalloids in the soil solution [20].

On the other hand, active assimilation necessitates energy and entails specific protein translocation. This process allows plants to absorb toxic metals/metalloids even when their concentrations in the soil solution are low. Active transport is essential to uptake critical toxic metals/metalloids, which plants require in trace amounts for various physiological functions [8, 21].

It is worth noting that the balance between passive and active uptake can be influenced by various factors, including the plant species, the type of toxic metals/metalloids, and the environmental conditions [8, 21]. For instance, some plants have developed mechanisms to limit the passive uptake of nonessential and potentially toxic metals/metalloids while enhancing the active uptake of essential ones [8].

The Casparian strip (CS) and endodermis constitute pivotal elements in the plant root system that act as barriers to metal uptake. The CS is a band-like structure of lignin impregnation in the primary cell wall between endodermal cells, preventing the apoplastic flow of substances from the soil solution to the vascular cylinder [22]. This barrier ensures that minerals, including toxic metals/metalloids, must pass through the selective symplastic pathway to enter the plant's vascular system, allowing for the selective uptake of essential mineral elements and restricting the movement of potentially harmful substances [22, 23].

The endodermis, the innermost cortex layer, surrounds the vascular cylinder and plays a pivotal role in controlling the radial movement of minerals from the soil solution to the stele [22]. The presence of suberin and lignin in the cell walls of the endodermis further contributes to the barrier function, limiting the radial transport of metals and other solutes [24]. This selective permeability is crucial for the plant's ability to regulate the internal concentrations of toxic metals/metalloids and other elements, thereby protecting the plant from potential toxicity [23, 24].

Moreover, the CS and suberin lamellae in the endodermis can adapt in response to toxic metal/metalloid stress, potentially modifying their structure and composition to fortify the plant's defense mechanisms against the toxicity of toxic metals/metalloids [23, 24]. These adaptations may include changes in the thickness and permeability of the endodermal cell walls, which can affect the plant's ability to absorb and translocate toxic metals/metalloids [23, 24].

2.2 Bioavailability of toxic metals/metalloids in soil

The bioavailability of toxic metals/metalloids in soil, which refers to the portion of toxic metals/metalloids that plants can absorb, is strongly influenced by various soil properties, including redox potential, pH, and organic matter content [19].

The pH of soil plays a vital role in determining the solubility of toxic metals/metalloids. Generally, toxic metals/metalloids are more soluble and, hence, more bioavailable in acidic soils than in neutral or alkaline soils. The increased hydrogen ion concentration in acidic soils can displace toxic metals/metalloids from soil particles, making them more available for plant uptake [20, 25].

Redox potential, often denoted as Eh, also affects the solubility of toxic metals/metalloids. Under reducing conditions (low Eh), some toxic metals/metalloids can be transformed into more soluble forms, increasing their bioavailability. Conversely, under oxidizing conditions (high Eh), toxic metals/metalloids may precipitate or adsorb onto soil particles, reducing their bioavailability [26, 27].

Organic matter in soil can influence toxic metal/metalloid bioavailability in several ways. It can bind toxic metals/metalloids, reducing their solubility and bioavailability. However, specific components of organic matter, known as chelating agents, can form soluble complexes with toxic metals/metalloids, enhancing their bioavailability. Moreover, organic matter can affect soil pH and redox potential, indirectly influencing toxic metal/metalloid bioavailability [19, 20, 28, 29].

A comprehensive elucidation of the determinants influencing the bioavailability of toxic metals/metalloids in the soil matrix and their subsequent implications for phytoaccumulation is paramount for the nuanced understanding of plant-soil interaction dynamics under contamination stress. **Table 1** delineates these determinants, their mechanisms of action, and the resultant implications for plant uptake of toxic metals/metalloids.

Other metals can significantly influence plants' bioavailability and soil uptake of toxic metals/metalloids. This phenomenon is known as competitive adsorption, where multiple toxic metals/metalloids may compete for the same sorption sites within the soil matrix [30, 31]. The outcome of this competition can affect the strength and

Determinant	Influence on Bioavailability	Implications for Phytoaccumulation
Soil pH	Acidic conditions facilitate the solubilization of metals, thereby augmenting their bioavailability due to proton-induced desorption from soil particulates.	Enhanced phytoaccumulation in acidic environments, potentially escalating toxicity risks
Redox Potential (Eh)	Reductive states increase the solubility and availability of certain metals/metalloids by promoting their transformation into more soluble forms. Oxidative states, conversely, may precipitate metals, thus diminishing their bioavailability.	Reduction conditions may amplify the uptake of more bioavailable metal/metalloid species, while oxidation could inhibit uptake due to the precipitation of metals.
Organic Matter Content	Organic matter modulates metal solubility and bioavailability through complexation. Chelating agents enhance bioavailability by forming soluble complexes, whereas other organic compounds may immobilize metals.	The impact on metal uptake is contingent upon the composition of organic matter, potentially either facilitating or impeding phytoaccumulation.
Metal Interactions	The phenomenon of competitive adsorption among metals at soil sorption sites and plant root interfaces alters each metal's bioavailability.	The presence of competing metals may reduce or enhance the uptake of specific metals, affecting the equilibrium of nutrient and metal assimilation.

Table 1.
Determinants of metals bioavailability in soil and implications for phytoaccumulation.

magnitude of metal retention by the soil, potentially increasing the mobility and availability of certain toxic metals/metalloids [30].

For example, a high concentration of one metal can reduce the adsorption of another metal as they compete for binding sites on soil particles or within the root system of plants [30, 31]. This can increase the concentration of the less strongly adsorbed metal in the soil solution, making it more available for plant uptake.

The specific interactions between metals can vary depending on several factors, including the types of metals involved, their relative concentrations, and the soil properties, such as pH and organic matter content [30, 31]. Some metals may have a higher affinity for adsorption sites, thereby out-competing others and influencing their bioavailability.

Understanding the competitive interactions between toxic metals/metalloids is crucial for predicting their soil behavior and potential plant uptake, particularly in areas with multi-metal contamination [32, 33]. This knowledge can inform management practices to reduce the risk of toxic metal/metalloid uptake by crops, thereby protecting food safety and human health.

The bioavailability of toxic metals/metalloids in soil and their plant uptake can be influenced by other metals, leading to synergistic or antagonistic effects. These interactions can either inhibit or optimize the uptake and toxicity of different metals, affecting plant growth and development of metals [34, 35].

Synergistic effects occur when one metal's presence optimizes another's uptake or toxicity. For instance, in a study on a plant, *Conocarpus lancifolius*, it was found that Pb) facilitated the accumulation of Cd without affecting the accumulation of nickel (Ni) [35]. On the other hand, antagonistic effects occur when one metal's presence inhibits the uptake or toxicity of another. For instance, applying Zn fertilizers reduced the values of all toxic metals/metalloids in groundnut seeds except Zn, indicating an antagonistic effect [36]. Similarly, it was found that phosphorus interferes with Zn uptake as Zn uptake by plants is reduced by increasing phosphorus in soil [36].

The toxicity of toxic metals/metalloids in plants can lead to several physiological and morphological changes responsible for the decline in growth. For instance, plants exposed to Cd demonstrated reduced water and nutrient uptake and a decrease in the rate of photosynthesis, leading to inhibition of growth and browning of root tips that ultimately lead to cell death [37].

The bioavailability and uptake of toxic metals/metalloids are also influenced by pollution level, soil pH, and exposure time [38]. Furthermore, plants' uptake of toxic metals/metalloids is concentration dependent and depends on whether the plant is exposed to single or multiple toxic metals/metalloids [35].

2.3 Role of transporters and chelators

Transport proteins are crucial in the uptake and movement of toxic metals/metalloids in plants. Critical families of transport proteins include toxic metal/metalloid ATPase transporter proteins (toxic metal/metalloid As), natural resistance-associated macrophage proteins (NRAMPs), cation diffusion facilitator (CDF) family, and ZRT, IRT-like Protein (ZIP) family. These proteins transport various toxic metals/metalloids, including Zn, Fe, Cd, and Co, and are essential for plants' uptake and detoxification processes [39]).

The regulation and expression of these transport proteins can be particular to the type of metal, the plant species, and the environmental conditions. Understanding the functional mechanisms of these transporters and their regulation is vital for developing strategies to enhance plant tolerance to toxic metal/metalloid stress and for phytoremediation applications [40].

Plants have developed various strategies to cope with toxic metal/metalloid toxicity, including metal sequestration. This process involves compartmentalizing or detoxifying toxic metals/metalloids, restricting them from interfering with vital metabolic pathways [41].

Toxic metals/metalloids can be transported out of the cell or restricted to the vacuole, a process known as sequestration or detoxification. This process allows plants to survive in metal-contaminated areas without experiencing toxic effects [41]. The toxic metal/metalloid transporters are critical in plant resistance to metal toxicity, managing uptake, and cellular regulation of metals. Critical transporter families include CPx-type ATPases, ZIP, and NRAMP, which handle essential metals such as Fe, Cu, Zn, and Cd. Intracellular transporters, especially from the HMA, ABC, and CDF families, are sequestering toxic metals/metalloids into vacuoles, aiding in detoxification and homeostasis [41].

In addition to compartmentalization, plants also employ chelation to cope with toxic metal/metalloid toxicity. Nonprotein thiol compounds, such as glutathione and phytochelatins (PCs), are directly involved in the fight against toxic metal/metalloid stress. Under exposure to toxic metals/metalloids, these compounds can chelate toxic metals/metalloids and then transport them into vacuoles *via* ABC-like transporters, transport proteins found in all kingdoms of life, including plants, thereby leading to a reduction in the number of metals present in the cytoplasm, which, in turn, shields the plants from the harmful consequences associated with these metals [42].

Another crucial defense mechanism in enhancing plant resistance to toxic metals/metalloids is the compartmentalization of toxic metals/metalloids in cell walls. This mechanism protects cellular organelles and components, such as cell membranes, mitochondria, endoplasmic reticulum, and nuclei, from the harmful effects of toxic metals/metalloids [43].

2.4 Influence of soil pH on metal uptake

The pH of the soil is a critical factor that influences the solubility and bioavailability of toxic metals/metalloids, affecting plant uptake. Lower pH increases toxic metal/metalloid solubility, facilitating plants' uptake of Mn, Cu, and Zn [44]. For example, Mn solubility is prompted by acidic soils, while Cu exists as hydrolysis products in soil with pH below 7, making it more bioavailable; Zn, being amphoteric, is highly mobile and bioavailable in acidic conditions; toxic metals/metalloids tend to form barely soluble phosphates and carbonates at higher pH levels, reducing their bioavailability [44].

The relationship between soil pH and toxic metal/metalloid solubility is complex and can be influenced by other soil properties and environmental conditions. However, the consensus is that acidic conditions increase the solubility and mobility of toxic metals/metalloids, making them more available for plant uptake, while alkaline conditions immobilize them [29, 44–46]. Comprehending these interactions is vital for effectively managing soil pollution and ensuring crop safety in impacted regions.

Acidic soil conditions enhance root expansion and growth in *Rhododendron*, potentially increasing the plant's uptake of toxic metals/metalloids, as evidenced by the more substantial root systems observed at a pH of 5.5 compared to 6.5 [47]. Likewise, *Lupinus angustifolius* demonstrates increased specific root length and a predominance of thinner roots in low pH environments, suggesting a possible alteration in toxic metal/metalloid absorption. At the same time, alkaline conditions notably suppress lateral root formation and favor thicker root development, influencing plant-metal interactions [48]. Furthermore, soil pH fluctuations, particularly at higher pH levels, induce changes in root surface traits, such as iron plaque formation, which correlates with the uptake of metals such as cadmium and zinc, indicating a pH-dependent modulation of toxic metal/metalloid sequestration [49].

Soil's buffering capacity and resistance to pH fluctuations are critical in controlling toxic metal/metalloid bioavailability and plant uptake. Calcareous soils, known for their higher buffering capacity, tend to immobilize toxic metals/metalloids more than acidic soils, even under similar contamination levels [45, 50]; such immobilization, characteristic of high pH environments, limit toxic metal/metalloids mobility [46]. Conversely, bioavailable toxic metals/metalloids increase with soil acidity due to their lower buffering capacity, enhancing mobility [45]. Enhancing soil buffering can mitigate acidification and reduce available toxic metal/metalloid content [51].

2.5 Influence of organic matter on metal uptake

Recent studies have highlighted the complex interplay between soil organic matter (SOM) decomposition and the release and availability of toxic metals/metalloids to plants. A variety of soil properties and environmental factors influence these interactions.

Soil pH is a pivotal factor affecting toxic metal/metalloid solubility and plant uptake, with lower pH increasing Mn, Cu, and Zn availability [44]. Mn solubility rises in acidic soils, Cu becomes more available as hydrolysis products below pH 7, and Zn's mobility and bioavailability are heightened in acidic conditions. In alkaline soils, toxic metal/metalloids are immobilized, forming insoluble phosphates and carbonates [29, 45, 46], with soil pH's influence on toxic metal/metalloid bioavailability being further complexified by other soil properties and environmental conditions.

Organic matter, especially dissolved organic matter (DOM), has intricate structures and interactions that can significantly impact environmental processes. One such process is the complexation and interface reactions of toxic metals/metalloids, which can be affected by the formation of complex molecules between organic matter and toxic metals/metalloids. This, in turn, can alter the metals' bioavailability [52].

Organic functional groups within soil organics play a pivotal role in the adsorption dynamics of metals, with carboxylate groups being notably influential, as demonstrated by Engel et al. [53]. Stefanowicz et al. [54] highlight that organic matter contributes to the vitality of soil microbiota, which indirectly impacts metal bioavailability. Yu et al. provide insights into how humic substances' diverse binding properties can sequester or mobilize metals, significantly affecting plant uptake. This collective research underscores the complexity of soil metal interactions and their significance in plant-metal assimilation.

In addition to other factors, organic matter can affect the bioavailability of toxic metals/metalloids through chelation. Chelation is a binding process in which a single metal ion is bound to two or more atoms of organic matter. This process can increase the solubility and mobility of toxic metals/metalloids, enhancing their bioavailability to plants [55].

However, the effect of organic matter on metal bioavailability can be more complex. For instance, some studies have found that applying organic amendments, such as compost, can sometimes increase the accumulation of certain toxic metals/metalloids in plants [56].

Humic and fulvic acids, significant components of soil organic matter, can significantly influence plants' uptake of toxic metals/metalloids. They can form complexes with metals, affecting their solubility and bioavailability [57–64].

Humic substances, including humic and fulvic acids, can significantly affect the environmental behavior of metal pollutants through processes, such as ion exchange and physical adsorption [60]. They can form complexes with toxic metals/metalloids, altering their mobility and bioavailability [57, 58, 61]. For instance, forming metal-humic complexes can decrease mobility and bioavailability [61]. However, in some cases, the formation of humic and fulvic acid complexes increased toxic metal/metalloid toxicity [58].

The binding of toxic metals/metalloids by humic and fulvic acids can be influenced by the number and identity of organic functional groups within these acids Steinberg & Hodge [58]. This binding can either immobilize metals, reducing their bioavailability, or keep them soluble, increasing their bioavailability [60].

Humic and fulvic acids can impact the bioavailability of toxic metals/metalloids by chelation. Chelation is a process where a single metal ion is bound to dual or more atoms of organic matter; this process can enhance the solubility and mobility of toxic metals/metalloids, which may increase their bioavailability to plants [62].

However, the effect of humic and fulvic acids on metal bioavailability can be more complex. For instance, some studies have found that humic acids can interrelate with metals to form metal-humate complexes, enhancing metal adsorption capacity and influencing plant growth and metal availability [57]. On the other hand, fulvic acids have been reported to inhibit metal availability and might be employed to decrease metal bioavailability [62].

The decomposition of organic matter is a vital process in the global carbon cycle and climate feedback as it redistributes energy and nutrients in ecosystems and determines the amount of soil organic carbon (SOC); this process involves the physical breakdown of substrate and biochemical transformation of complex organic compounds [65].

The relationship between the decomposition of organic matter and the release of toxic metals/metalloids is complex. The decomposition process can influence the environment's mobility and bioavailability of toxic metals/metalloids. Moreover, the decomposition of organic matter can lead to the release of toxic metals/metalloids, with the release rate increasing with increases in organic carbon mineralization [66].

Microbial activity related to organic matter breakdown can also transform toxic metals/metalloids into more bioavailable forms. For instance, some bacteria can convert less bioavailable forms of toxic metals/metalloids into more soluble forms that plants can take up [11]. Additionally, producing organic acids during decomposition can lead to the dissolution of metal complexes and increase the concentration of free metal ions in the soil solution [67].

However, releasing toxic metals/metalloids during organic matter decomposition can also pose risks, such as increased metal leaching to groundwater or uptake by nontarget organisms, which can have ecological and health implications [11, 31, 67, 68].

Microbial activity in organic-rich soils plays a pivotal role in influencing the bioavailability of toxic metals/metalloids [69–71] through various processes, including mineralization and immobilization [54, 72–75].

Microorganisms can initiate metal mobilization or immobilization by redox reactions, impacting bioremediation processes [76]. For example, certain bacteria can convert less bioavailable forms of toxic metals/metalloids into more soluble forms that plants can take up [54, 72]. Additionally, microbial-induced biomineralization can transform free v ions into less bioavailable forms, such as carbonates or phosphates, thus reducing their mobility and uptake by plants [77].

Microbial interventions in the bioremediation of toxic metal/metalloid contaminants in agroecosystems are crucial to prevent the leaching of toxic metals/metalloids or their mobilization and simplify toxic metal/metalloid extraction [69, 72]. Using genetically modified microbes and immobilized microbial cells for toxic metal/metalloid remediation is promising [72].

Furthermore, microbial communities can influence the bioavailability of soil pollutants, facilitating their entry into terrestrial food webs [78]. The application of nitrogen and sulfur fertilizers can regulate microbial communities to affect toxic metal/metalloid absorption, as seen in studies with *Salix integra Thunb*., where the bioavailability of toxic metals/metalloids was increased, and the microbial community was altered [71].

3. Physiological responses of plants to stress of toxic metals/metalloids

3.1 Perturbation of cellular mechanisms

Toxic metals/metalloids s can disrupt cellular membranes, leading to a cascade of detrimental effects on plant cells. The coherence of cellular membranes is paramount for maintaining homeostasis and controlling the movement of substances into and out of the cell. When toxic metals/metalloids, such as Cd and Pb, accumulate in plant tissues, they can interact with membrane lipids and proteins, causing lipid peroxidation, a critical oxidative process damaging cell membranes and modifying membrane viscosity and permeability [79–82].

Lipid peroxidation is when ROS, often induced by toxic metal/metalloid stress, attacks the unsaturated fatty acids in the lipid bilayer, forming malondialdehyde (MDA) and other toxic by-products. This oxidative damage to membranes can disrupt normal cellular functions, including nutrient transport, signal transduction, and energy transduction [80, 81].

Toxic metals/metalloids can also bind to membrane proteins, affecting their structure and function. This can inhibit the activity of membrane-bound enzymes, disrupt ion channels and transporters, and interfere with membrane receptors, leading to impaired cellular signaling and metabolic dysregulation [79, 80, 83, 84].

The disruption of membrane integrity by toxic metals/metalloids can have various physiological consequences, including decreased photosynthetic efficiency, altered water relations, and reduced nutrient uptake, ultimately leading to stunted plant growth and development [79–82].

Toxic metals/metalloids can inhibit various plant enzymes, leading to various detrimental effects on plant physiology and soil health. Understanding these interactions is critical for managing toxic metal/metalloid contamination and protecting plant and soil health [80, 85–88].

Toxic metals/metalloids can inhibit essential enzymatic reactions within plant cells, disrupting normal physiological processes. Enzymes are crucial for a wide range of metabolic activities, and their inhibition can lead to reduced plant growth, compromised pigment production, hormonal and nutrient imbalance, and even cell death [80].

The cytotoxic and genotoxic effects of toxic metals/metalloids are partly due to their interaction with enzymes, which can bind to the active sites or alter the enzyme's conformation, rendering them inactive [80]. For example, toxic metals/metalloids, such as Cd and Pb, have been shown to inhibit enzymes involved in photosynthesis and respiration, affecting the plant's ability to produce energy and synthesize important biomolecules [86, 88].

Toxic metals/metalloids can also affect soil enzymatic activity, which is crucial for nutrient cycling and availability. For instance, Zn and Cu have been reported to inhibit dehydrogenase and urease activities in soil, which are essential for organic matter decomposition and nitrogen cycling [85]. This inhibition can lead to changes in soil properties, productivity limitations, and altered ecosystem function [87].

The inhibition of enzymes by toxic metals/metalloids is not limited to those directly involved in metabolic pathways. Still, it can also extend to those responsible for DNA replication, gene expression, and cell division, leading to genome instability and affecting plant health and development [80].

3.2 Impact on photosynthesis and respiration

Toxic metals/metalloids can significantly impact the synthesis and function of chlorophyll, a crucial component of the photosynthesis process, in plants, leading to adverse effects on plant growth and productivity.

Cd stress has been observed to hinder the expression of crucial enzymes responsible for chlorophyll synthesis in tobacco leaves, reducing chlorophyll concentration. This inhibition of chlorophyll synthesis can significantly downregulate the expressions of photosynthesis-related proteins or subunits, suppress both the xanthophyll cycle and NDH-CEF process, and block the photosynthetic electron transport, resulting in severe photoinhibition of both photosystem II (PSII) and photosystem I (PSI) [89].

In contrast, Zn stress has little effect on tobacco leaves' photosynthetic function [89]. However, both Cd and Zn can disturb the energy migration from the antenna complexes to the chlorophyll of the reaction centers, leading to an increased inhibition of photosynthetic electron transport [88].

Another study found that chlorophyll synthesis products increased with increasing concentrations of Pb and Cd while the degradation products decreased. These findings indicate that toxic metals/metalloids indirectly impact photosynthesis by regulating the functions of enzymes involved in the manufacture and breakdown of chlorophyll [90].

Toxic metals/metalloids can significantly impact plants' stomatal conductance and gas exchange, reducing photosynthesis rates and other physiological changes. These influences can vary depending on the type and concentration of the toxic metals/metalloids and the specific plant species involved [79, 91, 92].

The impact of toxic metals/metalloids on stomatal conductance and gas exchange in plants is a well-documented area of research. Stomatal conductance refers to the degree of stomatal opening, which directly affects plant transpiration. Exposure to toxic metals/metalloids primarily induces a reduction in plant stomatal conductance, although there are also opposing findings [91]; for instance, Cd has been shown to cause a significant decrease in the stomatal conductance of various plant species [91].

Toxic metals/metalloids can also affect the hydraulic conductivity of plants. For example, high Zn concentrations significantly reduced the root hydraulic conductivity of *Quercus suber L.*, while As contamination reduces root water absorption in soybean (*Glycine max L.*) [79]. The photosynthetic metrics, including net photosynthesis

(Pn), transpiration rate (E), and stomatal conductance (Gs), exhibited a notable decrease in response to toxic metal/metalloid stress [79].

Different toxic metal/metalloid elements, including Cu, Cr, Cd, and Zn, have been found to lower the Pn rates, water-use efficiency (WUE), and stomatal conductance [92]. Toxic metals/metalloids cause the rate of CO_2 production to decrease and subsequently diminish, thus reducing the leaf's internal CO_2 concentration, which eventually affects the gaseous exchange process, especially the photosynthetic rate [92].

Understanding the mechanisms of toxic metal/metalloid-induced mitochondrial dysfunction is crucial for developing strategies to mitigate the harmful effects of toxic metals/metalloids on plants. Toxic metals/metalloids can significantly impact a plant's mitochondrial functions, altering ATP production. They can disrupt the normal functioning of mitochondria, the powerhouses of the cell, leading to ATP depletion and increased ROS production. This can result in oxidative stress, further damaging the mitochondria and other cellular components, ultimately leading to cell death [93–95].

Toxic metals/metalloids can significantly impact plants' mitochondrial functions, leading to ATP production alterations. Mitochondria, the primary sites of ATP generation, play a vital role in generating cellular energy. However, exposure to toxic metals/metalloids can disrupt their function, leading to ATP depletion and increased production of ROS, ultimately resulting in cell death [95].

Toxic metals/metalloids can impair mitochondrial function by increasing ROS production and decreasing antioxidant activity [93]. The overproduction of ROS can lead to oxidative stress, which can damage cellular components, including mitochondria; this damage can disrupt the mitochondrial respiratory chain, decouple oxidative phosphorylation, and lead to mitochondrial death [94].

For instance, Cd has been shown to interact directly or indirectly with different macromolecules, accumulating environmental pollutants and oxidative stress in plants; this interaction can affect pathological processes such as excess ROS production, ATP depletion, mitochondrial death, and mitochondrial respiratory chain damage [94].

Moreover, toxic metals/metalloids can also affect the efficiency of water flow in plants by reducing transpiration, which can further impact the plant's ability to produce ATP [79]. Reducing ATP production can decrease plant growth and development, as ATP is essential for various plant metabolic processes [82].

3.3 Oxidative stress and antioxidant responses

The production of ROS is a critical physiological response of plants to toxic metal/metalloid stress. This process can lead to oxidative stress, significantly destroying plant cells and disrupting their normal functioning [94, 96, 97].

Exposure of plants to toxic metals/metalloids leads to the overproduction of ROS, chemically reactive molecules containing oxygen. This overproduction is a part of the normal metabolism of plant cells. However, when plants are exposed to toxic concentrations of toxic metals/metalloids, the production of ROS can overwhelm the systems protecting the plants, resulting in oxidative stress [97].

One of the immediate effects when plant cells are exposed to toxic concentrations of toxic metals/metalloids is the production of ROS, such as superoxide (O–– 2) and hydroxyl radicals (-OH), as well as non-radicals, such as hydrogen peroxide (H_2O_2) and singlet oxygen (1O2) [97]; ROS can be produced either directly by ROS-active metals through the Haber-Weiss/Fenton reactions or indirectly through the disruption of cellular functions [97].

Toxic metals/metalloids can induce ROS production in two ways. First, they can interact directly or indirectly with different macromolecules and signaling pathways, leading to the accumulation of environmental contaminants and oxidative stress in plants [94]; second, toxic metals/metalloids can lead to the depletion of ATP, excessive formation of reactive oxygen species (ROS), damage to the mitochondrial respiratory chain, disruption of oxidative phosphorylation, and ultimately, mitochondrial death [94].

The overproduction of ROS can affect the redox status of cells, causing dramatic physiological changes leading to oxidative stress [96]. Toxic metal/metalloid-induced ROS can cause damage to plants, such as enzymatic activity suppression, protein oxidation, and lipid peroxidation. At the same time, previous studies demonstrated that ROS production is more toxic to plant macromolecules than toxic metals/metalloids themselves [96].

Activating plant antioxidant systems is a crucial physiological response to toxic metal/metalloid stress. These systems, composed of both enzymatic and nonenzymatic components, work together to neutralize ROS, mitigate oxidative damage, and help the plant manage the toxic effects of toxic metals/metalloids [80, 94, 98, 99].

When exposed to toxic metals/metalloids, plants experience oxidative stress due to the overproduction of ROS. This overproduction results from toxic metals/metalloids interacting directly or indirectly with different macromolecules and signaling pathways, accumulating environmental pollutants, and oxidative stress in plants [94].

To counteract this oxidative stress, plants activate their antioxidant systems. These systems are composed of various antioxidant enzymes, for instance, superoxide dismutase (SOD), catalase (CAT), and peroxidases, which work together to neutralize ROS and mitigate the damaging effects of toxic metals/metalloids [80].

For instance, *Brassica juncea*, a plant species known for its hyperaccumulation of toxic metals/metalloids, has been observed to activate efficient detoxification mechanisms, such as glutamylcysteine-γ synthetase (γECS) and glutathione reductase (GR1), to avoid metal toxicity. These mechanisms help the plant manage oxidative stress, manifested by excessive ROS production, lipid peroxidation increases, and oxidized protein levels [98].

In addition to these enzymatic defenses, plants also activate nonenzymatic antioxidant defenses. These include the production of antioxidant compounds, such as phenolic antioxidants, which play a crucial role in mitigating toxic metal/metalloid-induced oxidative stress [99].

Moreover, plants have developed strategies to regulate the activation of their antioxidant systems in response to toxic metal/metalloid stress. Various signaling molecules can activate the mitogen-activated protein kinase (MAPK) cascade, leading to the differential expression of genes and the subsequent activation of the antioxidant system [99].

3.4 Effects on nutrient uptake and metabolism

Toxic metals/metalloids can compete with essential nutrients for plant uptake, leading to deficiencies and disrupting plant growth and metabolism. For instance, Cd can interfere with the uptake of Zn and calcium (Ca), while Pb can inhibit the uptake of potassium (K) [100–102]. This competition occurs because toxic metals/metalloids can mimic essential nutrients, leading to their uptake by the same transporters or interfering with the transport process [100].

Cd is known to have a chemical similarity to Zn, which leads to competitive uptake and can affect the translocation of essential metals, such as Zn and Mn, within the plant [100]. The presence of Cd can also influence the uptake of Fe and Mg, which are critical for chlorophyll synthesis and photosynthesis [101].

The competition between toxic metals/metalloids and essential nutrients can significantly impact plant nutrient metabolism. For example, toxic metal/metalloid stress can decrease the concentration of vital nutrients in plant tissues, affecting photosynthesis, respiration, and enzymatic activities [100, 101]. This can result in stunted plant growth and reduced productivity and ultimately affect the nutritional quality of the plants [100–102].

The physiological responses of plants to toxic metal/metalloid stress can be diverse, affecting various aspects of plant function and metabolism. One such response is the disruption of nutrient transporters, which can lead to reduced nutrient uptake in plants.

This can significantly impact plant growth and development and make plants more susceptible to other forms of stress. Therefore, understanding the mechanisms of toxic metal/metalloid uptake and its effects on nutrient transporters is crucial for developing strategies to mitigate the impacts of toxic metal/metalloid contamination in soils [8, 39, 80, 101–103].

For instance, it has been suggested that no specific transporters exist for toxic metals/metalloids, such as Cd and Pb, since these elements do not have any direct biological function. Instead, these toxic metals/metalloids are often cotransported along with other soil nutrients across the root cell membrane, which can disrupt the normal nutrient uptake process [80].

Moreover, toxic metals/metalloids can also interfere with the molecular networks that maintain the homeostasis of essential elements in plants. This can further disrupt nutrient uptake and translocation, leading to nutrient lack and reduced plant growth [102, 103].

Furthermore, toxic metals/metalloids can generate ROS within plants, leading to oxidative harm and additional disruption of nutrient transporters [80, 102]. To counteract this, plants may bolster their defenses by activating antioxidant systems [102].

Furthermore, toxic metals/metalloids can inhibit the expression of auxin biosynthetic and catabolic genes, affecting root growth and further impacting nutrient uptake [8].

The physiological responses of plants to toxic metal/metalloid stress, particularly the inhibition of enzymes involved in nutrient metabolism, is a complex process that can significantly impact plant health and productivity. Toxic metals/metalloids, such as Cr, can inhibit enzymes crucial for nutrient metabolism, affecting ATP production and overall plant health [10, 80, 104].

Cr, for instance, has been found to negatively impact plant growth by impairing essential metabolic processes. This includes inhibiting enzymes involved in nutrient metabolism, such as phosphorus metabolism, which is crucial for ATP production [104]. Cr toxicity can lead to the generation of ROS, causing oxidative stress in plants and further disrupting metabolic processes [104, 105].

Furthermore, toxic metals/metalloids can impede the activity of vital enzymes involved in the metabolism and assimilation of nitrate and ammonia, such as nitrate reductase and nitrite reductase; this can decrease photosynthetic rate and CO_2 assimilation, further impacting plant health and productivity [80].

In addition to Cr, other toxic metals/metalloids, such as Cd and Pb, can also inhibit enzymes involved in nutrient metabolism. For example, Cd can significantly reduce the activity of glucose-6-phosphate dehydrogenase, an enzyme part of the oxidative pentose phosphate pathway. This pathway has essential functions, including the production of building blocks for nucleic acids and fatty acids and the generation of NADPH, which is crucial for basic metabolic processes and combating oxidative stress [106].

Toxic metals/metalloids can lead to significant alterations in root morphology, which can further impact the plant's ability to absorb nutrients from the soil. This can have important implications for plant health and productivity, particularly in areas with high soil toxic metal/metalloid contamination levels [24, 37, 79, 107].

Toxic metals/metalloids can significantly alter the root morphology of plants, which can further impact the plant's ability to absorb nutrients from the soil. The absorption efficiency of toxic metals/metalloids by plant roots directly affects their metal enrichment ability [107].

Exposure to toxic metals/metalloids can reduce root diameter due to decreased cell division and size. This is often due to reduced growth and lower elasticity of cell walls, along with a reduction in vessel size, contributing to a reduced vascular cylinder area [24]. For instance, plants exposed to Cd have demonstrated reduced water and nutrient uptake, inhibition of growth, and browning of root tips, which ultimately lead to cell death [37].

Toxic metals/metalloids can also cause significant damage to the root and shoot cells and internal organelles, such as chloroplasts and mitochondria, reducing energy production, imposing oxidative stress, and ultimately affecting plant morphology and survival rate [24].

Moreover, toxic metals/metalloids can cause changes in plant physiology and hydraulics and induce anatomical changes; for instance, toxic metals/metalloids in tailings have been demonstrated to cause a decrease in plant growth, which may be related to plant physiological changes [79].

Toxic metals/metalloids can disrupt the internal balance of nutrients within plant cells, leading to deficiencies or toxicities. This disruption can significantly impact plant health and productivity, affecting various physiological and biochemical processes within the plant [79, 91, 94, 108].

Toxic metals/metalloids, including Cu, can induce toxicity in biological systems by interacting with sulfhydryl groups, generating reactive oxygen species (ROS), inactivating enzymes, and causing oxidative stress. While these mechanisms broadly impact cellular functions and can lead to nutrient imbalances, the specific interaction between Cu and Fe in plants, such as competition for transport mechanisms leading to Fe deficiency, underscores the complex dynamics of toxic metals/metalloids uptake and its potential to disrupt plant nutrient homeostasis [108].

Furthermore, toxic metals/metalloids have the potential to disrupt plant metabolism, leading to the suppression of plant development and reduced crop productivity. Plant productivity is contingent upon their growth and development, which toxic metal/metalloid-induced phytotoxicity can negatively impact. This toxin hinders the operation and effectiveness of the photosynthetic system, stomatal function, and cambium activity [91].

Furthermore, toxic metals/metalloids can cause damage to the stomatal morphology and structure and ultimately interfere with several physiological metabolic processes in plants. For example, Cd can inhibit photosynthesis in plants by decreasing stomatal conductance and initiating a reduction in the availability of intracellular CO_2, leading to a decrease in the photosynthetic rate [91].

Furthermore, toxic metals/metalloids can induce poisoning and turn off enzyme systems, significantly impacting vital physiological and biochemical activities such as photosynthesis, respiration, protein synthesis, and the creation of cellular organic matter. This may result in impaired information transfer, thereby causing changes in structural and physiological aspects of the plant such as leaf area, stomatal density, and stomatal conductance [91].

While mycorrhizal fungi and nitrogen-fixing bacteria can mitigate the effects of toxic metals/metalloids on plants, high concentrations of these metals can disrupt the symbiotic relationships, reducing plant growth and nutrient uptake [24, 109–111].

Toxic metals and metalloids can significantly affect plants' symbiotic relationships with mycorrhizal fungi and nitrogen-fixing bacteria, which are crucial for nutrient acquisition and plant growth. However, these symbiotic microorganisms can also enhance plant tolerance to toxic metals and metalloids and improve plant growth under metal stress conditions [109, 111].

Mycorrhizal fungi, for example, can sequester large quantities of toxic metals/metalloids, reducing their bioavailability in the soil and, thus, their plant uptake. They can detoxify metal ions by chelating metallothioneins (MTs) or conjugating toxic metals/metalloids with organic molecules, such as glutathione or organic acids [109]. This reduces the toxicity of toxic metals/metalloids to the plant and improves the plant's nutrient uptake and growth [109, 111].

Furthermore, the diversity, nitrogen-fixing capacity, and toxic metal/metalloid tolerance of culturable rhizobia associated with leguminous plants, such as *Pongamia pinnata,* have been studied, demonstrating that these bacteria can survive in toxic

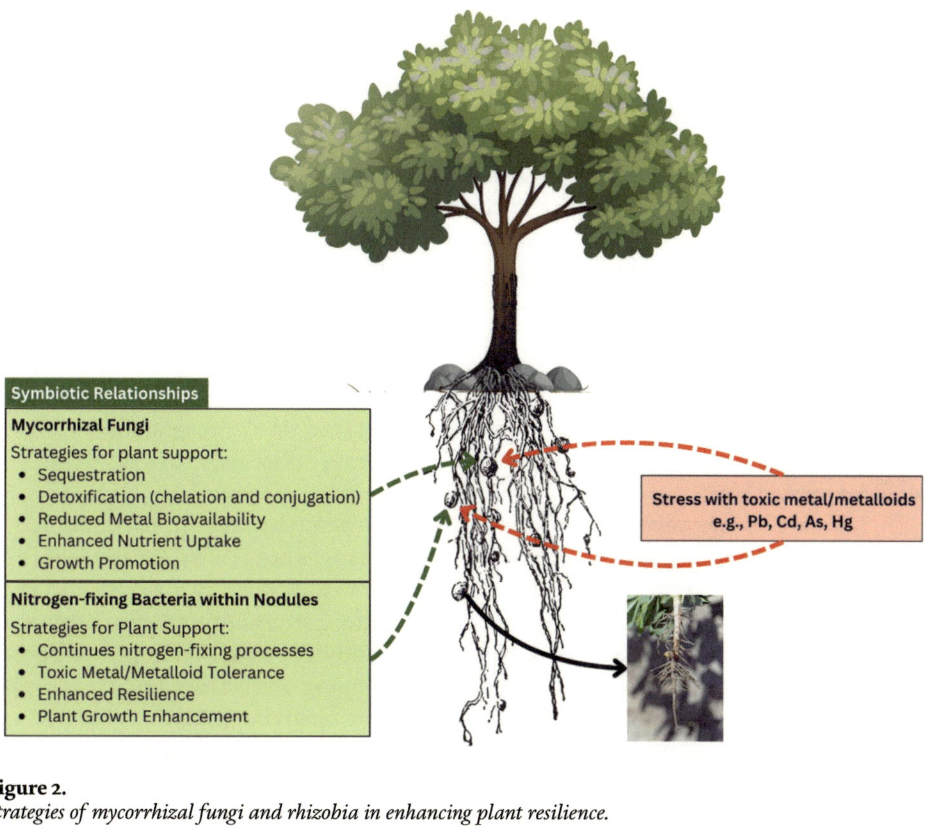

Figure 2.
Strategies of mycorrhizal fungi and rhizobia in enhancing plant resilience.

metal/metalloid-contaminated environments and continue to provide the essential service of nitrogen fixation to their host plants [110]. **Figure 2** summarizes the strategies of these microorganisms for enhancing plant resilience and mitigating the effects of toxic metals/metalloids.

However, toxic metal/metalloid stress can negatively impact these beneficial plant-microbe interactions. For instance, increased bioaccumulation of toxic metals/metalloids beyond the threshold level can damage the root and shoot cells and internal organelles, such as chloroplasts and mitochondria, and affect symbiotic relationships. This can reduce plant growth and productivity due to impaired nutrient acquisition and metabolism [24].

4. Molecular and genetic responses

4.1 Alterations in gene expression owing to toxic metal/metalloid stress

Plants' responses to toxic metal/metalloid stress involve significant changes in gene expression, with specific genes being upregulated (turned on) or downregulated (turned off) depending on the particular stress conditions. These changes in gene expression are part of the plant's adaptive response to mitigate the harmful effects of toxic metals/metalloids [80, 112, 113].

Upregulated genes are those whose expression is increased in response to toxic metal/metalloid stress. Toxic metal/metalloid ATPase (HMA) gene expression in rice was upregulated under toxic metal/metalloid stress [112]. These genes transport toxic metals/metalloids across cell membranes, which is crucial in metal detoxification. Similarly, genes that regulate antioxidant activity and stress responses are upregulated when plants are exposed to toxic metals/metalloids, as observed in *Lolium perenne L.* exposed to Cd^{2+} and Hg^{2+} [113]. This upregulation enhances the plant's defense mechanisms against the oxidative stress of toxic metals/metalloids.

Conversely, downregulated genes are those whose expression is reduced under toxic metal/metalloid stress. While specific examples of downregulated genes were not provided in the search results, toxic metal/metalloid stress can lead to the downregulation of genes involved in average plant growth and the development of stress response mechanisms [80].

It is important to note that the specific genes that are upregulated or downregulated can vary depending on the type of toxic metals/metalloids, the plant species, and the specific environmental conditions. Furthermore, these changes in gene expression can be heritable, indicating they can be passed on to subsequent generations, allowing plants to "remember" and respond more effectively to toxic metal/metalloid stress [112].

Under toxic metal/metalloid stress, plants change gene expression, particularly those involving metal-responsive elements. These changes are crucial for the plant's adaptive response to stress and can affect various processes, such as the transport of metals and the overall tolerance to toxic metal/metalloid stress. Alterations in gene expression can result in the increased or decreased activity of particular genes. These changes are part of the plant's adaptive response to coping with stress and can even be passed down to subsequent generations, making them heritable [112].

One of the critical components of this response involves metal-responsive elements, which are specific DNA sequences that can bind to metals, thereby influencing gene expression. For instance, research on the Cd-tolerant cultivar of Chinese flowing cabbage and *Sedum plumbizincicola* revealed elevated expression of specific genes in response to Cd stress [112]. Similarly, in *Lycopersicum esculentum*, the toxic metal/metalloid transporters COPT1 and COPT2 could be prompted to express under Cu stress [112].

A study investigated the expression patterns of genes in the metal tolerance protein family in *Eucalyptus grandis* exposed to metal stress conditions. These proteins, which are membrane-bound transporters, play a crucial role in the selective transport of various metals. Variations in their transcript levels may reflect the plant's adaptive response to metal-induced stress [114].

Moreover, plants respond to toxic metal/metalloid stress-induced oxidative and genotoxic damage *via* rapid gene expression changes. Various families of transcription factors play a crucial role in triggering such responses [80].

4.2 Function of noncoding RNAs

MicroRNAs (miRNAs) are essential in the plant's reaction to toxic metal/metalloid stress, regulating a wide array of physiological processes to mitigate the harmful effects of these metals. As our understanding of these small but powerful molecules continues to grow, they may offer promising avenues for enhancing plant tolerance to toxic metals/metalloids and improving phytoremediation strategies [115–117].

The miRNAs are small, noncoding RNAs that play a fundamental role in controlling gene expression in response to toxic metal/metalloid stress in plants. These miRNAs, usually 20-24 nucleotides long, regulate gene expression following transcription. They mainly achieve this by inhibiting translation or cleaving their target [116, 117].

Rapid industrialization and large-scale chemical fertilizers and pesticides have increased toxic metals/metalloids and soil contamination. This has necessitated the evolution of complex plant mechanisms to cope with toxic metal/metalloid stress. One such mechanism involves using miRNAs to combat metal-induced toxicity [115, 117].

miRNAs regulate plant responses to metal stress through several mechanisms, including antioxidant functions, root growth, hormone signals, transcription factors (TF), and metal transporters [115]. For instance, they can specifically target and reduce the activation of genes responsible for accumulating toxic metals/metalloids in plants, thus reducing the level of metal present and mitigating its detrimental effects on the plant [117].

Next-generation sequencing (NGS) has significantly enhanced our understanding of the population and function of miRNAs in the context of metal stress [115, 116]. These technologies have enabled the discovery of several known and unknown miRNAs that respond to metal stress, revealing their diverse roles in several plants [115].

For example, miRNAs regulate antioxidant systems, which help mitigate the oxidative stress caused by toxic metals/metalloids [115]. They also regulate root growth, which can be adversely affected by toxic metal/metalloid toxicity [116].

Furthermore, miRNAs can regulate hormone signals and transcription factors, which are critical components of the plant's response to stress; they can also influence the activity of metal transporters, which are responsible for the uptake and distribution of metals within the plant [115].

Noncoding RNAs (lncRNAs) are RNA molecules that do not code for proteins; however, they play a vital role in regulating gene expression in response to toxic metal/metalloid stress in plants [118, 119]. They are "biological regulators" for various developmental processes and plant stress responses [120]. These lncRNAs can modify gene expression at the posttranscriptional and epigenetic levels by targeting various stress-responsive mRNAs, regulatory genes encoding transcription factors, and numerous miRNAs controlling gene expression [120].

In the context of toxic metal/metalloid stress, lncRNAs have been found to respond to toxic metals/metalloids such as Pb, Fe, Cu, Mn, and aluminum (Al) [118]. For instance, a genome-wide analysis of lncRNAs in rice under Cd stress revealed differentially expressed lncRNAs between Cd stress and normal conditions. A total of 69 lncRNAs were found to be upregulated, while 75 lncRNAs were downregulated in the Cd stress group [119]. This suggests that lncRNAs play a role in the plant's response to toxic metal/metalloid stress, potentially by modulating the expression of genes in stress response pathways.

Nevertheless, the precise mechanisms *via* which lncRNAs control gene expression in the existence of toxic metal/metalloid stress remain incompletely comprehended. Some studies suggest that lncRNAs may act as a nucleator, a scaffold for numerous complexes, a template, a decoy, or a signal [118]. For example, lncRNAs may bind to specific DNA sequences (metal-responsive elements), influencing gene expression [118]. They may also interact with other noncoding RNAs, such as miRNAs, to regulate gene expression [121].

Despite these advances, our understanding of the role of lncRNAs in toxic metal/metalloid stress responses is still in its infancy. Many lncRNAs have been detected in plants, but their specific functions and the target genes they regulate remain primarily unknown [120]. Further research is needed to elucidate the complex molecular mechanisms by which lncRNAs contribute to plant adaptation to toxic metal/metalloid stress [118, 120]. This could lead to developing new strategies for enhancing plant tolerance to toxic metals/metalloids, such as manipulating the expression of specific lncRNAs [121].

Small interfering RNAs (siRNAs) play a crucial role in plants by mediating epigenetic modifications, particularly DNA methylation, under toxic metal/metalloid stress. These modifications can affect gene expression and potentially enhance the plant's tolerance to toxic metals/metalloids, leading to transgenerational stress memory [112, 122, 123].

The siRNAs significantly impact the plant's response to toxic metal/metalloid stress, especially their influence on DNA methylation, a crucial epigenetic modification. DNA methylation is the process of introducing a methyl group to the DNA molecule, resulting in the modification of gene expression without any changes to the underlying DNA sequence. This process is essential for plant adaptation to various environmental stresses, including toxic metal/metalloid toxicity [122, 123].

siRNAs are known to mediate DNA methylation in plants. This process, known as RNA-directed DNA methylation (RdDM), involves the generation of 24-nucleotide siRNAs that guide the *de novo* DNA methyltransferase to specific genomic regions, leading to methylation [123]. This methylation can suppress the transcription of specific genes, thereby influencing the plant's response to stress [122, 123].

Several plant species have observed changes in DNA methylation patterns in toxic metal/metalloid stress. For instance, exposure to toxic metals/metalloids, such as Cd and Zn, has been demonstrated to alter global DNA methylation patterns in wheat, potentially conferring tolerance to these metals. Similarly, a study on rice revealed that changes in DNA methylation potentially mediate toxic metal/metalloid stress-induced locus-specific changes in gene expression [112].

Noncoding RNAs (ncRNAs) could engineer plants that are more resistant to toxic metals/metalloids. However, more research is needed to understand how these ncRNAs control gene expression at a molecular level in response to toxic metal/metalloid stress. As research progresses, engineering ncRNAs, particularly miRNAs and lncRNAs, could become an effective tool to develop crop varieties that can tolerate toxic metals/metalloids. These ncRNAs have emerged as critical regulators in plant responses to toxic metal/metalloid stress [115, 118, 120, 121, 124].

miRNAs are tiny, noncoding RNA molecules that regulate gene expression post-transcriptionally, often by silencing genes through mRNA degradation or translational repression. They have been implicated in various plant processes, including development, nutrient homeostasis, and stress responses. In toxic metal/metalloid stress, miRNAs can target metal uptake, transport, and sequestration genes, thereby modulating the plant's tolerance to toxic metals/metalloids [115, 124].

lncRNAs, exceeding 200 nucleotides in length, can regulate gene expression by many mechanisms such as modifying chromatin, controlling transcription, and influencing posttranscriptional processing. They have been demonstrated to respond to toxic metals/metalloids such as Pb, Fe, Cu, and Mn, suggesting their involvement in toxic metal/metalloid stress responses [118, 120].

The engineering of ncRNAs offers a novel approach to developing plants with enhanced tolerance to toxic metals/metalloids. For example, overexpressing specific miRNAs that confer stress tolerance or silencing miRNAs that negatively regulate stress response genes could be strategies to improve plant resilience to toxic metals/metalloids [115, 121]. Also, manipulating lncRNAs involved in the stress response could be a viable strategy for enhancing tolerance [118, 120].

4.3 Role of metallothioneins and PCs

The MTs act as metal detectors in toxic metal/metalloid stress, binding to excess metal ions and reducing their toxicity. This makes them essential to the plant's defense against toxic metal/metalloid stress [125]. MTs are small proteins with a molecular weight of 6-7 kilodaltons (kDa) crucial in mitigating the effects caused by excess metal ions. They are rich in cysteine, a sulfur-containing amino acid, which allows them to bind with metals and detoxify them by reducing their presence [59, 125].

MTs involve cellular processes such as cell growth regulation, ROS management, and DNA repair. The structure of MTs consists of two domains, each containing a cluster of metal atoms. These proteins have several isoforms coded by different alleles on chromosome 16 [125].

The induction of MTs is constant with metals such as Cd, Cu, and Zn. The cysteine residues in MTs are essential for their function and are required for life. The architectural integrity, evolutionary conservation, pervasive presence, gene duplication, and regulated expression of MTs during living organisms' development, regeneration, and reproduction present compelling evidence for their essential functions within biological systems [125, 126].

PCs are essential for detoxifying toxic metals and metalloids in plants, and their synthesis is regulated in response to metal exposure.

PCs are small, cysteine-rich peptides crucial in detoxifying metal/metalloids in plants. Their synthesis is mediated by the enzyme phytochelatins synthase (PCs), which is activated in response to metal exposure. This activation leads to the chelation of toxic metal ions, such as Cd, Cu, and Zn, thus mitigating their toxicity to plants [127–129].

PCs are synthesized from glutathione, a tripeptide composed of glutamate, cysteine, and glycine, through an enzymatic reaction catalyzed by PCs. The reaction involves the gamma-glutamylcysteine moiety of glutathione being polymerized to form PCs, which can bind to and detoxify metals/metalloids within the plant cells. This process not only plays a significant role in the detoxification and tolerance of metal/metalloids but also highlights the potential for phytoremediation applications, where plants are used to remove pollutants from the environment [127, 129].

Gene regulation of PCS plays a pivotal role in this process. The presence of metal/metalloids highly regulates the expression of PCS genes, and the activation of these genes leads to the synthesis of PCs, which chelate and sequester metals/metalloids in the vacuoles of plant cells. This mechanism ensures the detoxification of metals/metalloids and the protection of cellular functions from metal toxicity [127].

Furthermore, glutathione's implications in synthesizing PCs highlight the interconnectedness of plant detoxification pathways. Glutathione is a substrate for PCS and acts as an antioxidant, reducing oxidative stress induced by heavy metal exposure. The synthesis of PCs and their role in heavy metal chelation exemplify the plant's comprehensive strategy to cope with environmental stressors, ensuring survival and maintaining homeostasis in the presence of toxic metals [127, 128].

To encapsulate, the enzyme PCs, regulated by metal/metalloids, catalyzes the formation of PCs from glutathione. These PCs then bind to toxic metals, facilitating their detoxification and sequestration within the plant. This intricate mechanism underlines plants' adaptive responses to environmental stresses and has significant implications for enhancing phytoremediation strategies and understanding plant resilience to heavy metal toxicity.

5. Conclusions

In summary, as presented in this chapter, the comprehensive examination of toxic metal/metalloid contamination in soil and its implications for crop resilience and abiotic stress management underscores the multifaceted challenges and innovative strategies in sustainable agriculture. The detailed exploration of the sources of natural and anthropogenic soil contamination, alongside the mechanisms of plant absorption and response to these toxicants, provides a foundational understanding of the complexities involved in plant-soil interactions under stress conditions. The physiological and molecular responses of plants to toxic metal/metalloid stress, particularly the impact on photosynthesis, nutrient uptake, and oxidative stress management, highlight the critical areas where plant resilience needs bolstering.

The discussion on phytoremediation and genetic engineering as strategies to alleviate toxic metal/metalloid stress showcases the potential of biotechnological advancements in enhancing plant growth and resilience in contaminated soils. The role of PCs in detoxifying plants and the promising results from the overexpression of PCs genes in *Arabidopsis thaliana* for improved tolerance and accumulation of toxic metals/metalloids point toward a genetic manipulation approach in developing crops with enhanced resistance to environmental stressors.

Furthermore, the interaction between plants and soil microbes offers an intriguing insight into the bioavailability of toxic metals/metalloids and the support of plant health under adverse conditions. This symbiotic relationship underscores the importance of microbial communities in the soil ecosystem and their potential to support plant resilience and health.

The chapter contributes significantly to the current body of knowledge by presenting advanced mitigation strategies and research findings that can inform and guide sustainable agricultural practices. The insights gained from this chapter are invaluable for researchers, agronomists, and policymakers, aiming to enhance crop resilience against toxic metal/metalloid contamination and to ensure food security in the face of increasing environmental challenges. The innovative approaches discussed herein, including phytoremediation, genetic engineering, and the exploitation of plant-microbe interactions, offer promising pathways toward achieving sustainable agricultural systems resilient to abiotic stresses.

Conflict of interest

The authors declare no conflict of interest.

Disclaimer

The views expressed in this paper are those of the author(s) and do not necessarily reflect those of the SFDA or its stakeholders. Guaranteeing the accuracy and validity of the data is the sole responsibility of the research author(s).

Author details

Hany Almotairy
Suadi Food and Drug Authority (SFDA), Riyadh, Saudi Arabia

*Address all correspondence to: hmmotairy@sfda.gov.sa

References

[1] Li C, Zhou K, Qin W, Tian C, Qi M, Yan X, et al. A review on heavy metals contamination in soil: Effects, sources, and remediation techniques. Soil and Sediment Contamination: An International Journal. 2019;**28**(4):380-394. DOI: 10.1080/15320383.2019.1592108

[2] Das S, Sultana KW, Ndhlala AR, Mondal M, Chandra I. Heavy metal pollution in the environment and its impact on health: Exploring green technology for remediation. Environmental Health Insights. 2023;**17**:1-10. DOI: 10.1177%2F11786302231201259 17

[3] She W, Guo L, Gao J, Zhang C, Wu S, Jiao Y, et al. Spatial distribution of soil heavy metals and associated environmental risks near major roads in southern Tibet, China. International Journal of Environmental Research and Public Health. 2022;**19**(14):8380. DOI: 10.3390%2Fijerph19148380

[4] Shiyi Y, Xiaonuo L, Weiping C. High-resolution risk mapping of heavy metals in soil with an integrated static-dynamic interaction model: A case study in an industrial agglomeration area in China. Journal of Hazardous Materials. 2023;**455**:131650. DOI: 10.1016/j.jhazmat.2023.131650

[5] Pourret O, Hursthouse A. It's time to replace the term "heavy metals" with "potentially toxic elements" when reporting environmental research. International Journal of Environmental Research and Public Health. 2019;**16**(22):4446. DOI: 10.3390/ijerph16224446

[6] Abd Elnabi MK, Elkaliny NE, Elyazied MM, Azab SH, Elkhalifa SA, Elmasry S, et al. Toxicity of heavy metals and recent advances in their removal: A review. Toxics. 2023;**11**(7):580. DOI: 10.3390/toxics11070580

[7] Slobodian MR, Petahtegoose JD, Wallis AL, Levesque DC, Merritt TJ. The effects of essential and non-essential metal toxicity in the Drosophila melanogaster insect model: A review. Toxics. 2021;**9**(10):269. DOI: 10.3390%2Ftoxics9100269

[8] Angulo-Bejarano PI, Puente-Rivera J, Cruz-Ortega R. Metal and metalloid toxicity in plants: An overview on molecular aspects. Plants. 2021;**10**(4):635. DOI: 10.3390%2Fplants10040635

[9] Riyazuddin R, Nisha N, Ejaz B, Khan MIR, Kumar M, Ramteke PW, et al. A comprehensive review on the heavy metal toxicity and sequestration in plants. Biomolecules. 2021;**12**(1):43. DOI: 10.3390%2Fbiom12010043

[10] Alengebawy A, Abdelkhalek ST, Qureshi SR, Wang MQ. Heavy metals and pesticides toxicity in agricultural soil and plants: Ecological risks and human health implications. Toxics. 2021;**9**(3):42. DOI: 10.3390%2Ftoxics9030042

[11] Priya AK, Muruganandam M, Ali SS, Kornaros M. Clean-up of heavy metals from contaminated soil by phytoremediation: A multidisciplinary and eco-friendly approach. Toxics. 2023;**11**(5):422. DOI: 10.3390%2Ftoxics11050422

[12] Eugenio RN, McLaughlin M, Pennock D. Soil Pollution: A Hidden Reality [Internet]. Rome, Italy: FAO; 2018. ISBN: 978-92-5-130505-8. Available from: https://www.fao.org/documents/card/en/c/I9183EN

[13] Ma T, Zhang Y, Hu Q, Han M, Li X, Zhang Y, et al. Accumulation

characteristics and pollution evaluation of soil heavy metals in different land use types: Study on the whole region of Tianjin. International Journal of Environmental Research and Public Health. 2022;**19**(16):10013. DOI: 10.3390%2Fijerph191610013

[14] Al-Swadi HA, Usman AR, Al-Farraj AS, Al-Wabel MI, Ahmad M, Al-Faraj A. Sources, toxicity potential, and human health risk assessment of heavy metals-laden soil and dust of urban and suburban areas as affected by industrial and mining activities. Scientific Reports. 2022;**12**(1):8972. DOI: 10.1038/s41598-022-12345-8

[15] Sosa D, Hilber I, Buerge-Weirich D, Faure R, Escobar A, Bucheli TD. Heavy metals in soils of Mayabeque, Cuba: Multifaceted and hardly discernable contributions from pedogenic and anthropogenic sources. Environmental Monitoring and Assessment. 2022;**194**(6):441. DOI: 10.1007/s10661-022-10097-6

[16] Shimod KP, Vineethkumar V, Prasad TK, Jayapal G. Effect of urbanization on heavy metal contamination: A study on major townships of Kannur District in Kerala, India. Bulletin of the National Research Centre. 2022;**46**(1):1-14. DOI: 10.1186/s42269-021-00691-y

[17] Qasem NA, Mohammed RH, Lawal DU. Removal of heavy metal ions from wastewater: A comprehensive and critical review. npj Clean Water. 2021;**4**(1):36. DOI: 10.1038/s41545-021-00127-0

[18] Peñalver R, Jacobs MR, Hegarty S, Regan F. Assessment of anthropogenic pollution by monitoring occurrence and distribution of chemicals in the river Liffey in Dublin. Environmental Science and Pollution Research. 2021;**28**(38):53754-53766. DOI: 10.1007/s11356-021-14508-y

[19] Rashid A, Schutte BJ, Ulery A, Deyholos MK, Sanogo S, Lehnhoff EA, et al. Heavy metal contamination in agricultural soil: Environmental pollutants affecting crop health. Agronomy. 2023;**13**(6):1521. DOI: 10.3390/agronomy13061521

[20] Yan A, Wang Y, Tan SN, Mohd Yusof ML, Ghosh S, Chen Z. Phytoremediation: A promising approach for revegetation of heavy metal-polluted land. Frontiers in Plant Science. 2020;**11**:359. DOI: 10.3389%2Ffpls.2020.00359

[21] Tan HW, Pang YL, Lim S, Chong WC. A state-of-the-art of phytoremediation approach for sustainable management of heavy metals recovery. Environmental Technology & Innovation. 2023;**30**:103043. DOI: 10.1016/j.eti.2023.103043

[22] Wang Z, Yamaji N, Huang S, Zhang X, Shi M, Fu S, et al. OsCASP1 is required for Casparian strip formation at endodermal cells of rice roots for selective uptake of mineral elements. The Plant Cell. 2019;**31**(11):2636-2648. DOI: 10.1105%2Ftpc.19.00296

[23] Ricachenevsky Wen J, Wu C, Bi X, Zhang S, Ouyang H, Ye J, et al. Soil pH change induced by smelting activities affects secondary carbonate production and long-term Cd activity in subsoils. Applied Geochemistry. 2023;**152**:105663. DOI: 10.1016/j.apgeochem.2023.105663

[24] Pandey AK, Zorić L, Sun T, Karanović D, Fang P, Borišev M, et al. The anatomical basis of heavy metal responses in legumes and their impact on plant–rhizosphere interactions. Plants. 2022;**11**(19):2554. DOI: 10.3390%2Fplants11192554

[25] Li Q, Wang Y, Li Y, Li L, Tang M, Hu W, et al. Speciation of heavy metals in soils and their immobilization at micro-scale interfaces among diverse soil components. Science of the Total Environment. 2022;**825**:153862. DOI: 10.1016/j.scitotenv.2022.153862

[26] Meng D, Li J, Liu T, Liu Y, Yan M, Hu J, et al. Effects of redox potential on soil cadmium solubility: Insight into microbial community. Journal of Environmental Sciences. 2019;**75**:224-232. DOI: 10.1016/j.jes.2018.03.032

[27] Gao Y, Jia J, Xi B, Cui D, Tan W. Divergent response of heavy metal bioavailability in soil rhizosphere to agricultural land use change from paddy fields to various drylands. Environmental Science: Processes & Impacts. 2021;**23**(3):417-428. DOI: 10.1039/D0EM00501K

[28] Zhu Q, Ji J, Tang X, Wang C, Sun H. Bioavailability assessment of heavy metals and organic pollutants in water and soil using DGT: A review. Applied Sciences. 2023;**13**(17):9760. DOI: 10.3390/app13179760

[29] Xu D, Shen Z, Dou C, Dou Z, Li Y, Gao Y, et al. Effects of soil properties on heavy metal bioavailability and accumulation in crop grains under different farmland use patterns. Scientific Reports. 2022;**12**(1):9211. DOI: 10.1038/s41598-022-13140-1

[30] Liu J, Li Y, Wang Y, Wang Y, Xu J, Liu X. Competitive adsorption of lead and cadmium on soil aggregate at micro-interfaces: Multi-surface modeling and spectroscopic studies. Journal of Hazardous Materials. 2023;**448**:130915. DOI: 10.1016/j.jhazmat.2023.130915

[31] Yu H, Li C, Yan J, Ma Y, Zhou X, Yu W, et al. A review on adsorption characteristics and influencing mechanism of heavy metals in farmland soil. RSC Advances. 2023;**13**(6):3505-3519. DOI: 10.1039%2Fd2ra07095b

[32] Campillo-Cora C, Conde-Cid M, Arias-Estévez M, Fernández-Calviño D, Alonso-Vega F. Specific adsorption of heavy metals in soils: Individual and competitive experiments. Agronomy. 2020;**10**(8):1113. DOI: 10.3390/agronomy10081113

[33] Li Y, Liu J, Wang Y, Tang X, Xu J, Liu X. Contribution of components in natural soil to Cd and Pb competitive adsorption: Semi-quantitative to quantitative analysis. Journal of Hazardous Materials. 2023;**441**:129883. DOI: 10.1016/j.jhazmat.2022.129883

[34] Sharma A, Pandey H, Devadas VS, Kartha BD, Vashishth A. Beneficial effect of heavy metals, antagonistic effect and gene regulations in nutrient translocation through soilless culture. Journal of Agriculture and Food Research. 2023;**12**:100600. DOI: 10.1016/j.jafr.2023.100600

[35] Redha A, Al-Hasan R, Afzal M. Synergistic and concentration-dependent toxicity of multiple heavy metals compared with single heavy metals in Conocarpus lancifolius. Environmental Science and Pollution Research. 2021;**28**:23258-23272. DOI: 10.1007/s11356-020-12271-0

[36] Aboyeji CM, Dunsin O, Adekiya AO, Suleiman KO, Chinedum C, Okunlola FO, et al. Synergistic and antagonistic effects of soil applied P and Zn fertilizers on the performance, minerals and heavy metal composition of groundnut. Open Agriculture. 2020;**5**(1):1-9. DOI: 10.1515/opag-2020-0002

[37] Pande A, Mun BG, Methela NJ, Rahim W, Lee DS, Lee GM, et al. Heavy

metal toxicity in plants and the potential NO-releasing novel techniques as the impending mitigation alternatives. Frontiers. Plant Science. 2022;**13**:1019647. DOI: 10.3389/fpls.2022.1019647

[38] Tőzsér D, Horváth R, Simon E, Magura T. Heavy metal uptake by plant parts of Populus species: A meta-analysis. Environmental Science and Pollution Research. 2023;**30**(26):69416-69430. DOI: 10.1007/s11356-023-27244-2

[39] Jain S, Muneer S, Guerriero G, Liu S, Vishwakarma K, Chauhan DK, et al. Tracing the role of plant proteins in the response to metal toxicity: A comprehensive review. Plant Signaling & Behavior. 2018;**13**(9):e1507401. DOI: 10.1080/15592324.2018.1507401

[40] Yang Z, Yang F, Liu JL, Wu HT, Yang H, Shi Y, et al. Heavy metal transporters: Functional mechanisms, regulation, and application in phytoremediation. Science of the Total Environment. 2022;**809**:151099. DOI: 10.1016/j.scitotenv.2021.151099

[41] Ejaz U, Khan SM, Khalid N, Ahmad Z, Jehangir S, Fatima Rizvi Z, et al. Detoxifying the heavy metals: A multipronged study of tolerance strategies against heavy metals toxicity in plants. Frontiers in Plant Science. 2023;**14**:1154571. DOI: 10.3389/fpls.2023.1154571

[42] Salbitani G, Maresca V, Cianciullo P, Bossa R, Carfagna S, Basile A. Non-protein thiol compounds and antioxidant responses involved in bryophyte heavy-metal tolerance. International Journal of Molecular Sciences. 2023;**24**(6):5302. DOI: 10.3390/ijms24065302

[43] Lao R, Guo Y, Hao W, Fang W, Li H, Zhao Z, et al. The role of lignin in the compartmentalization of cadmium in maize roots is enhanced by mycorrhiza. Journal of Fungi. 2023;**9**(8):852. DOI: 10.3390/jof9080852

[44] Adamczyk-Szabela D, Wolf WM. The impact of soil pH on heavy metals uptake and photosynthesis efficiency in Melissa officinalis, Taraxacum officinalis, Ocimum basilicum. Molecules. 2022;**27**(15):4671. DOI: 10.3390/molecules27154671

[45] Kicińska A, Pomykała R, Izquierdo-Diaz M. Changes in soil pH and mobility of heavy metals in contaminated soils. European Journal of Soil Science. 2022;**73**(1):e13203. DOI: 10.1111/ejss.13203

[46] Wen J, Wu C, Bi X, Zhang S, Ouyang H, Ye J, et al. Soil pH change induced by smelting activities affects secondary carbonate production and long-term Cd activity in subsoils. Applied Geochemistry. 2023;**152**:105663. DOI: 10.1016/j.apgeochem.2023.105663

[47] Turner AJ, Arzola CI, Nunez GH. High pH stress affects root morphology and nutritional status of hydroponically grown Rhododendron (*Rhododendron* spp.). Plants. 2020;**9**(8):1019. DOI: 10.3390/plants9081019

[48] Robles-Aguilar AA, Pang J, Postma JA, Schrey SD, Lambers H, Jablonowski ND. The effect of pH on morphological and physiological root traits of Lupinus angustifolius treated with struvite as a recycled phosphorus source. Plant and Soil. 2019;**434**:65-78. DOI: 10.1007/s11104-018-3787-2

[49] Zhang Q, Chen H, Xu C, Zhu H, Zhu Q. Heavy metal uptake in rice is regulated by pH-dependent iron plaque formation and the expression of the metal transporter genes. Environmental and Experimental Botany. 2019;**162**:392-398. DOI: 10.1016/j.envexpbot.2019.03.004

[50] Wang Z, Jia M, Li Z, Liu H, Christie P, Wu L. Acid buffering capacity of four contrasting metal-contaminated calcareous soil types: Changes in soil metals and relevance to phytoextraction. Chemosphere. 2020;**256**:127045. DOI: 10.1016/j.chemosphere.2020.127045

[51] Lu HL, Li KW, Nkoh JN, He X, Hong ZN, Xu RK. Effects of the increases in soil pH and pH buffering capacity induced by crop residue biochars on available Cd contents in acidic paddy soils. Chemosphere. 2022;**301**:134674. DOI: 10.1016/j.chemosphere.2022.134674

[52] Li Y, Gong X. Effects of dissolved organic matter on the bioavailability of heavy metals during microbial dissimilatory iron reduction: A review. Reviews of Environmental Contamination and Toxicology. 2021;**257**:69-92. DOI: 10.1007/398_2020_63

[53] Engel M, Pacheco JSL, Noël V, Boye K, Fendorf S. Organic compounds alter the preference and rates of heavy metal adsorption on ferrihydrite. Science of the Total Environment. 2021;**750**:141485. DOI: 10.1016/j.scitotenv.2020.141485

[54] Stefanowicz AM, Kapusta P, Zubek S, Stanek M, Woch MW. Soil organic matter prevails over heavy metal pollution and vegetation as a factor shaping soil microbial communities at historical Zn–Pb mining sites. Chemosphere. 2020;**240**:124922. DOI: 10.1016/j.chemosphere.2019.124922

[55] Nurchi VM, Cappai R, Crisponi G, Sanna G, Alberti G, Biesuz R, et al. Chelating agents in soil remediation: A new method for a pragmatic choice of the right chelator. Frontiers in Chemistry. 2020;**8**:597400. DOI: 10.3389/fchem.2020.597400

[56] Medyńska-Juraszek A, Bednik M, Chohura P. Assessing the influence of compost and biochar amendments on the mobility and uptake of heavy metals by green leafy vegetables. International Journal of Environmental Research and Public Health. 2020;**17**(21):7861. DOI: 10.3390/ijerph17217861

[57] Rashid I, Murtaza G, Dar AA, Wang Z. The influence of humic and fulvic acids on Cd bioavailability to wheat cultivars grown on sewage irrigated Cd-contaminated soils. Ecotoxicology and Environmental Safety. 2020;**205**:111347. DOI: 10.1016/j.ecoenv.2020.111347

[58] Steinberg SM, Hodge VF. Measurement of lead complexation by humic acids and humic acid analogues using competitive ligand exchange. Heliyon. 2022;**8**(12):e12437. DOI: 10.1016/j.heliyon.2022.e12437

[59] Gao C, Gao K, Yang H, Ju T, Zhu J, Tang Z, et al. Genome-wide analysis of metallothionein gene family in maize to reveal its role in development and stress resistance to heavy metal. Biological Research. 2022;**55**(1):1-13. DOI: 10.1186/s40659-021-00368-w

[60] Song C, Sun S, Wang J, Gao Y, Yu G, Li Y, et al. Applying fulvic acid for sediment metals remediation: Mechanism, factors, and prospect. Frontiers in Microbiology. 2023;**13**:1084097. DOI: 10.3389/fmicb.2022.1084097

[61] Klučáková M. The effect of supramolecular humic acids on the diffusivity of metal ions in agarose hydrogel. Molecules. 2022;**27**(3):1019. DOI: 10.3390/molecules27031019

[62] Yildirim E, Ekinci M, Turan M, Ağar G, Dursun A, Kul R, et al. Humic+ fulvic acid mitigated Cd adverse effects

on plant growth, physiology and biochemical properties of garden cress. Scientific Reports. 2021;**11**(1):8040. DOI: 10.1038/s41598-021-86991-9

[63] Rong Q, Zhong K, Huang H, Li C, Zhang C, Nong X. Humic acid reduces the available cadmium, copper, lead, and zinc in soil and their uptake by tobacco. Applied Sciences. 2020;**10**(3):1077. DOI: 10.3390/app10031077

[64] Zhang Y, Liu G, Gao S, Zhang Z, Huang L. Effect of humic acid on phytoremediation of heavy metal contaminated sediment. Journal of Hazardous Materials Advances. 2023;**9**:100235. DOI: 10.1016/j.hazadv.2023.100235

[65] Ravn NR, Michelsen A, Reboleira ASP. Decomposition of organic matter in caves. Frontiers in Ecology and Evolution. 2020;**8**:554651. DOI: 10.3389/fevo.2020.554651

[66] Li C, Zhang Y, Chen R, Wang N, Liu J, Liu F. Influence of mineralized organic carbon in marine sediments on ecological heavy metal risk: Bohai Bay case study. Environmental Research. 2024;**240**:117542. DOI: 10.1016/j.envres.2023.117542

[67] Tunlid A, Floudas D, Op De Beeck M, Wang T, Persson P. Decomposition of soil organic matter by ectomycorrhizal fungi: Mechanisms and consequences for organic nitrogen uptake and soil carbon stabilization. Frontiers in Forests and Global Change. 2022;**5**:934409. DOI: 10.3389/ffgc.2022.934409

[68] Wydro U, Jabłońska-Trypuć A, Hawrylik E, Butarewicz A, Rodziewicz J, Janczukowicz W, et al. Heavy metals behavior in soil/plant system after sewage sludge application. Energies. 2021;**14**(6):1584. DOI: 10.3390/en14061584

[69] Wróbel M, Śliwakowski W, Kowalczyk P, Kramkowski K, Dobrzyński J. Bioremediation of heavy metals by the genus bacillus. International Journal of Environmental Research and Public Health. 2023;**20**(6):4964. DOI: 10.3390/ijerph20064964

[70] Henao GS, Ghneim-Herrera T. Heavy metals in soils and the remediation potential of bacteria associated with the plant microbiome. Frontiers in Environmental Science. 2021;**9**:604216. DOI: 10.3389/fenvs.2021.604216

[71] Wang S, Niu X, Di D, Huang D. Nitrogen and sulfur fertilizers promote the absorption of lead and cadmium with Salix integra Thunb. By increasing the bioavailability of heavy metals and regulating rhizosphere microbes. Frontiers in Microbiology. 2022;**13**:945847. DOI: 10.3389/fmicb.2022.945847

[72] Pande V, Pandey SC, Sati D, Bhatt P, Samant M. Microbial interventions in bioremediation of heavy metal contaminants in agroecosystem. Frontiers in Microbiology. 2022;**13**:824084. DOI: 10.3389/fmicb.2022.824084

[73] Joshi S, Gangola S, Bhandari G, Bhandari NS, Nainwal D, Rani A, et al. Rhizospheric bacteria: The key to sustainable heavy metal detoxification strategies. Frontiers in Microbiology. 2023;**14**:1229828. DOI: 10.3389/fmicb.2023.1229828

[74] Newsome L, Falagán C. The microbiology of metal mine waste: Bioremediation applications and implications for planetary health. GeoHealth. 2021;**5**(10):e2020GH000380. DOI: 10.1029/2020GH000380

[75] Nnaji ND, Onyeaka H, Miri T, Ugwa C. Bioaccumulation for heavy

metal removal: A review. SN Applied Sciences. 2023;**5**(5):125. DOI: 10.1007/s42452-023-05351-6

[76] Kapahi M, Sachdeva S. Bioremediation options for heavy metal pollution. Journal of Health and Pollution. 2019;**9**(24):191203. DOI: 10.5696/2156-9614-9.24.191203

[77] Lai HJ, Ding XZ, Cui MJ, Zheng JJ, Chen ZB, Pei JL, et al. Mechanisms and influencing factors of biomineralization based heavy metal remediation: A review. Biogeotechnics. 2023;**1**(3):100039. DOI: 10.1016/j.bgtech.2023.100039

[78] Madawala HMSP. Role of microorganisms in bioavailability of soil pollutants. One Health: Human, Animal, and Environment Triad. 2023;**9**:113-132. DOI: 10.1002/9781119867333.ch9

[79] Gao T, Wang H, Li C, Zuo M, Wang X, Liu Y, et al. Effects of heavy metal stress on physiology, hydraulics, and anatomy of three desert plants in the Jinchang mining area, China. International Journal of Environmental Research and Public Health. 2022;**19**(23):15873. DOI: 10.3390/ijerph192315873

[80] Dutta S, Mitra M, Agarwal P, Mahapatra K, De S, Sett U, et al. Oxidative and genotoxic damages in plants in response to heavy metal stress and maintenance of genome stability. Plant Signaling & Behavior. 2018;**13**(8):e1460048. DOI: 10.1080/15592324.2018.1460048

[81] Keyster M, Niekerk LA, Basson G, Carelse M, Bakare O, Ludidi N, et al. Decoding heavy metal stress signalling in plants: Towards improved food security and safety. Plants. 2020;**9**(12):1781. DOI: 10.3390/plants9121781

[82] Vasilachi IC, Stoleru V, Gavrilescu M. Analysis of heavy metal impacts on cereal crop growth and development in contaminated soils. Agriculture. 2023;**13**(10):1983. DOI: 10.3390/agriculture13101983

[83] De Caroli M, Furini A, DalCorso G, Rojas M, Di Sansebastiano GP. Endomembrane reorganization induced by heavy metals. Plants. 2020;**9**(4):482. DOI: 10.3390/plants9040482

[84] Tiwari S, Lata C. Heavy metal stress, signaling, and tolerance due to plant-associated microbes: An overview. Frontiers in Plant Science. 2018;**9**:452. DOI: 10.3389/fpls.2018.00452

[85] Lukowski A, Dec D. Influence of Zn, Cd, and Cu fractions on enzymatic activity of arable soils. Environmental Monitoring and Assessment. 2018;**190**:1-12. DOI: 10.1007/s10661-018-6651-1

[86] Houri T, Khairallah Y, Al Zahab A, Osta B, Romanos D, Haddad G. Heavy metals accumulation effects on the photosynthetic performance of geophytes in Mediterranean reserve. Journal of King Saud University-Science. 2020;**32**(1):874-880. DOI: 10.1016/j.jksus.2019.04.005

[87] Maphuhla NG, Lewu FB, Oyedeji OO. Enzyme activities in reduction of heavy metal pollution from Alice landfill site in eastern cape, South Africa. International Journal of Environmental Research and Public Health. 2022;**19**(19):12054. DOI: 10.3390/ijerph191912054

[88] Paunov M, Koleva L, Vassilev A, Vangronsveld J, Goltsev V. Effects of different metals on photosynthesis: Cadmium and zinc affect chlorophyll fluorescence in durum wheat. International Journal of Molecular Sciences. 2018;**19**(3):787. DOI: 10.3390/ijms19030787

[89] Zhang H, Xu Z, Guo K, Huo Y, He G, Sun H, et al. Toxic effects of heavy metal Cd and Zn on chlorophyll, carotenoid metabolism and photosynthetic function in tobacco leaves revealed by physiological and proteomics analysis. Ecotoxicology and Environmental Safety. 2020;**202**:110856. DOI: 10.1016/j.ecoenv.2020.110856

[90] Yang Y, Zhang L, Huang X, Zhou Y, Quan Q, Li Y, et al. Response of photosynthesis to different concentrations of heavy metals in Davidia involucrata. PLoS One. 2020;**15**(3):e0228563. DOI: 10.1371/journal.pone.0228563

[91] Guo Z, Gao Y, Yuan X, Yuan M, Huang L, Wang S, et al. Effects of heavy metals on stomata in plants: A review. International Journal of Molecular Sciences. 2023;**24**(11):9302. DOI: 10.3390/ijms24119302

[92] Farooq TH, Rafay M, Basit H, Shakoor A, Shabbir R, Riaz MU, et al. Morpho-physiological growth performance and phytoremediation capabilities of selected xerophyte grass species toward Cr and Pb stress. Frontiers. Plant Science. 2022;**13**:997120. DOI: 10.3389/fpls.2022.997120

[93] Koyama H, Kamogashira T, Yamasoba T. Heavy metal exposure: Molecular pathways, clinical implications, and protective strategies. Antioxidants. 2024;**13**(1):76. DOI: 10.3390/antiox13010076

[94] Mansoor S, Ali A, Kour N, Bornhorst J, AlHarbi K, Rinklebe J, et al. Heavy metal induced oxidative stress mitigation and ROS scavenging in plants. Plants. 2023;**12**(16):3003. DOI: 10.3390/plants12163003

[95] Cheng H, Yang B, Ke T, Li S, Yang X, Aschner M, et al. Mechanisms of metal-induced mitochondrial dysfunction in neurological disorders. Toxics. 2021;**9**(6):142. DOI: 10.3390/toxics9060142

[96] Georgiadou EC, Kowalska E, Patla K, Kulbat K, Smolińska B, Leszczyńska J, et al. Influence of heavy metals (Ni, Cu, and Zn) on nitro-oxidative stress responses, proteome regulation and allergen production in basil (Ocimum basilicum L.) plants. Frontiers in Plant Science. 2018;**9**:862. DOI: 10.3389/fpls.2018.00862

[97] Berni R, Luyckx M, Xu X, Legay S, Sergeant K, Hausman JF, et al. Reactive oxygen species and heavy metal stress in plants: Impact on the cell wall and secondary metabolism. Environmental and Experimental Botany. 2019;**161**:98-106. DOI: 10.1016/j.envexpbot.2018.10.017

[98] Małecka A, Konkolewska A, Hanć A, Ciszewska L, Staszak AM, Jarmuszkiewicz W, et al. Activation of antioxidative and detoxificative systems in Brassica juncea L. plants against the toxicity of heavy metals. Scientific Reports. 2021;**11**(1):22345. DOI: 10.1038/s41598-021-01827-w

[99] Goncharuk EA, Zagoskina NV. Heavy metals, their phytotoxicity, and the role of phenolic antioxidants in plant stress responses with focus on cadmium. Molecules. 2023;**28**(9):3921. DOI: 10.3390/molecules28093921

[100] Vasile GG, Tenea AG, Dinu C, Iordache AMM, Gheorghe S, Mureseanu M, et al. Bioavailability, accumulation and distribution of toxic metals (As, Cd, Ni and pb) and their impact on sinapis alba plant nutrient metabolism. International Journal of Environmental Research and Public Health. 2021;**18**(24):12947. DOI: 10.3390/ijerph182412947

[101] Sperdouli I. Heavy metal toxicity effects on plants. Toxics. 2022;**10**(12):715. DOI: 10.3390/toxics10120715

[102] Ali B, Gill RA. Heavy metal toxicity in plants: Recent insights on physiological and molecular aspects, volume II. Frontiers in Plant Science. 2022;**13**:1016257. DOI: 10.3389/fpls.2022.1016257

[103] Clemens S, Eroglu S, Grillet L, Nozoye T. Metal transport in plants. Frontiers in Plant Science. 2021;**12**:644960. DOI: 10.3389/fpls.2021.644960

[104] Sharma A, Kapoor D, Wang J, Shahzad B, Kumar V, Bali AS, et al. Chromium bioaccumulation and its impacts on plants: An overview. Plants. 2020;**9**(1):100. DOI: 10.3390/plants9010100

[105] Ali S, Mir RA, Tyagi A, Manzar N, Kashyap AS, Mushtaq M, et al. Chromium toxicity in plants: Signaling, mitigation, and future perspectives. Plants. 2023;**12**(7):1502. DOI: 10.3390/plants12071502

[106] De Lillo A, Cardi M, Landi S, Esposito S. Mechanism(s) of action of heavy metals to investigate the regulation of plastidic glucose-6-phosphate dehydrogenase. Scientific Reports. 2018;**8**(1):13481. DOI: 10.1038/s41598-018-31348-y

[107] Hu Z, Zhao C, Li Q, Feng Y, Zhang X, Lu Y, et al. Heavy metals can affect plant morphology and limit plant growth and photosynthesis processes. Agronomy. 2023;**13**(10):2601. DOI: 10.3390/agronomy13102601

[108] Balali-Mood M, Naseri K, Tahergorabi Z, Khazdair MR, Sadeghi M. Toxic mechanisms of five heavy metals: Mercury, lead, chromium, cadmium, and arsenic. Frontiers in Pharmacology. 2021;**12**:643972. DOI: 10.3389/fphar.2021.643972

[109] Hachani C, Lamhamedi MS, Cameselle C, Gouveia S, Zine El Abidine A, Khasa DP, et al. Effects of ectomycorrhizal fungi and heavy metals (Pb, Zn, and Cd) on growth and mineral nutrition of Pinus halepensis seedlings in North Africa. Microorganisms. 2020;**8**(12):2033. DOI: 10.3390/microorganisms8122033

[110] Shen T, Jin R, Yan J, Cheng X, Zeng L, Chen Q, et al. Study on diversity, nitrogen-fixing capacity, and heavy metal tolerance of culturable Pongamia pinnata rhizobia in the vanadium-titanium magnetite tailings. Frontiers in Microbiology. 2023;**14**:1078333. DOI: 10.3389/fmicb.2023.1078333

[111] Boorboori MR, Zhang HY. Arbuscular mycorrhizal fungi are an influential factor in improving the phytoremediation of arsenic, cadmium, lead, and chromium. Journal of Fungi. 2022;**8**(2):176. DOI: 10.3390/jof8020176

[112] Cong W, Miao Y, Xu L, Zhang Y, Yuan C, Wang J, et al. Transgenerational memory of gene expression changes induced by heavy metal stress in rice (Oryza sativa L.). BMC Plant Biology. 2019;**19**(1):1-14. DOI: 10.1186/s12870-019-1887-7

[113] Cruz Y, Villar S, Gutiérrez K, Montoya-Ruiz C, Gallego JL, Delgado MDP, et al. Gene expression and morphological responses of Lolium perenne L. exposed to cadmium (Cd2+) and mercury (Hg2+). Scientific Reports. 2021;**11**(1):11257. DOI: 10.1038/s41598-021-90826-y

[114] Shirazi Z, Khakdan F, Rafiei F, Balalami MY, Ranjbar M. Genome-wide identification and expression profile

analysis of metal tolerance protein gene family in Eucalyptus grandis under metal stresses. BMC Plant Biology. 2023;**23**(1):240. DOI: 10.1186/s12870-023-04240-9

[115] Yang Y, Huang J, Sun Q, Wang J, Huang L, Fu S, et al. microRNAs: Key players in plant response to metal toxicity. International Journal of Molecular Sciences. 2022;**23**(15):8642. DOI: 10.3390/ijms23158642

[116] Dubey S, Shri M, Chakrabarty D. MicroRNA mediated regulation of gene expression in response to heavy metals in plants. Journal of Plant Biochemistry and Biotechnology. 2021;**30**:744-755. DOI: 10.1007/s13562-021-00718-5

[117] Talukder P, Saha A, Roy S, Ghosh G, Roy DD, Barua S. Role of miRNA in phytoremediation of heavy metals and metal induced stress alleviation. Applied Biochemistry and Biotechnology. 2023;**195**(9):5712-5729. DOI: 10.1007/s12010-023-04599-3

[118] Yang H, Cui Y, Feng Y, Hu Y, Liu L, Duan L. Long non-coding RNAs of plants in response to abiotic stresses and their regulating roles in promoting environmental adaption. Cells. 2023;**12**(5):729. DOI: 10.3390/cells12050729

[119] Chen L, Shi S, Jiang N, Khanzada H, Wassan GM, Zhu C, et al. Genome-wide analysis of long non-coding RNAs affecting roots development at an early stage in the rice response to cadmium stress. BMC Genomics. 2018;**19**:1-10. DOI: 10.1186/s12864-018-4807-6

[120] Jha UC, Nayyar H, Jha R, Khurshid M, Zhou M, Mantri N, et al. Long non-coding RNAs: Emerging players regulating plant abiotic stress response and adaptation. BMC Plant Biology. 2020;**20**(1):1-20. DOI: 10.1186/s12870-020-02595-x

[121] Zhang Y, Zhou Y, Zhu W, Liu J, Cheng F. Non-coding RNAs fine-tune the balance between plant growth and abiotic stress tolerance. Frontiers in Plant Science. 2022;**13**:965745. DOI: 10.3389/fpls.2022.965745

[122] Liu J, He Z. Small DNA methylation, big player in plant abiotic stress responses and memory. Frontiers in Plant Science. 2020;**11**:595603. DOI: 10.3389/fpls.2020.595603

[123] Akhter Z, Bi Z, Ali K, Sun C, Fiaz S, Haider FU, et al. In response to abiotic stress, DNA methylation confers epigenetic changes in plants. Plants. 2021;**10**(6):1096. DOI: 10.3390/plants10061096

[124] Raza A, Tabassum J, Zahid Z, Charagh S, Bashir S, Barmukh R, et al. Advances in "omics" approaches for improving toxic metals/metalloids tolerance in plants. Frontiers in Plant Science. 2022;**12**:794373. DOI: 10.3389/fpls.2021.794373

[125] Parameswari E, Ilakiya T, Davamani V, Kalaiselvi P, Sebastian SP. Metallothioneins: Diverse protein family to bind metallic ions. In: Heavy Metals-their Environmental Impacts and Mitigation. London, UK: IntechOpen; 2021. DOI: 10.5772/intechopen.97658

[126] Luo M, Finet C, Cong H, Wei HY, Chung H. The evolution of insect metallothioneins. Proceedings of the Royal Society B. 2020;**287**(1937):20202189. DOI: 10.1098/rspb.2020.2189

[127] Faizan M, Alam P, Hussain A, Karabulut F, Tonny SH, Cheng SH, et al. Phytochelatins: A key regulator against heavy metal toxicity in plants. Plant Stress. 2024;**11**:100355. DOI: 10.1016/j.stress.2024.100355

[128] Chia JC. Phytochelatin synthase in heavy metal detoxification and xenobiotic metabolism. In: Mendes KF, de Sousa RN, Mielke KC, editors. Biodegradation Technology of Organic and Inorganic Pollutants. Vol. 18. 2021. pp. 1-18. DOI: 10.5772/intechopen.99077

[129] Zhu S, Shi W, Jie Y. Overexpression of BnPCS1, a novel phytochelatin synthase gene from ramie (Boehmeria nivea), enhanced Cd tolerance, accumulation, and translocation in Arabidopsis thaliana. Frontiers in Plant Science. 2021;**12**:639189. DOI: 10.3389/fpls.2021.639189

Chapter 6

Increasing Mango Production Efficiency under the Fast-Changing Climate

Augustine Antwi-Boasiako, Priscilla Amponsah, Jacinta Adoma Opoku, Daouda Coulibaly and Paul Mintah

Abstract

Mango (*Mangifera indica*) is an economically and nutritionally significant fruit crop in the tropical and subtropical regions. The demand for mango fruits and processed products has been high both in the internal and external markets due to its taste and its associated health benefits. In meeting the demands requirement, reengineering the various tools that are available to improve mango production in response to the varied stresses in their environment, especially in the era of climate change, is crucial. The prime goal is to demonstrate the effects of climate change on mango growth, yield and quality while showing interventions that have been deployed to combat it. The chapter focuses on the role climate variability plays in the growth and yield parameters of mango trees, as well as the improvement objectives and approaches employed in enhancing their production and quality. It offers the diverse progress made in overcoming the biotic stresses that hinder mango production as well as the intervention made in improving the nutritional and quality traits of mangoes while offering directions for future research works on mango trees.

Keywords: mango production, climate change, biotic stress, traditional breeding techniques, multi-omics, yield, fruit quality, nutrition

1. Introduction

High interest shown by farmers including the smallholders in diversifying their farm produce to include fruits crops such as mango and with profit-making goal accounts assist in attaining food security in the fast-changing climate era. Fruit crops such as mango (*Mangifera indica* L.) plays an essential role in achieving the prime goal of food security and hunger reduction. Mangoes are among the most popular fruits due to changes in consumer preferences [1]. This is partly attributed to its special aroma, colour, taste, juicy pulp and nutritional significance [2, 3]. Mango contains nutrients such as iron, calcium, phosphorus and vitamins A and C [4]. It is recommended that fruit nutrients are required for physical and mental people especially for children [5, 6]. These attributes term mango as the king of fruits. Mangoes as a tropical fruit crops ranks fifth globally in terms of production among major fruit crops [7, 8]. The total volume of mangoes produced globally in 2021 is 58.3 million metric tons, which is projected to increase by 1.9%, reaching 65 million metric tons by 2026 [9]. Mangoes are produced in more than 90 countries globally, with India as the world's leading producer [10]. Asia is the largest region in terms of mango production, contributing to over 70% of the total mango produced yearly. Africa contributes to 10% of the total mangoes produced yearly [11]. In Ghana, mango is ranked third after pineapple and orange in terms of its fresh juice consumption [12]. Mango production provides an opportunity as an alternative source of livelihood, especially in rural areas and offers high profit [13, 14]. The increase in demand for high-quality mango to meet standards for local, national and international markets has resulted in the designing of new production strategies and techniques that are essential elements for its production in a fast-changing climate. For instance, the amount of rainfall in most producing regions is not adequate to meet mango water requirements, demanding the application of irrigation, biostimulants and paclobutrazol to promote flowering, fruit yield and quality [15–17]. Unfortunately, there are a number of bottlenecks that underpin the production of mango. Limiting factors implicated in the yield gap of mango production may include pests, diseases, insects, weeds, changes in rainfall pattern, seeds, soil fertility and irregular flowering, among others. These form a larger component of the core production and productivity constraints of mango. A better understanding of these challenges positions the mango industry for higher output through the designing of effective and efficient management systems. However, a number of efforts are geared towards nutritional management by employing a number of fertilisation strategies to limit physiological disorders of mangoes while promoting fruit qualities [18–20]. Given the wide range of research on mango, the current chapter focuses on climate change and mango production and its effects on the mango tree growth, development, yield and quality. It further explores the farmers' reactions and adaptation strategies used in combating climate change. Also, general mango improvement objectives, challenges and methods are employed. Lastly, we present post-harvest and value-addition improvement strategies for mango improvement.

2. Climate change and mango production

The actions of humans have increased emissions of greenhouse gases, causing global warming and climate change. The effects of climate change may be helpful, harmful or unbiased, depending on the area [21]. Climate remains a key component in regulating the farming potential of a locality and classifying areas for cultivating certain crops [21–23]. Climate variability may lead to changes in temperatures, rainfall amount, atmospheric CO2 concentration, drought and soil salinity, among others. Therefore, the growth, development, flowering and yields of mango trees are influenced by prevailing environmental conditions. A shift in the required climatic condition may negatively impact mangoes in terms of their fruit set, quality and yields [24].

2.1 Impact of climate change on mango tree growth and development

The vegetative and phenological periods in mangoes are mainly controlled by temperature and rainfall. For instance, the best temperature for a mango tree's vegetative growth is within the range of 20–29°C, and below 15°C; this incites blossoming [25]. Temperatures ranging from 20 to 28°C record a significant effect on leaf size [26], possibly attributed to translocation of dry matter to the root during high CO_2 concentrations [25]. Mango responds differently to rises in temperatures compared to annual crops [27]. Hence, it stays alive under desiccating conditions, and this trait is advantageous for yielding during succeeding growth seasons. However, under stress conditions, the development of the plant may be delayed and may lead to high yield loss [28]. Mango trees experience negative effects under drought and flood stresses by limiting their growth. Matching crop varieties to climate is an imperative action for lucrative mango production in conventional and newer areas. The cold period (wet season) and the dry periods are critical for flowering in mangoes, and wet seasons favour vegetative flow emissions [29–31]. Usually, the onset of rains during mango flowering can be harmful and may cause total crop failure [32]. Mango trees grow continuously under high temperatures, rainfall and humidity without a distinct pause in growth, thereby impeding flowering. However, extreme drought and vapour pressure deficit negatively impact photosynthesis due to the rapid closure of the stomatal of the mango plant as a product of variations in climatic situations [33]. Flower bud development happens in the shoot of most varieties with termination of growth. Growth chamber studies on mango trees confirm that low night temperatures within 8–15°C, coupled with day temperatures below 20°C, lead to flowering once the shoot initiation happens [34].

2.2 Impact of climate change on mango tree yield and quality

The variation in the climatic elements imposes stress, which causes anatomical, biochemical, morphological and physiological changes in the mango plant, affecting mango tree growth, yield and quality. For instance, drastic change in temperatures coupled with unseasonal rain leads to yield and quality reduction in mango fruits. For instance, the quality and the number of mangoes are lowered as the result of the changes in the plant's physical features (vigour and canopy), reproductive traits (fruiting potential & fruit size) and quality traits (juice content, shelf life, pest & disease attack). Exposure of mango plants to high temperatures and low humidity lowers the efficiency of transpiration, photosynthesis and water potential in mango leaves [35, 36]. In an attempt by the mango plant to thwart the negative effects of these stresses, it reduces canopy area, inhibiting stomatal conductance and may, in the end, affect its yield [37]. Yields of mangoes are reduced when pre-flowering and flowering time coincides with rains. Moreover, it impedes fruit set and pollination activities. Similarly, storms cause major harvest loss in the later stage of fruit development. Soil water content in the soil has an effect on the fruit size, fruit yield as well as fruit quality parameters (soluble solids and sugar, starch and vitamin C quantity) of mangoes [38]. A number of studies confirm the vital role played by irrigation in supplementing rainfed in a fast-changing climate in attaining higher yield and quality of mango fruits [38–40]. Variability in climate influences fruit quality and the appearance of matured mango fruits. This is partly attributed to high incidences of pests and diseases such as anthracnose under high humidity and rainfall [32, 41]. During these periods, mango fruits may lose their aesthetic value due to the blackening of the peel, among others [32]. On the contrary, low humidity and rainfall support fruit setting and harvesting and the reduction of biotic stresses [42]. The huge variety in the genomic resources of mango is a benefit for collection and breeding plans to face climatic fluctuations. Fortunately, varietal disparities exist among mango genotypes in terms of their productivity under wet and dry conditions. Thus, these differences can be exploited and mango yields and qualities will be less affected by climate variability when utilised. Adaptation strategies are crucial in flight against climate change and contribute to mango yields and quality. For instance, matching crop varieties to climate results in yield in mangoes in both traditional and newer production areas.

2.3 Mango farmer' reactions and adaptation strategies towards climate change

Strategies adopted in mango seedling production against climate change include creating drainage channels, applying agrochemicals, grafting, better-quality varieties, establishing trees as shades, irrigation and soil nutrition improvement methods have been applied by farmers to diminish climate inconsistency and variation outcomes [43]. Mango producers' adaptation behaviour to climate change arises for the purposes of maintaining their farms and also their lives [44]. The provision of resources and knowledge on mitigating the negative consequences of climate alteration is key in mango production. With these, farmers are able to reduce these effects by adhering to weather forecasts while adjusting their time dates and supplying water. Strategies such as substitution and complementation of low temperature via the use of growth retardant coupled with the imposition of moisture stress for about 4 months preceding blossoming, post-harvest reduction of sizeable branches of vigorous and late varieties grown in heavy soils to boost soil moistness loss through winter periods [45]. Similarly, creating trenches along the drip line enhances drainage, thus preventing moisture stress conditions during rain peaks. Traditional measures for managing water deficits are unrealistic in the medium or long term. Mango producers are recommended to adopt deficit irrigation methods to meet drought conditions to save water, improve water use efficiency and maintain yield and quality [42]. Climate change acclimatisation tactics recommended by other researchers embrace drip irrigation approaches, high plant population and cover belts [46]. Mango farmers' acclimatisation to the changing climate is challenged by some factors which may arise from technological, socioeconomic and institutional barriers [46].

3. Mango improvement in fast-changing climate

3.1 General mango tree improvement objectives

Genetic improvements in most fruit crops like mango are underdeveloped. Most of the improvements in mangoes have arisen from natural selection through several generations. The upsurge in consumer and market demands coupled with a fast-changing climate demands mangos of high yield and quality. Unfortunately, mango breeding takes several years, going beyond 20 years. This is ascribed to its generation period, long juvenile stage, high heterozygosity, one seed in a fruit, recalcitrant seeds, incompatibility among cultivars, polyembryony, large area for evaluating hybrids and high prevalence of fruit shedding naturally. These challenges require a smart solution for improving mangoes. Nevertheless, there is a high range of genetic diversity in mango germplasm in addition to its asexual propagation, facilitating selection at any period along the breeding cycle [47].

The goals of mango improvement are centred on the needs of society in relation to agriculture. Thus, leads to three main pillars, namely, producing fruits of high quality, reduction in environmental impact and high returns on the production. These pillars must form an integral component in any mango improvement programmes to avoid trade-offs among them. However, a number of works also focus on the usage of post-harvest strategies and value-addition development to enhance the safety and quality of harvested fruits to their final consumer while reducing mango fruit wastage [48, 49]. This has resulted in the release of hybrid mangoes around the globe [50, 51]. This is achieved via designed, initiated and implemented well-defined breeding goals, and new varieties are expected to be introduced that are well suited to the changing climate and its associated biotic and abiotic stresses. Preferred traits in mango cultivars vary among producers, consumers and industries. However, a number of the traits such as early and regular bearing, dwarfness, nice appearance (attractive), good size, lack of physiological disorders, firmness, resistance to pests and diseases, and long shelf life, among others, are desirable traits of mango cultivar. A large number of commercial cultivars of mango on the market have been improved in terms of their biennial bearing potential. The combined application of multi-omics techniques has enhanced and limited uncertainty in mango improvement [52, 53].

3.2 Mango tree breeding challenges in fast challenging climate

Most of the improvements in mangoes arise from natural selection over several generations. The upsurge in consumer and market demands coupled with a fast-changing climate. Unfortunately, mango breeding takes several years, going beyond 20 years. This is ascribed to its generation period, long juvenile stage, high heterozygosity, one seed in a fruit, recalcitrant seeds, incompatibility among cultivars, polyembryony, large area for evaluating hybrids, and high prevalence of fruit shedding naturally. Factors that limit progress in traditional fruit tree breeding are the long juvenile phase, long generation time and large resource requirements in field areas and personnel for maintaining and evaluating hybrid populations. In meeting customers' and processors' demands for quality traits, traditional breeding approaches render this objective difficult to achieve [54]. In addition to these limitations, mango breeders are faced with high heterozygosity, polyembryony, low crossing rates (0.1%), a high incidence of fruitlet fall and a single seed per fruit [55]. Moreover, mango breeding programmes are challenged with long juvenile periods affecting the evaluation, selection and release of hybrid cultivars. Additionally, large areas are needed to evaluate the hybrid seedlings, coupled with the cost of maintaining them.

4. Approaches used in mango improvement

Mango tree has witnessed improvement via the introduction of new varieties and evaluation, selection (clones, monoembryomic and polyembryomic seedlings), pollination (assisted and controlled), in vitro selection, embryo rescue, molecular markers and multi-omics techniques. Applications of omics technologies such as proteomics, transcriptomics and metabolomics facilitate the development and identification of traits governing key genes of interest, supporting reverse and forward breeding. Genetic transformation in mango via *Agrobacterium tumefaciens* using genes regulating fruit ripening is reported and markers such as SNPs are efficient in revealing the selfing rate and pollinator behaviour of mangoes at the seedling stage [56]. Both the main approaches (traditional and conventional) have their pros and cons. Similarly, rootstocks have been utilised to improve mango tree cultivation and overcome stresses arising from several factors including drought, salinity and pathogens. Specifically, nanoparticles (nano-zinc oxide and nano-silicon) enhanced mango productivity by improving salinity tolerance [57]. To improve biotic stress tolerance in mango breeding programmes, researchers have used a variety of cutting-edge strategies that combine both the most recent techniques with conventional breeding techniques.

4.1 Traditional breeding techniques for mango improvement

For many years, traditional breeding methods have been the cornerstone of mango advancement against biotic stresses (**Table 1**). Cross-pollination of various mango types is used in these techniques to combine beneficial features, such as disease resistance and pest tolerance. Breeders can find and enhance mango cultivars with increased biotic stress tolerance by repeatedly selecting and evaluating progenies [55]. Similarly, breeders can create mango varieties with increased biotic stress tolerance by using conventional breeding methods in conjunction with thorough screening and evaluation procedures. To speed up the breeding process and improve the overall effectiveness of creating stress-tolerant mango varieties, it is crucial to combine these old methods with contemporary tools like molecular markers and biotechnological developments [65]. Traditional breeding, however, takes time and could take several generations to succeed.

Technique	Description	Examples	Reference
Germplasm selection	Accessions resistant or tolerant to biotic stressors are advanced to the next stage.	Anthracnose and powdery mildew	[58]
Cross-Pollination and Recurrent Selection	Controlled cross-pollination among mango varieties to combine desirable genes for biotic stress tolerance. Repeatedly choosing the most successful offspring for additional breeding cycles.	Increased disease and pest resistance in hybrid mango varieties.	[59]
Backcrossing	Transferring particular biotic stress tolerance genes from a donor parent to a recipient parent.	Mango variety with improved resistance to a specific disease through backcrossing.	[60]
Mass selection	Choosing mango trees from a wide population that has desired characteristics, like tolerance to biotic stressors.	Local landraces with inherent resistance to specific pests or diseases.	[61]
Line breeding	Developing genetically homogenous lines by choosing and breeding mango plants with favourable features over several generations.	Mango lines with improved biotic stress.	[62]
Inbreeding and selection	Controlled inbreeding to concentrate desirable traits for biotic stress tolerance.	Mango varieties with increased resistance to specific diseases.	[63]
Open pollination and selection	Promoting genetic variety by allowing mango trees to undergo natural pollination.	Mango trees with unique biotic stress tolerance traits from open pollination.	[64]

Table 1.
Traditional mango improvement technique.

4.2 Current approaches

Smart solutions are required for improving mangoes. The wide range of genetic diversity in mango germplasm in addition to its asexual propagation, facilitates selection at any period along the breeding cycle [47]. A number of strategies involving marker-assisted selection (MAS), transgenic approaches and genome editing technologies are employed in improving mango production for high yield and quality.

Marker-assisted selection (MAS): To improve mangoes, MAS serves as a potent tool used in addition to conventional breeding methods. Finding molecular markers connected to traits of interest (disease or insect resistance) is the goal of MAS. The time and resources required for phenotypic evaluations are decreased because these markers allow breeders to choose plants with desirable features at the early seedling stage. Mango cultivars with improved resistance to diseases like anthracnose and powdery mildew have been developed successfully using MAS [62]. For instance, to choose mango offspring with favourable alleles for desirable qualities, [54] used MAS to determine which progeny carry the required alleles; the procedure involved genotyping the progeny. These variants can be located using a collection of trait-associated SNP markers that the researchers designed. Thus cutting down on the time and expense needed for field use, maintenance and long-term evaluation of material. In conclusion, MAS is a helpful tool for choosing mango progenies with desirable features and the trait-associated SNP markers created in the study can assist breeders in more effectively identifying these progenies.

Transgenic approaches: A direct and targeted method of introducing particular genes into mango plants to provide biotic stress resistance through genetic engineering. Genes from other organisms, such as those from disease-resistant plants or insecticidal proteins, can be inserted into the mango genome using transgenic techniques [66]. The development of mango cultivars with enhanced resistance to pests like fruit flies has shown potential thanks to this technique. Transgenic methods do, however, also spark worries about potential environmental and regulatory problems.

The fastest-growing agricultural technology is transgenic technology [67]. It refers to a group of strategies for introducing desired genes into a particular plant by unconventional means from any source (including plants, animals, microbes and even genes that have been artificially created). The primary benefit of transgenic technology is the ability to change plants by using the genes governing a variety of crucial agronomically significant features that may be obtained from any organism, including plants, bacteria, etc. As a result, the target plant can easily adopt unique features from any background.

The defensin *J1* gene was discovered to have been used to create transgenic mango embryos (*Mangifera indica*) cv. "*Ataulfo*." Defensins are tiny cysteine-rich peptides with antibacterial activity against a variety of pathogens, including bacteria, fungi and viruses. The transgenic mango plants exhibited enhanced resistance to anthracnose, a fungal disease caused by *Colletotrichum gloeosporioides* [68]. According to the study, creating disease-resistant mango types may be facilitated through genetic engineering.

Genome editing technologies: The development of genome editing technologies, particularly CRISPR-Cas9, has completely changed how mangoes can be improved to withstand biotic stress. Without adding foreign DNA, genome editing enables precise alterations of particular genes inside the mango genome. Breeders can now use this technology to increase the expression of genes related to pest tolerance or disease resistance [69]. While resolving the ethical issues with transgenic methods, genome editing has enormous potential for accelerating the production of stress-tolerant mango varieties [70]. Key genes linked to disease resistance in mango have been edited using CRISPR-Cas9. The possibility of creating disease-resistant mango varieties has been demonstrated by targeting genes involved in plant-pathogen interactions, such as those in charge of resistance to anthracnose and powdery mildew [71]. CRISPR-Cas9 may be used by scientists to improve mango's resistance to environmental challenges including salinity and drought. The goal is to create mango varieties that are more tolerant to harsh environmental conditions by focusing on genes implicated in stress response pathways [72]. CRISPR-Cas9 can be used to change the genes responsible for the size, flavour and ripening of fruit. Mango cultivars with enhanced flavour and longer shelf lives may be developed by focusing on specific genes involved in fruit scent, sugar content and texture [72]. Additionally, CRISPR-Cas9 is being investigated as a way to improve resistance to viral infections that damage mangoes and cause diseases including mango malformation disease and mango leaf distortions. Researchers want to develop mango cultivars with improved virus resistance by introducing particular mutations in genes that are sensitive [73].

Multi-omics approaches: Understanding the molecular pathways underpinning mango's ability to withstand biotic stress is now substantially easier due to multi-omics technologies (genomics, transcriptomics, proteomics and metabolomics) by deepening our understanding of how mango genes and stressors interact [74]. Breeders can design tailored breeding strategies for biotic stress tolerance by understanding these relationships of key genes and their metabolic processes involved in stress responses. Recent genome sequences on mango varieties (Carabao, Tommy Atkins, Alphonso & Irwin) facilitate the study of mangoes at the transcriptomics, comparative genomics and genotyping to provide reliable gene sets and references [75–77]. A number of molecular markers are also used to explore genetic diversity, generate genetic linkage maps and identify markers for key traits (pulp colour, fruit colour, taste and flavour) [54, 78–80]. These investigations aid in the identification of the genes and proteins involved in early and late flowering, various stages of fruit development and biotic stress situations. The regulation of post-harvest and biannual flowering is possible through metabolomic research on the mango, which sheds light on the metabolites, hormones and pathways involved in fruit softening and flower induction [81].

Public and private collaboration: Mango enhancement for biotic stress tolerance depends critically on collaboration between governmental research institutions and private businesses [82]. Public institutions frequently have significant collections of germplasm, the ability to conduct fundamental research and knowledge of breeding methods. On the other side, private businesses can make investments in cutting-edge biotechnological equipment and market the stress-tolerant mango cultivars that have been created. Collaboration between these industries can hasten the creation and spread of enhanced mango cultivars that are resistant to a variety of biotic stressors. An evaluation of the extensive collection of germplasm in mango breeding is urgently needed [61]. Germplasm is employed to find sources of resistance to different biotic stresses in the context of mango breeding for biotic stress. It is important to screen the polyembryonic cultivars for these stresses because their variability is lower, and they are typically found in areas with significant rainfall. To improve the quality of these cultivars, it is also necessary to introduce diversity into them. Using markers to identify zygotic and nucellar embryony might aid in the selection of new recombinants.

Due to the fusion of conventional breeding techniques with cutting-edge biotechnological methodologies, mango development for biotic stress resistance has advanced dramatically. The way mango breeders create stress-tolerant cultivars has been transformed by the use of omics technology, transgenic methods, genome editing technologies and marker-assisted selection (MAS). The effectiveness and efficiency of mango improvement programmes have been further improved by collaboration between the public and commercial sectors. The future of mango breeding for biotic stress tolerance is bright as research advances; it has the potential to provide sustainable mango production and availability, protecting this beloved tropical fruit for future generations.

5. Improving mangoes against biotic stresses tolerance

Biotic stressors present a number of difficulties in mango farming. A wide range of diseases caused by fungi, bacteria, viruses and nematodes pose a serious danger to crop productivity [83]. Additionally, a variety of pests, including insects and mites, can seriously harm mango plants and fruits. Mango growers around the world continue to face chronic challenges from common diseases and pests like the mango hopper and fruit fly, as well as diseases like anthracnose, powdery mildew and bacterial black spot [84]. For instance, yield loss attributed to fruit fly (*Diptera: Tephritidae*) is billions of dollars annually globally, with country-specific losses of around USD 242 million from Brazil and 50% losses from the entire mango produced in Mali [85, 86]. Recent studies demonstrate that penetration of sunlight in mango plantation activates the defence signals in mango fruits. For instance, mango fruit develops tolerance to fungal pathogen *Collectotrichum gloeosporioides* by activating defence-related pathways and hormones (ethylene, brassinosteroid and jasmonic acid), which triggers phenylpropanoid biosynthesis [87]. Rootstocks from polyembryonic are used to improve mango tree resistance to fruit fly occurrences and mango wilt disease [88, 89]. Current research has designed mango fruit trap detection to monitor their population on the field and to establish an economic threshold for managing them [90]. Scientists are keen on improving the genetic base of mango cultivars to enhance their resistance to varied biotic stress (**Table 2**). However, current management of these stresses largely uses integrated management approaches.

Stress	Key Management Approach	Reference
Anthracnose	*MiPR1A* genes enhanced resistance to *Colletotrichum gloeosporioides*	[91]
Powdery mildew	Resistant cultivars and sulphur-based fungicides application	[92]
Bacterial black spot	Copper-based sprays and pruning of affected tissues	[93, 94]
Fruit fly	Fruit fly integrated pest management, trap-based monitoring and protein bait sprays	[95, 96]
Anthracnose	Cultural practices (pruning and sanitation) and fungicide application	[97]
Mango hopper	Biological control (parasitoids) and neem-based insecticides	[98]

Table 2.
Prevalent major biotic stresses in mango production.

6. Post-harvest and value addition improvement strategies

6.1 Post-harvest losses in mango

The promotion of mango production among smallholder mango farmers contributed to the development and reduction of poverty. However, post-harvest losses and challenges in value addition have hindered its full potential in the market [99]. Post-harvest losses in mango occur due to factors such as physiological changes, mechanical damage, fungal and bacterial infections, and inadequate storage and transportation [100]. Climate change-induced factors like extreme temperatures and unpredictable rainfall patterns also contribute to losses. However, the short shelf life of mango fruit, the difficulty in regulating the post-harvest quality, and the lack of preservation technology are the main problems that need to be solved in the mango industry. Post-harvest practices encompass all the activities that take place after a crop is harvested until it reaches the ultimate market or consumer. These essential steps include harvesting, packing, storing, transporting, processing and marketing of the mango. Implementing effective post-harvest practices is crucial as it not only enhances the overall quality and value of the mango but also minimises the extent of deterioration. By gaining insights into the ideal conditions for handling your specific product, you can significantly reduce losses and optimise the outcome.

6.2 Post-harvest handling and management

Mango is a climacteric fruit with a relatively short shelf life, making post-harvest management crucial for maintaining quality and reducing losses. In recent years, climate change has further exacerbated post-harvest challenges, necessitating innovative strategies to ensure food security and economic sustainability. Post-harvest management is the careful processing of agricultural products to preserve their quality and freshness and lengthen the shelf life before marketing in order to increase market value [99]. The first step in post-harvest handling is to ensure that the mangoes are harvested at the optimum maturity stage and subsequently ripened to obtain the desired flavour and texture [99]. For instance, mangoes harvested too early will not have developed their full flavour and aroma, while those harvested too late will be prone to decay. Once the mangoes have been harvested, they should be handled with care (**Figure 1**). During and after harvesting, mangoes are critically susceptible to damage; once the fruit is damaged, all the biological processes, including ethylene production and respiration, proceed ahead swiftly, producing/leading to its rapid deterioration. After mangoes are harvested, procedures like sorting, grading, pre-cooling/washing, packaging, storage, transportation and marketing, among others, are said to be crucial [100]. Spreading fruits on a paddy straw cushion floor in the orchard's yard for nearly 24 hours, followed by washing the fruit to remove dirt, has been described as being a very important step in extending the shelf life. They should be packed in clean, well-ventilated containers and stored at a cool temperature. The ideal storage temperature for mangoes is 10–12°C [101].

6.3 Strategies for enhancing mango shelf life

A number of methods such as waxing, modified atmosphere packaging (MAP), controlled atmosphere storage and hot water treatment are used to prolong the shelf life of mangoes.

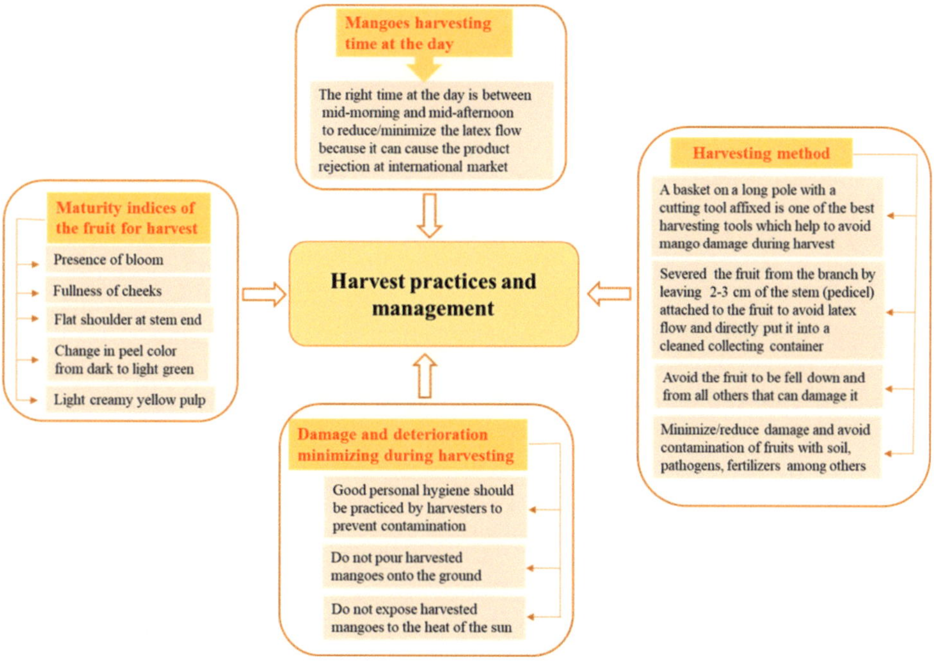

Figure 1.
Technical practices and management of harvest.

- Waxing: Waxing/Coatings can be effectively used to increase the shelf life and marketable period of mango [99]. Beeswax-emulsified chitosan–*Aloe vera* has been used to coat mangoes to extend the shelf life.

- Modified atmosphere packaging (MAP): Modified atmosphere packaging (MAP) involves altering the atmospheric composition around the fruit to slow down respiration and reduce ethylene production. Several studies have shown the efficacy of MAP in extending mango shelf life and reducing post-harvest losses.

- Controlled atmosphere (CA) storage: This involves maintaining specific gas concentrations (oxygen and carbon dioxide) and temperature levels. Research has indicated that CA storage can effectively delay ripening and extend the storage life of mangoes [101].

- Hot water treatment: Hot water treatment is an alternative quarantine method that is effective in controlling fruit fly infestation. Studies have demonstrated that hot water treatment significantly reduces fruit fly infestation without compromising fruit quality [102]. The mangoes are dipped in hot water for a short period of time, which kills the bacteria and fungi that can cause decay.

6.4 Value addition improvement strategies

Value addition in agriculture is the process of enhancing the quality, quantity and marketability of agricultural products. It entails transforming raw materials into more valuable forms. Value addition not only raises the revenue of farmers and agribusinesses but also creates employment opportunities and reduces post-harvest losses, therefore improving food security. Value addition can also contribute to environmental sustainability by reducing waste and greenhouse gas emissions. Value addition entails converting a raw product into a mango and processing it into puree and concentrates. Processing mangoes into a variety of products, such as juice, pulp, chutney, pickle, jam, puree and concentrates, not only extends their shelf life but also adds value to the fruit than fresh mangoes, and they can be exported to other countries. These products are used as ingredients in various food products and beverages, increasing the marketability of mangoes [103]. Value-addition improvement strategies include:

- Drying and dehydration: These are traditional preservation methods that reduce moisture content, thereby inhibiting microbial growth and enzymatic reactions. Solar drying and vacuum drying are among the commonly used techniques for mango preservation [104].

- Mango by-product utilisation: By-products of mango processing, such as peels and seeds, can be utilised to produce value-added products like pectin, mango kernel oil and animal feed [105].

In the era of climate change, post-harvest and value-addition improvement strategies are essential for enhancing mango production and reducing losses. The adoption of innovative technologies and processing techniques can lead to increased market opportunities, significant income generation and overall sustainability in mango production.

7. Conclusion

Mango has an established export market and poses bright opportunities for export in the international market, whether in fresh or processed forms. Similarly, the mango industry has provided livelihood opportunities to its growers and those involved in its marketing channel. The development of mango varieties that are climate-smart with improved qualities is desired. Also, current mango improvement programmes should focus on the development of pests (fruit flies) and diseases (black spot disease) resistance, early maturing, wide adaptation and identification of key traits controlling yield, pulp colour and fruit shape. There should be a strong linkage between research institutions and farmers in order to combat the key challenges confronting the mango industries. In addition, value addition along the mango production chain should be strengthened especially in African countries where post-harvest losses in handling and marketing are higher. Similarly, since mangoes are seasonal fruits, there is scope to establish mango preservation factories. Considerable amounts of waste material, e.g., mango stones and peels, remain unutilized and can be used properly by processors to earn more profit. This will also help to improve sanitary conditions around factory premises. High-quality commercial cultivation of mango trees by using improved planting and drip irrigation leads to multiple benefits, viz.

Conflict of interest

The authors declare no conflict of interest.

Author details

Augustine Antwi-Boasiako[1,2]*, Priscilla Amponsah[3], Jacinta Adoma Opoku[1], Daouda Coulibaly[4,5] and Paul Mintah[1]

1 Council for Scientific and Industrial Research, Crops Research Institute CSRI-CRI), Fumesua, Kumasi, Ghana

2 National Center for Soybean Improvement, Key Laboratory of Biology and Genetics and Breeding for Soybean, Ministry of Agriculture, State Key Laboratory of Crop Genetics and Germplasm Enhancement, Nanjing Agricultural University, Nanjing, China

3 Key Laboratory of Integrated Pest Management on Crops in East China, Ministry of Agriculture, Nanjing Agricultural University, Nanjing, China

4 Department of Agricultural Sciences and Techniques-Horticulture, Rural Polytechnic Institute for Training and Applied Research (IPR/IFRA) of Katibougou, Koulikoro, BP, Mali

5 Laboratory of Fruit Tree Biotechnology, College of Horticulture, Nanjing Agricultural University, Nanjing, China

*Address all correspondence to: bbbantwi@cropsresearch.org

References

[1] Altendorf S. Global prospects for major tropical fruits: Short-term outlook, challenges and opportunities in a vibrant global marketplace. Food Outlook Special Feature. 2017:69-81

[2] Tharanathan RN, Yashoda HM, Prabha TN. Mango (*Mangifera indica* L.) "The king of fruits"—An overview. Food Reviews International. 2006;**22**(2):95-123

[3] Ntsoane ML, Zude-Sasse M, Mahajan P, Sivakumar D. Quality assessment and postharvest technology of mango: A review of its current status and future perspectives. Scientia Horticulturae. 2019;**249**:77-85. DOI: 10.1016/j.scienta.2019.01.033

[4] Lebaka VR, Wee YJ, Ye W, Korivi M. Nutritional composition and bioactive compounds in three different parts of mango fruit. International Journal of Environmental Research in Public Health. 2021;**18**:741. DOI: 10.3390/ijerph18020741

[5] Carrillo JÁ, Zafrilla MP, Marhuenda J. Cognitive function and consumption of fruit and vegetable polyphenols in a young population: Is there a relationship? Food. 2019;**8**(10):507. DOI: 10.3390/foods8100507

[6] Lamprecht M, Obermayer G, Steinbauer K, Cvirn G, Hofmann L, Ledinski G, et al. Supplementation with a juice powder concentrate and exercise decrease oxidation and inflammation, and improve the microcirculation in obese women: Randomised controlled trial data. British Journal of Nutrition. 2013;**110**:1685-1695. DOI: 10.1017/S0007114513001001

[7] Okino-Delgado CH, Prado DZ, Pereira MS, Camargo DA, Koike A, Fleuri LF. Mango. In: Valorization of Fruit Processing by-Products. United States: Academic Press; 2020. pp. 167-181. DOI: 10.1016/B978-0-12-817106-6.00008-3

[8] FAO. Major Tropical Fruits: Preliminary Results 2021. Rome; 2022

[9] Mango Market Outlook. 2023. Available from: https://www.reportlinker.com/clp/global/2974#introduction [Accessed: 30 May, 2023]

[10] FAO. FAOSTAT crop production [Online]. 2023. Available from: http://www.fao.org/faostat/en/#home

[11] FAO. FAOSTAT crop production [Online]. 2021. Available from: http://www.fao.org/faostat/en/#home

[12] Broek J, Apenteng-Sackey N, Arnoldus M, Keita S, Waardenburg R. West Africa fruit-scoping study. Netherlands Enterprise Agency. 2016. Publication number: RVO-007-1701/RP-INT

[13] Testa R, Tudisca S, Schifani G, Di Trapani AM, Migliore G. Tropical fruits as an opportunity for sustainable development in rural areas: The case of mango in small-sized Sicilian farms. Sustainability. 2018;**10**(5):1436

[14] Yiran GA, Ablo AD, Asem FE. Is mango farming sustainable? An integrated analysis of remote sensing techniques and smallholder farmers' perception in mango farming communities in Ghana. Ghana Social Science Journal. 2023;**20**(1):71-96

[15] Lobo JT, De Sousa KDSM, Neto VBDP, Pereira RN, Silva LDS, et al.

Biostimulants on fruit yield and quality of mango cv. Kent grown in semiarid. Journal American Pomological Society. 2019;**73**(3):152-160

[16] Simões WL, do Carmo Mouco MA, de Andrade VPM, Bezerra PP, Coelho EF. Fruit yield and quality of palmer mango trees under different irrigation systems. Comunicata Scientiae. 2020;**11**:e3254-e3254

[17] Ferreira KM, Simões WL, do Carmo Mouco MA, da Silva JL, da Silva JS, Mesquita AC. Efficient management of the application of paclobutrazol for the production and quality of 'Tommy Atkins' mango. Research, Society and Development. 2020;**9**(8):e34898489 4-e348984894

[18] Taiz L, Zeiger E, Møller IM, Murphy A. Fisiologia Edesenvolvimento Vegetal. 6th ed. Porto Alegre: ArtMed; 2017. p. 858

[19] Zhang D, Chong W, Li XL. Yield gap and production constraints of mango (*Mangifera indica*) cropping systems in Tianyang County, China. Journal of Integrative Agriculture. 2019;**18**(8):1726-1736

[20] Simões WL, Silva JSD, Mouco MADC, de Oliveira CP, Silva DJ, Oliveira FFD. Marine calcium application on 'Palmer' mango production. Revista Brasileira de Engenharia Agrícola e Ambiental. 2022;**26**:618-623

[21] Janjua PZ, Samad G, Khan N. Climate change and wheat production in Pakistan: An autoregressive distributed lag approach. NJAS-Wageningen Journal of Life Sciences. 2014;**68**:13-19. DOI: 10.1016/j.njas.2013.11.002

[22] Fain SJ, Quiñones M, Álvarez-Berríos NL, Parés-Ramos IK, Gould WA. Climate change and coffee: Assessing vulnerability by modeling future climate suitability in the Caribbean island of Puerto Rico. Climatic Change. 2018;**146**(1-2):175-186

[23] Drebenstedt I, Marhan S, Poll C, Kandeler E, Högy P. Annual cumulative ambient precipitation determines the effects of climate change on biomass and yield of three important field crops. Field Crops Research. 2023;**290**:108766

[24] Shahraki AS, Caloiero T, Bazrafshan O. Influence of climatic factors on yields of pistachio, mango, and bananas in Iran. Sustainability. 2023;**15**(11):8962

[25] Whiley AW, Rasmussen TS, Saranah JB, Wolstenholme BN. Effect of temperature on growth, dry matter production and starch accumulation in ten mango (*Mangifera indica* L.) cultivars. Journal of Horticultural Science. 1989;**64**:753-765

[26] Dambreville A, Normand F, Lauri PE. Plant growth co-ordination in natura: A unique temperature-controlled law among vegetative and reproductive organs in mango. Functional Plant Biology. 2013;**40**:280-291

[27] Litz RE. The mango. In: Botany, Production and Uses. 2nd ed. UK: CABI International; 2009

[28] Morison IL, Morecroft M. Pant Growth and Climate Change. UK: Blackwell Publishing Ltd; 2006

[29] Sukhvibul N, Whiley AW, Smith MK, Hetherington SE, Vithanage V. Effect of temperature on inflorescence and floral development in four mango (*Mangifera indica* L.) cultivars. Scientia Horticulturae. 1999;**82**(1-2):67-84. DOI: 10.1016/S0304-4238(99)00041-2

[30] Spreer W, Ongprasert S, Hegele M, Wünsche JN, Müller J. Yield and fruit

development in mango (*Mangifera indica* L. cv. Chok Anan) under different irrigation regimes. Agricultural Water Management. 2009;**96**(4):574-584. DOI: 10.1016/j.agwat.2008.09.020

[31] Ramírez F, Davenport TL. Mango (*Mangifera indica* L.) flowering physiology. Scientia Horticulturae. 2010;**126**(2):65-72. DOI: 10.1016/j.scienta.2010.06.024

[32] Makhmale S, Bhutada P, Yadav L, Yadav BK. Impact of climate change on phenology of mango-the case study. Ecology, Environment and Conservation. 2016;**22**(9):S127-S132

[33] Normand F, Lauri PE, Legave JM. Climate change and its probable effects on mango production and cultivation. In: Espinal JJ et al., editors. Proceeding in X International Mango Symposium. Vol. 1075. ACTA Horticulture, ISHS; 2015

[34] Shu ZH, Sheen TF. Floral induction in axillary buds of mango (*Mangifera indica* L.) as affected by temperature. Scientia Horticulture. 1987;**31**:81-87

[35] Faria LN, Donato SL, Santos MRD, Castro LG. The effects of irrigation management on floral induction of 'Tommy Atkins' mango in Bahia semiarid. Energia na Agricultura. 2016;**36**:387-398. DOI: 10.1590/1809-4430-Eng.Agric.v36n3p387-398/2016

[36] Khanum Z, Tiznado-Hernández ME, Ali A, Musharraf SG, Shakeel M, Khan IA. Adaptation mechanism of mango fruit (*Mangifera indica* L. cv. Chaunsa white) to heat suggest modulation in several metabolic pathways. RSC Advances. 2020;**10**:35531-35544. DOI: 10.1039/D0RA01223H

[37] Shanthala L, Ramachandra RK, Mohan KS, Swamy GSK, Honnabyraiah MK, Pallavi HM. Evaluation of mango (*Mangifera indica* L.) rootstocks under induced moisture stress condition for various Morpho-physiological traits. International Journal of Current Microbiology and Applied Sciences. 2021;**10**:143-148. DOI: 10.20546/ijcmas.2021.1005.019

[38] Wei J, Liu G, Liu D, Chen Y. Influence of irrigation during the growth stage on yield and quality in mango (*Mangifera indica* L). PLoS One. 2017;**12**(4):e0174498. DOI: 10.1371/journal.pone.0174498

[39] Durán Zuazo VH, Rodríguez Pleguezuelo CR, Gálvez Ruiz B, Gutiérrez Gordillo S, García-Tejero IF. Water use and fruit yield of mango (*Mangifera indica* L.) grown in a subtropical Mediterranean climate. International Journal of Fruit Science. 2019;**19**(2):136-150. DOI: 10.1080/15538362.2018.1493960

[40] Hahn F, García JA. Mango stem response under different irrigation regimes. International Journal of Fruit Science. 2022;**22**(1):35-56. DOI: 10.1080/15538362.2021.1970080

[41] Ploetz R, Ploetz R. Diseases of Tropical Fruit Crops. Wallingford: CABI Publishing; 2003. pp. 327-363

[42] Zuazo VHD, García-Tejero IF, Rodríguez BC, Tarifa DF, Ruiz BG, Sacristán PC. Deficit irrigation strategies for subtropical mango farming. A review. Agronomy for Sustainable Development. 2021;**41**:1-13. DOI: 10.1007/s13593-021-00671-6

[43] Asare-Nuamah P, Antwi-Agyei P, Dick-Sagoe C. Mitigating the risks of climate variability and change on mango seedlings in Ghana: Evidence from mango seedlings producers in the Yilo Krobo municipality. Environmental Challenges. 2022;**8**:100594

[44] Esperanza D, Putri G, Rasmikayati E, Qanti S, Yuwariah Y. Adaptation behavior of mango farmers to climate change. Jurnal Manajemen & Agribisnis. 2018;**15**(3):268-268

[45] Halder S, Hasan MA. Climate change and mango production. Chemical Science Review Letters. 2020;**9**(33):55-57. DOI: 10.37273/chesci.cs122050121

[46] Singh NP, Ashok A, Pavithra S, Balaji SJ, Anand B, Khan MA. Mainstreaming climate change, adaptation into development planning. In: Policy Paper 32, ICAR National Institute of Agricultural Economics and Policy Research. New Delhi; 2017

[47] Iyer CPA, Dinesh MR. Advances in classical breeding and genetics in mango. Acta Horticulturae. 1997;**455**:252-267. DOI: 10.17660/ActaHortic.1997.455.34

[48] Arinloye ADD, Degrande A, Fassinou Hotegni VN, Asaah E, Bockarie R, Nyemeck JB, et al. Value chain development for mango (*Mangifera indica*) around Outamba Kilimi National Park in Sierra Leone: Constraints and opportunities for smallholders. Agriculture & Food Security. 2017;**2017**(6):1-13

[49] Esguerra EB, Rolle R. Post-Harvest Management of Mango for Quality and Safety Assurance. Rome: Food and Agriculture Organization of the United Nations; 2018

[50] Jha SK, Sethi S, Srivastav M, Dubey AK, Sharma RR, Samuel DVK, et al. Firmness characteristics of mango hybrids under ambient storage. Journal of Food Engineering. 2010;**97**(2):208-212. DOI: 10.1016/j.jfoodeng.2009.10.011

[51] Rajwana IA, Malik AU, Khan AS, Saleem BA, Malik SA. A new mango hybrid shows better shelf life and fruit quality. Pakistan Journal of Botany. 2010;**42**(4):2503-2512

[52] Gómez-Ollé A, Bullones A, Hormaza JI, Mueller LA, Fernandez-Pozo N. MangoBase: A genomics portal and gene expression atlas for *Mangifera indica*. Plants. 2023;**12**(6):1273. DOI: 10.3390/plants12061273

[53] Srivastav M, Radadiya N, Ramachandra S, Jayaswal PK, Singh N, Singh S. High resolution mapping of QTLs for fruit color and firmness in Amrapali/sensation mango hybrids. Frontiers in Plant Science. 2023;**14**:1135285. DOI: 10.3389/fpls.2023.1135285

[54] Kuhn DN, Bally IS, Dillon NL, Innes D, Groh AM, Rahaman, et al. Genetic map of mango: A tool for mango breeding. Frontiers in Plant Science;**8**:577, 10.3389/fpls.2017.00577

[55] Bally IS, Lu P, Johnson PR. Mango breeding. In: Breeding Plantation Tree Crops: Tropical Species. New York: Springer; 2009. pp. 51-82

[56] Michael VN, Crane J, Freeman B, Kuhn D, Chambers AH. Mango seedling genotyping reveals potential self-incompatibility and pollinator behavior. Scientia Horticulturae. 2023;**308**:111599. DOI: 10.1016/j.scienta.2022.111599

[57] Elsheery NI, Helaly MN, El-Hoseiny HM, Alam-Eldein SM. Zinc oxide. Silicone nanoparticles to improve the resistance mechanism and annual productivity of salt stressed mango trees. Agronomy. 2020;**10**(4):558. DOI: 10.3390/agronomy10040558

[58] Dangana MC, Auta Y, Mailafiya SC, Moses BE, Oluwafemi OKJ, Ebyna KC. Determination of heavy metals in four mango fruit varieties sold in Minna

modern market, Niger state, Nigeria. International Journal of Biological and Environmental Engineering. 2018;**1**(1):24-29

[59] Kumar N. Breeding of Horticultural Crops: Principles and Practices. New Delhi, India: New India Publishing; 2006

[60] Dijkman M, Soule M Jr. A tentative method of mango selection. In: Paper Presented at the Proceedings of the Florida State Horticultural Society. 1951

[61] Dinesh M, Vasugi C, Ravishankar K. Morphological, molecular characterization and breeding for biotic and abiotic stress in mango (*Mangifera indica* L.). In: Paper Presented at the Global Conference on Augmenting Production and Utilization of Mango: Biotic and Abiotic Stresses. 2011. p. 1066

[62] Parmar N, Singh KH, Sharma D, Singh L, Kumar P, Nanjundan J, et al. Genetic engineering strategies for biotic and abiotic stress tolerance and quality enhancement in horticultural crops: A comprehensive review. 3 Biotech. 2017;7:1-35. DOI: 10.1007/s13205-017-0870-y

[63] Litz RE. Biotechnology and mango improvement. Acta Horticulturae. 2004;**645**:85-92

[64] Shamili M, Fatahi R, Hormaza J. Characterization and evaluation of genetic diversity of Iranian mango (*Mangifera indica* L., Anacardiaceae) genotypes using microsatellites. Scientia Horticulturae. 2012;**148**:230-234

[65] Tongyoo P, Duangjit J, Sharma N, Chunwongse J. Molecular mapping and breeding in mango. The Mango Genome. 2021:153-163

[66] Singh Z, Singh RK, Sane VA, Nath P. Mango-postharvest biology and biotechnology. Critical Reviews in Plant Sciences. 2013;**32**(4):217-236

[67] Penna S, Jain SM. Fruit crop improvement with genome editing, in vitro and transgenic approaches. Horticulturae. 2023;**9**(1):58. DOI: 10.3390/horticulturae9010058

[68] Rivera Domínguez M, Astorga Cienfuegos KR, Vallejo Cohen S, Vargas Arispuro I, Sánchez Sánchez E. Transgenic mango embryos (*Mangifera indica*) cv.'Ataulfo' with the Defensin J1 gen. Revista Mexicana de Fitopatología. 2011;**29**(1):78-80

[69] Dongariyal A, Chandra AK, Dongriyal A, Kumar A, Sharma P. Tending genome editing via CRISPR/Cas9-induced mutagenesis: Opportunity and challenges for yield, quality and nutritional improvement of fruit crops. Scientia Horticulturae. 2023;**311**:111790

[70] Kaur N, Awasthi P, Tiwari S. Fruit crops improvement using CRISPR/Cas9 system. In: Genome Engineering Via CRISPR-Cas9 System. Amsterdam, The Netherlands: Elsevier; 2020. pp. 131-145

[71] Ahmad S, Tang L, Shahzad R, Mawia AM, Rao GS, Jamil S, et al. CRISPR-based crop improvements: A way forward to achieve zero hunger. Journal of Agricultural and Food Chemistry. 2021;**69**(30):8307-8323

[72] Mathiazhagan M, Chidambara B, Hunashikatti LR, Ravishankar KV. Genomic approaches for improvement of tropical fruits: Fruit quality, shelf life and nutrient content. Genes. 2021;**12**(12):1881. DOI: 10.3390/genes12121881

[73] Goulin EH, Galdeano DM, Granato LM, Matsumura EE, Dalio RJD, Machado MA. RNA interference and CRISPR: Promising approaches to better

understand and control citrus pathogens. Microbiological Research. 2019;**226**:1-9. DOI: 10.1016/j.micres.2019.03.006

[74] Yang Y, Saand MA, Huang L, Abdelaal WB, Zhang J, Wu Y. Applications of multi-omics technologies for crop improvement. Frontiers in Plant Science. 2021;**12**:563953

[75] Wang P, Luo Y, Huang J, Gao S, Zhu G, Dang Z. The genome evolution and domestication of tropical fruit mango. Genome Biology. 2020;**21**(1):1-17. DOI: 10.1186/s13059-020-01959-8

[76] Bally IS, Bombarely A, Chambers AH, Cohen Y, Dillon NL, et al. The 'Tommy Atkins' mango genome reveals candidate genes for fruit quality. BMC Plant Biology. 2021;**21**:1-18. DOI: 10.1186/s12870-021-02858-1

[77] Cortaga CQ, Lachica JAP, Lantican DV, Ocampo ETM. Genome-wide SNP and InDel analysis of three Philippine mango species inferred from whole-genome sequencing. Journal of Genetic Engineering and Biotechnology. 2022;**20**(1):46. DOI: 10.1186/s43141-022-00326-3

[78] Jena RC, Chand PK. Multiple DNA marker-assisted diversity analysis of Indian mango (*Mangifera indica* L.) populations. Scientific Reports. 2021;**11**(1):10345. DOI: 10.1038/s41598-021-89470-3

[79] Arogundade O, Matthew JO, Akinyoola OI, Akin-Idowu PE, Akinyemi SOS. Phenotypic and molecular characterization of mango cultivars in Southwest Nigeria. International Journal of Fruit Science. 2022;**22**(1):151-159. DOI: 10.1080/15538362.2021.2019652

[80] Song M, Wang H, Fan Z, Huang H, Ma H. Advances in sequencing and key character analysis of mango (*Mangifera indica* L.). Horticulture Research. 2023;**10**(2):259. DOI: 10.1093/hr/uhac259

[81] Mathiazhagan M, Padala S, Doddahejjaji SGC, Murugan S, Makki DR, Kundapura RV. Omics of mango: A tropical fruit tree. In: Omics in Horticultural Crops. United States: Academic Press; 2022. pp. 427-448

[82] Jain SM. Major mutation-assisted plant breeding programs supported by FAO/IAEA. Plant Cell, Tissue and Organ Culture. 2005;**82**:113-123. DOI: 10.1007/s11240-004-7095-6

[83] Suzuki N, Rivero RM, Shulaev V, Blumwald E, Mittler R. Abiotic and biotic stress combinations. New Phytologist. 2014;**203**(1):32-43

[84] Pena JE, Mohyuddin AI, Wysoki M. A review of the pest management situation in mango agroecosystems. Phytoparasitica. 1998;**26**:129-148

[85] Vayssières JF, Fatogoma S, Moussa N. Inventory of species of fruit flies (Diptera: Tephritidae) linked to the mangotree in Mali and testing of rational control. Fruits. 2004;**59**:1-14. DOI: 10.1051/fruits:2007029

[86] Oliveira CM, Auad AM, Mendes SM, Frizzas MR. Economic impact of insect pests in Brazilian agriculture. Journal of Applied Entomology. 2012;**137**:1-15

[87] Sudheeran PK, Sela N, Carmeli-Weissberg M, Ovadia R, Panda S, Feygenberg O, et al. Induced defense response in red mango fruit against Colletotrichum gloeosporioides. Horticulture Research. 2021;**8**. DOI: 10.1111/jen.12018

[88] Rossetto CJ, Ribeiro IJA, Gallo PB, Soares NB, Sabino JC, Martins ALM, et al. Mango breeding for resistance to

diseases and pests. In: V International Mango Symposium. Vol. 455. Acta Horticulturae; 1996. pp. 299-304

[89] Ribeiro IJA, Rossetto CJ, Donadio LC, Sabino JC, Martins AIM, Gallo PB, et al. Mango Wilt. XIV selection of mango (*Mangifera indica* L.) rootstocks resistant to the mango wilt fungus Ceratocystis fimbriata ell & Halst. Acta Horticulturae. 1993;**370**:159-166. DOI: 10.17660/ActaHortic.1995.370.26

[90] Hahn F, Valle S, Rendón R, Oyorzabal O, Astudillo A. Mango fruit fly trap detection using different wireless communications. Agronomy. 2023;**13**(7):1736. DOI: 10.3390/agronomy13071736

[91] Li JJ, Luo C, Yang XZ, Peng LH, Lu TT, Yang JH, et al. Genome-wide identification of the mango pathogenesis-related 1 (PR1) gene family and functional analysis of MiPR1A genes in transgenic Arabidopsis. Scientia Horticulturae. 2023;**321**:112254. DOI: 10.1016/j.scienta.2023.112254

[92] Rajan S, Srivastav M, Rymbai H. Genetic resources in mango. The Mango Genome. 2021:45-73

[93] Pruvost O, Roumagnac P, Gaube C, Chiroleu F, Gagnevin L. New media for the semiselective isolation and enumeration of Xanthomonas campestris pv. Mangiferaeindicae, the causal agent of mango bacterial black spot. Journal of Applied Microbiology. 2005;**99**(4):803-815. DOI: 10.1111/j.1365-2672.2005.02681.x

[94] Jean-martial KKF, N'Goran N, Gaston KK, Seydou T, Dramane DD, Brahima C, et al. Current situation of mango bacterial black spot caused by Xanthomonas campestris pv mangiferaeindicae in the Poro and Tchologo regions in Côte d'Ivoire. American Journal of Food and Nutrition. 2021;**9**(3):117-121. DOI: 10.12691/ajfn-9-3-4

[95] Muriithi BW, Gathogo NG, Diiro GM, Mohamed SA, Ekesi S. Potential adoption of integrated pest management strategy for suppression of mango fruit flies in East Africa: An ex ante and ex post analysis in Ethiopia and Kenya. Agriculture. 2020;**10**(7):278. DOI: 10.3390/agriculture10070278

[96] Kibira M, Affognon H, Njehia B, Muriithi B, Mohamed S, Ekesi S. Economic evaluation of integrated management of fruit fly in mango production in Embu County, Kenya. African Journal of Agricultural and Resource Economics. 2015;**10**(311):343-353

[97] Zainab M, Shinkafi S. Isolation and identification of fungi responsible for leaf spots disease of mango (*Mangifera indica linneaus*) in Sokoto state, Nigeria. Bayero Journal of Pure and Applied Sciences. 2016;**9**(2):166-173

[98] Bokonon-Ganta AH, de Groote H, Neuenschwander P. Socio-economic impact of biological control of mango mealybug in Benin. Agriculture, Ecosystems & Environment. 2002;**93**(1):367-378

[99] Amin U, Khan MKI, Khan MU, Ehtasham AM, Pateiro M, Lorenzo JM, et al. Improvement of the performance of chitosan—Aloe vera coatings by adding beeswax on postharvest quality of mango fruit. Food. 2021;**10**:2240. DOI: 10.3390/foods10102240

[100] Liu B, Xin Q, Zhang M, Chen J, Lu Q, Zhou X, et al. Research progress on mango post-harvest ripening physiology and the regulatory technologies. Food. 2022;**12**(1):173. DOI: 10.3390/foods12010173

[101] Khan MA. Postharvest management of mango: A review. Scientia Horticulturae. 2011;**129**(1):1-12

[102] Narang RS, Kataria AK. Postharvest management of mango: A global perspective. Postharvest Biology and Technology. 2014;**84**:1-16

[103] Malik AU, Amin M, Mazhar MS, Johnson P, Hofma PJ, Campbell J, et al. Mango value chain improvement through postharvest research and development: A developing country case study. Acta Horticulturae. 2017:410-419. DOI: 10.17660/ActaHortic.2017.1183.58

[104] Sharma SK, Thakur NK. Postharvest management of mango: A review. Indian Journal of Horticulture. 2012;**69**(1):1-16

[105] Sudheeran PK, Love C, Feygenberg O, Maurer D, Ovadia R, Oren-Shamir M, et al. Induction of red skin and improvement of fruit quality in 'Kent', "Shelly" and 'Maya' mangoes by preharvest spraying of prohydrojasmon at the orchard. Postharvest Biology and Technology. 2019;**149**:8-26

Section 2

Approaches in Conferring Abiotic Stress Tolerance

Chapter 7

Mitigating Metal/Metalloid Stress in Crops: Strategies for Sustainable Agricultural Resilience

Hany Almotairy

Abstract

In the face of escalating soil contamination, this chapter meticulously examines the multifaceted strategies employed to mitigate metal/metalloid stress in crops, an imperative endeavor for maintaining agricultural productivity and ensuring food security. Central to the discussion is exploring advanced phytoremediation techniques alongside the strategic use of soil amendments, highlighting their efficacy in decontaminating metal/metalloid-laden soils. The narrative further extends to the crucial role of mycorrhizal fungi in enhancing plant resilience against metal/metalloid toxicity and the innovative application of genetic engineering and breeding techniques aimed at cultivating metal/metalloid-tolerant crop varieties. Moreover, the chapter sheds light on integrating cutting-edge soil remediation technologies, including electrokinetic and nanotechnology, showcasing their potential to revolutionize conventional remediation practices. The synthesis of these strategies underscores the importance of adopting an interdisciplinary approach, blending traditional methods with technological innovations to develop sustainable and effective solutions for metal/metalloid stress in agriculture. Additionally, the chapter emphasizes the need for robust policy frameworks and sophisticated monitoring tools to manage soil health comprehensively, advocating for a holistic strategy to safeguard agricultural landscapes against metal/metalloid contamination.

Keywords: phytoremediation, soil amendments, crop metal resistance, sustainable agriculture, environmental remediation, metal/metalloid stress in crops

1. Introduction to metal/metalloid stress in sustainable agriculture

The escalating incidence of soil contamination with metal/metalloids poses a formidable global challenge to sustainable agriculture and food security [1]. These contaminants, including lead (Pb), cadmium (Cd), mercury (Hg), arsenic (As), and aluminum (Al), emanate from various anthropogenic sources such as industrial discharges, pesticide use, and mining activities, accumulating in agricultural soils and consequently in the food chain, thereby impairing plant growth, reducing crop yield and quality, and threatening human health through bioaccumulation and biomagnification processes [2, 3].

The imperative for research in this domain stems from the dual need to ensure agricultural resilience and food safety in an era of burgeoning environmental stressors. Traditional remediation techniques often fail to address the scale and complexity of metal/metalloid pollution sustainably and cost-effectively. This underscores a critical gap in the existing literature and practice: developing and implementing innovative, eco-friendly strategies that can effectively mitigate metal and metalloid stress in crops while bolstering sustainable agricultural practices [4, 5].

This study endeavors to bridge this gap by exploring a holistic suite of strategies encompassing advanced phytoremediation techniques, genetic engineering, and the strategic application of soil amendments. The study aims to elucidate the efficacy of these approaches in detoxifying metal/metalloid-laden soils, enhancing plant resilience to metal toxicity, and ultimately fostering sustainable agricultural productivity. This research contributes to burgeoning environmental and sustainable agriculture by integrating cutting-edge soil remediation technologies with traditional agronomic practices. Moreover, it aligns with the pressing global mandate to safeguard agricultural landscapes against contamination and to ensure security of food in the face of environmental challenges [6, 7].

This chapter addresses these concerns by filling a significant gap in the current scientific discourse on soil decontamination and plant resilience mechanisms and proposing actionable insights for deploying these strategies at a scale. Consequently, it contributes to the foundational knowledge for developing robust policy frameworks and sophisticated monitoring tools to manage soil health comprehensively. This work advocates for a paradigm shift toward sustainable and resilient agricultural systems, as supported by the research of the authors in [4, 5].

2. Impact of metal/metalloid accumulation on crop yield and quality

Soil contamination with metals and metalloids significantly threatens agricultural productivity and food safety. These contaminants from industrial, agricultural, and natural sources accumulate in soils, leading to crop toxicity, diminished productivity, and heightened health risks along the food chain [1, 8].

Metal/metalloid accumulation impairs plant productivity by stunting growth, inhibiting seed germination rates, and reducing biomass through disrupted cell division and elongation [2, 3, 9]. These adverse effects stem from plant physiological and biochemical alterations [3].

Furthermore, metal/metalloid toxicity affects a plant's reproductive capabilities, diminishing its ability to flower and bear fruit and impacting crop yield and quality. This reduction is linked to disruptions in photosynthesis, nutrient absorption, and enzyme activity [8, 10].

The absorption and translocation of metals/metalloids by plant roots can extend to the edible parts of crops (fruits and grains), affecting human and animal health throughout the food chain. The degree of translocation varies depending on the metal/metalloid type, plant species, and environmental conditions. For example, Typha domingensis primarily accumulates contaminants in its roots with minimal translocation to aerial parts, whereas hyperaccumulators have a translocation factor (TF) greater than 1, indicating efficient transport from roots to edible parts [7, 11]. Some metals also act as gerontogens, accelerating aging and increasing disease risks associated with aging in organisms [12, 13].

Consuming vegetables and fruits contaminated with metals such as Pb, Hg, Cd, and As is linked to severe health effects, including cancer and organ damage, particularly in developing regions where industrial and agricultural activities exacerbate soil contamination [14–16].

Metal/metalloid exposure can also induce premature senescence in harvested crops, adversely affecting their shelf life and marketability by accelerating tissue deterioration through oxidative stress. This impacts the economic value of agricultural produce and raises significant health concerns [17]. Furthermore, crops cultivated in metal-contaminated soils may require specialized storage conditions to mitigate further contamination risks during storage [16].

Addressing metal/metalloid pollution is crucial for ensuring the quality and safety of agricultural products, underscoring the need for effective soil decontamination strategies.

3. Comprehensive strategies for mitigating metal/metalloid stress in crops

Addressing metal/metalloid contamination in agricultural settings requires a comprehensive, multifaceted approach that integrates various remediation strategies. The selection of these strategies is influenced by factors such as the types of metals/metalloids present, the extent of contamination, soil properties, and the cost-effectiveness of the methods. Tailoring a combination of methodologies to meet each site's specific conditions and challenges is essential for optimizing decontamination outcomes.

3.1 Advanced phytoremediation techniques for soil decontamination

Phytoremediation is a leading eco-friendly technique to mitigate metal/metalloid stress in agricultural lands. This strategy involves using plants with exceptional abilities to absorb, stabilize, and detoxify metals/metalloids, thereby rejuvenating contaminated soils. Among these, hyperaccumulators play a critical role due to their unique capacity to absorb and sequester high concentrations of metals/metalloids in their tissues, making them instrumental in soil remediation efforts [4–6, 18, 19].

Hyperaccumulators are characterized by their ability to thrive in metal-contaminated environments, accumulating metals/metalloids to levels far exceeding those typically found in common plants. This trait makes them particularly effective for phytoextraction, where plant uptake removes metals from soils. Ongoing research is dedicated to identifying new hyperaccumulating species and enhancing the efficiency of these plants in remediation processes.

However, the use of natural hyperaccumulators is often limited by their slow growth rates and low biomass production. To overcome these challenges, researchers are exploring the potential of high-biomass non-hyperaccumulators and ensuring the use of non-edible plants to prevent metal/metalloid entry into the food chain [4, 18].

Phytoremediation, particularly when leveraging hyperaccumulators, offers a sustainable, cost-effective, and in-situ remediation strategy that contrasts sharply with traditional methods. It aligns with environmental sustainability goals and provides potential economic advantages when integrated into standard agricultural practices [6, 20].

Phytoremediation employs plants to address soil contamination in various ways:

3.1.1 Phytoextraction

Phytoextraction, a core strategy within the phytoremediation spectrum, leverages the natural capabilities of plants to absorb, translocate, and sequester metals/metalloids from contaminated soils into their biomass (**Figure 1**). This environmentally friendly remediation method offers a sustainable and economically viable alternative to traditional soil cleanup methods, contributing significantly to the restoration of soil health and agricultural productivity.

The efficacy of phytoextraction largely depends on the plant's ability to uptake metals/metalloids, move them to the aerial parts, and accumulate them in harvestable tissues. The root system plays a crucial role in metal uptake, influenced by the availability of metals in the soil, the presence of root exudates that modify metal bioavailability, and the expression of specific metal transporter proteins [21].

Selecting appropriate plant species is critical for effective phytoextraction. Hyperaccumulators, which can concentrate metals/metalloids in their tissues at levels 100–1000 times greater than typical plants, are particularly suitable for this process. However, their slow growth rate and low biomass output can restrict their practical

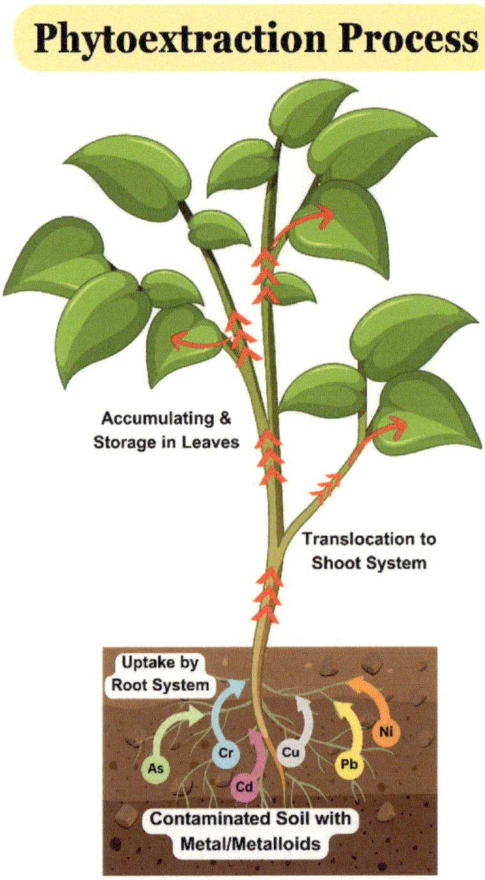

Figure 1.
The phytoextraction process illustrates how hyperaccumulator plants can uptake and accumulate metal/metalloids from contaminated soil. The roots absorb contaminants, translocated to the shoots, and stored in the leaves.

use. Consequently, research increasingly focuses on high-biomass, non-hyperaccumulator plants that still exhibit significant metal accumulation capabilities [21].

Strategic cultivation of hyperaccumulator plants is essential for removing metals/metalloids from soils. Advances in genetic engineering can potentially enhance these plants' efficiency by enabling them to express genes that improve metal uptake, translocation, and sequestration [22].

Despite its potential, phytoextraction faces challenges such as the slow removal rate of metals, potential food chain contamination, and the need for safe disposal of contaminated plant biomass. To optimize the phytoextraction process, overcoming these challenges requires a multidisciplinary approach that integrates plant physiology, soil science, and environmental engineering.

3.1.2 Phytostabilization

Phytostabilization offers a practical approach for managing soils contaminated with metals/metalloids, particularly where complete pollutant removal is impractical. By immobilizing contaminants within the soil matrix, phytostabilization prevents their migration to groundwater and entry into the food chain, making it an effective strategy for in-situ soil remediation.

The principal mechanisms of phytostabilization include the adsorption of contaminants onto plant roots, precipitation within the rhizosphere, and metal complexation with root exudates. These processes effectively reduce the bioavailability and mobility of pollutants, thereby mitigating their ecological and health impacts. The success of this method depends significantly on the choice of plant species, which must have a robust tolerance to the toxic effects of metals/metalloids, an extensive root system, and the ability to thrive in contaminated environments [23].

Non-edible plants are preferred in phytostabilization to avoid transferring contaminants into the food chain. Enhancing this method's effectiveness often involves integrating soil amendments such as biochar or phosphate fertilizers, which help immobilize metals and improve soil quality [24].

3.1.3 Phytovolatilization

Phytovolatilization utilizes the unique capabilities of certain plants to absorb contaminants from soil or water and release them into the atmosphere. This method is particularly suited for dealing with volatile organic compounds and metal/metalloids that can be transformed into less harmful volatile forms.

In phytovolatilization, plants take up contaminants through their roots, transport them to aerial parts, and release them as gases through transpiration. This process involves enzymatic reactions that convert metals like Hg and selenium (Se) into volatile forms such as elemental mercury vapor and dimethylselenide [25].

While offering a sustainable means of remediating contaminated sites, phytovolatilization requires careful selection of plant species with high transpiration rates and specialized enzymatic systems capable of converting contaminants into gaseous forms. Challenges include potential air quality impacts and the need for strategies to manage the released contaminants effectively [26].

By refining these sections, the manuscript will present a more polished, comprehensive, and scholarly discussion of phytoremediation strategies suitable for publication in a high-quality academic journal. Each strategy is contextualized within its practical application, challenges, and future research directions, providing a robust overview of contemporary approaches to managing metal/metalloid stress in crops.

3.2 Utilizing soil amendments and conditioners to mitigate metal/metalloid stress

3.2.1 Organic matter addition

Adding organic matter to soils contaminated with metals/metalloids is a well-established strategy for mitigating metal/metalloid stress in crops. Organic amendments such as compost, biochar, and manure can immobilize metals/metalloids, reducing their bioavailability to plants and decreasing the potential for uptake and accumulation in edible plant parts [4, 27–29].

These amendments enhance soil structure, increase water-holding capacity, and boost nutrient content, promoting healthier plant growth even in contaminated environments. Furthermore, they stimulate microbial activity in the soil, which is crucial in stabilizing and further immobilizing metals/metalloids [27, 28].

The strategic application of organic amendments improves soil properties and facilitates a conducive environment for microbial proliferation. This contributes significantly to detoxifying and rejuvenating metal/metalloid-contaminated soils, fostering sustainable agricultural practices [30].

3.2.2 Assisted phytoremediation

Assisted phytoremediation is a promising strategy for mitigating metal/metalloid stress in crops. This approach involves adding agents to the soil to enhance plants' ability to remediate metal/metalloid contamination. These agents can include plant growth-promoting bacteria (PGPB), organic amendments like compost and biochar, and symbiotic fungi such as arbuscular mycorrhizal fungi (AMF), a type of fungi that form a mutualistic relationship with the roots of plants [25, 31–35].

Plant growth-promoting bacteria (PGPB) can enhance phytoremediation by promoting plant growth and health, thereby increasing the plant's capacity to tolerate and accumulate metal/metalloids [25]. These bacteria can also directly bind to metal/metalloids, reducing their bioavailability and toxicity to plants.

Organic amendments like compost and biochar can immobilize metal/metalloids in the soil, reducing their bioavailability and plant uptake [33, 35]. These amendments can also improve soil health and fertility, promoting plant growth and metal/metalloid stress resilience. For instance, a study found that compost was constructed with a lower bioavailability of metal/metalloids, providing theoretical and technical support for the application of compost in metal/metalloid-polluted soils [32].

Biochar, recognized for its carbon-rich composition derived from biomass, exhibits a high surface area and abundant binding sites, making it an effective material for metal/metalloid remediation. This characteristic not only facilitates the immobilization of metal/metalloids, thereby preventing their uptake by plants and leaching into groundwater but also contributes to enhancing soil fertility. Applying biochar in contaminated soils is a significant sink for metal/metalloids, leveraging its adsorption capabilities to reduce the bioavailability and mobility of these pollutants. This dual function of biochar underscores its potential as a sustainable solution for soil remediation while improving soil health [27, 36, 37].

Arbuscular mycorrhizal fungi (AMF) form symbiotic relationships with plant roots, which make it easier for plants to take in nutrients and make them more resistant to abiotic stress factors like metal/metalloid contamination [33, 35]. These fungi can extend their hyphae into the soil, accessing a larger volume of soil than the plant roots alone. This allows them to capture nutrients with low mobility, such as phosphorus (P) and zinc (Zn), making them available to plants. In the context of metal/metalloid stress, AMF can help to immobilize metal/metalloids in the root zone, reducing their uptake and toxicity to plants [35].

In addition to these biological and organic amendments, assisted phytoremediation can also involve chemical amendments to alter soil properties and reduce metal/metalloid bioavailability. For instance, lime can be used to adjust soil pH, influencing the solubility and bioavailability of metal/metalloids [34].

3.2.3 Lime and pH adjustment

Soil pH profoundly influences the bioavailability of metals/metalloids, affecting their solubility and the extent of their binding to soil particles. Adjusting soil pH, typically by adding lime, can significantly alter the chemical forms of metals/metalloids, making them less available to plants [4, 38–40].

Metals tend to form insoluble phosphates and carbonates at higher pH levels, whereas they are more likely to exist in bioavailable ionic forms at lower pH levels. However, the effectiveness of pH adjustment strategies can vary depending on the specific metal/metalloids involved and the initial soil conditions [38, 41].

It is essential to tailor pH adjustment strategies to the specific soil characteristics and metal/metalloid contamination types present, as their effectiveness can differ significantly under different environmental conditions [4, 38, 41].

3.3 Enhancing crop resilience to metal/metalloids through mycorrhizal fungi symbiosis

The symbiotic relationships between AMF and plants enhance resilience against metal/metalloid stress. This mutualistic interaction is pivotal for improving plant performance under contaminated conditions through enhanced nutrient uptake, reduced metal toxicity, and facilitated phytostabilization.

3.3.1 Symbiotic relationships

As a symbiotic relationship, the mutualistic relationship between mycorrhizal fungi and plants is vital in improving plant resistance to metal/metalloids. Most plant species, including most crops, are associated with mycorrhizal fungi, and these symbioses can improve plant performance through enhanced soil nutrient uptake [42]. The relationship between mycorrhizal fungi and plants, specifically AMF, plays a crucial role in alleviating the adverse effects of metal/metalloid stress on crops. These fungi can enhance plant performance, improve phytoremediation of metal/metalloids, and induce responses in plants that help ease the effects of metal/metalloids [42–46].

The AMF is particularly influential in improving the phytoremediation of metal/metalloids, such as, Cd, Pb, and Cr; these fungi can reduce the toxicity of these elements in contaminated environments, thus preventing their bioavailability and environmental migration in the soil [43].

The AMF-assisted alleviation of metal/metalloids phytotoxicity is a cost-effective and environment-friendly strategy; it can induce both molecular and physiological responses in plants that help mitigate the effects of metal/metalloids [44].

Fungi are often found in metal/metalloid-polluted soils; most plants form symbiotic relationships [45]. They can play an essential role in plant growth in metal-contaminated soils by acting as bio-alleviators and/or biofertilizers [46].

3.3.2 Improved nutrient uptake

The role of AMF in promoting nutrient absorption in metal-contaminated soils is critical to improving plant growth and health. This natural symbiotic relationship presents a promising solution for reducing metal/metalloid stress in crops and can be a crucial aspect of sustainable agricultural practices in contaminated environments [33, 44, 47–51].

The AMF enhances plant nutrient uptake, particularly in soils contaminated with metal/metalloids. These fungi establish symbiotic associations with plant roots by extending their hyphal networks into the soil, enhancing the surface area available for nutrient uptake. This symbiosis is especially beneficial in metal-contaminated soils, where nutrient availability may be compromised [46, 49, 50].

The AMF improves the uptake of essential nutrients such as nitrogen, P, and potassium (K) and micronutrients like Zn and Cu, which are often less mobile in the soil [50]. Doing so supports better plant growth and health, even under the stress of metal/metalloid contamination [44, 47–49]. The fungi's ability to bind metal/metalloids can also reduce the toxicity of these metals to the plant, further aiding in nutrient uptake and overall plant performance [44, 51].

3.3.3 Facilitation of metal/metalloids phytostabilization

Moreover, AMF can alter the rhizosphere's chemistry, leading to the immobilization or chelation of metal/metalloids, thus reducing their bioavailability to plants [44, 52]. This mechanism protects the plant from metal toxicity and minimizes the translocation of metals to the edible parts of crops, therefore minimizing the potential health hazards linked to the consumption of vegetables tainted with metals [33, 44].

4. Innovations in soil remediation technologies for metal/metalloid removal

4.1 Electrokinetic remediation

Electrokinetic remediation (EKR) is an advanced technology used to clean contaminated soil and groundwater, and it is especially effective in addressing soil metal/metalloid contamination. This versatile approach utilizes electrical currents to mobilize and extract a broad spectrum of metal/metalloids, making it a crucial component in the suite of modern soil decontamination techniques.

EKR is particularly adept at treating fine-grained soils with low permeability, where conventional remediation methods struggle to achieve efficiency. The technology operates on principles including electromigration, electroosmosis, and electrophoresis, which collectively enable the movement of contaminants toward strategically placed electrodes in the soil [53–55].

Recent enhancements in EKR involve nanotechnology, which involves developing nanoelectrodes and nanoscale amendments, significantly improving the efficiency and selectivity of metal/metalloid extraction from soils [20].

4.1.1 Mechanisms and applications

The primary mechanism of EKR, electromigration, facilitates the movement of charged metal ions through the soil toward oppositely charged electrodes. This movement is supported by electroosmosis—the migration of water and solutes toward the cathode-and electrophoresis, the transport of charged particles through a fluid under a uniform electric field [55, 56].

Recent innovations such as interval power breaking have been adopted to enhance EKR's efficacy. This technique, which involves alternating cycles of power-on and power-off, improves the desorption and migration of ions, significantly increasing the removal efficiency of contaminants like Cd from soils [53]. Furthermore, introducing organic acids as facilitators has bolstered the solubilization and extraction of metal/metalloids, enhancing the overall effectiveness of the EKR process [57].

4.1.2 Challenges and innovations in EKR

Despite its many advantages, EKR faces challenges, including pH fluctuations within the soil that can impact the mobility and speciation of metals and metalloids. Acid generation at the anode and base formation at the cathode can create pH gradients that may precipitate metals near the electrodes, thereby reducing their mobility and extraction efficiency [55].

Innovative solutions, such as chelating agents and surfactants, have been explored to mitigate these issues by enhancing metal mobility and preventing precipitation. These adjustments help maintain a consistent soil environment conducive to effective metal/metalloid removal [58].

Additionally, ongoing advancements focus on improving the energy efficiency of EKR by integrating renewable energy sources and optimizing electrode configurations. For instance, the application of pulsed electric fields has significantly reduced energy consumption during the remediation process, thus minimizing this technology's environmental footprint [59].

4.2 Nanotechnology

Nanotechnology offers a versatile approach to mitigating metal/metalloid pollution in soil and water environments. By developing specialized nanomaterials and nanosensors, this technology provides efficient, selective, and environmentally friendly solutions for remediating contaminated sites.

4.2.1 Nanomaterials for soil remediation

Nanotechnology harnesses nanomaterials' unique properties to immobilize or effectively remove metal/metalloids from contaminated soils. Notably, nanoscale zero-valent iron particles are employed due to their high reactivity in degrading various pollutants, including metal/metalloids. These particles are engineered to target specific contaminants, enhancing soil quality and offering a sustainable soil remediation method [60].

Other nanomaterials, such as titanium dioxide (TiO_2) and carbon-based nanoparticles, have significantly succeeded in adsorbing and transforming pollutants. This capability is crucial for mitigating the adverse impacts of agricultural practices on soil health, marking a significant advancement in soil remediation [60].

Remediation technology	Mechanism	Efficacy	Applications	Challenges	References
Phytoremediation	Utilization of plants to absorb, stabilize, and detoxify metal/metalloids from the soil	Varies by plant species and metal type	Soil decontamination, especially for agricultural lands	Slow growth rate of hyperaccumulators, potential food chain contamination	[4–6, 18, 19]
Soil amendments	Addition of materials (organic matter, biochar, lime) to soil to improve health and reduce metal/metalloid bioavailability	Depends on the type of amendment and metal/metalloids	Soil health improvement, reduction of metal/metalloid bioavailability	Long-term effectiveness and potential adverse effects on soil chemistry	[27, 28, 31–34]
Electrokinetic remediation	Application of electrical currents to mobilize and extract metal/metalloids from soils	Effective in fine-grained, low-permeability soils	Removal of metal/metalloids from contaminated sites	Potential for pH changes in soil, energy-intensive	[53–55]
Nanotechnology	Use of nanomaterials for metal/metalloid immobilization or removal	High adsorption capacity, operates over a wide pH range	Soil and water remediation	Potential ecological risks of nanomaterials	[60–63]
Mycorrhizal fungi	Symbiotic relationships with plants enhancing resilience to metal/metalloid stress	Facilitates nutrient uptake and metal/metalloid phytostabilization	Improving plant performance in contaminated soils	Specificity to certain plant species and metals	[42–44, 47–52]

Table 1.
Comparative overview of soil remediation technologies for metal/metalloid stress mitigation in crops. The table presents a detailed comparison of various remediation methods, highlighting their mechanisms of action, overall efficacy, practical applications, and associated challenges.

4.2.2 Nanosensors for water remediation

In the context of treating metal/metalloid-contaminated water, nanosensors play a critical role. Due to their high surface area and excellent adsorption properties, these devices can effectively detect metal/metalloids. Nanosensors are tailored with specific nanomaterials to enhance their sensitivity and selectivity toward target metal/metalloid ions, making them invaluable in monitoring and managing water quality [60].

4.2.3 Nano-adsorbents

The application of nano-adsorbents in water remediation is a prominent use of nanotechnology. Iron oxide nanoparticles (Fe_2O_3) and various metal sulfides have been extensively researched for their capacity to remove divalent metal ions, such as copper (Cu), nickel (Ni), and cobalt (Co), from water. These nano-adsorbents are preferred due to their high adsorption capacity and effectiveness across a broad pH range, making them highly effective in water remediation scenarios [61, 62].

4.2.4 Challenges and future directions

While nanotechnology provides innovative solutions for remediation, it is essential to consider the potential risks associated with deploying nanomaterials in environmental settings. Understanding nanoparticles' behavior, movement, and potential hazards are crucial for ensuring nanotechnology applications' long-term safety and viability in soil and water decontamination [60, 63].

Future research should focus on developing strategies to mitigate the ecological risks associated with nanomaterials. This includes enhancing their biocompatibility and degradability to prevent adverse environmental impacts.

Table 1 synthesizes critical aspects such as mechanism, efficacy, applications, and challenges of each technology, drawn from an extensive review of current literature. By juxtaposing these methodologies, the table serves as a crucial reference point for researchers and practitioners alike, facilitating an informed selection of remediation strategies based on specific environmental conditions and contamination profiles.

5. Advanced tools for monitoring and assessing soil metal/metalloid contamination

5.1 Remote sensing and geographic information systems

Remote sensing (RS), gathering information about an object or phenomenon from a distance, typically using aircraft or satellite-based sensors, and geographical information systems (GIS) have emerged as indispensable tools in monitoring and strategically managing metal/metalloid stress in agriculture. By offering detailed insights into contamination levels and their spatial distribution, RS and GIS enhance the development, implementation, and evaluation of targeted mitigation strategies [64, 65].

5.1.1 Applications and advantages

RS technology facilitates the early detection of physiological alterations in crops attributable to metal/metalloid exposure, thereby allowing for targeted mitigation measures [64]. Concurrently, GIS's spatial analytical capabilities are pivotal in identifying contamination hotspots and assessing the spread of metal/metalloids within agricultural landscapes, which is crucial for formulating and monitoring effective mitigation strategies [65].

5.1.2 GIS for strategy development

Geographic Information Systems (GIS), computer systems that capture, store, analyze, and display geographical data, play a critical role in strategy development and impact assessment by mapping metal/metalloid contamination patterns and evaluating the success of mitigation strategies over time. This information guides the refinement of interventions, ensuring their efficacy in reducing metal/metalloid stress [65]. Furthermore, integrating RS and GIS data with agronomic models predicts the impact of metal/metalloid stress on crop yields, thereby informing the optimization of agricultural practices to mitigate such stress [66]. Predictive analytics provided by RS and GIS enable a proactive approach to managing metal/metalloid contamination, facilitating the implementation of preventive measures before crops are adversely affected [67]. **Table 2** offers a concise overview of case studies demonstrating their application. This table succinctly encapsulates the geographical areas investigated, pollutants assessed, methods utilized, key findings, and the impactful role of GIS in enhancing remediation strategies. It underscores the significance of RS and GIS in enabling precise, informed interventions for metal/metalloid contamination across diverse landscapes.

5.2 Biosensors

Biosensors represent a significant advancement in detecting and monitoring environmental pollutants, including metal/metalloids and other hazardous substances. Their integration into environmental management strategies offers a real-time, sensitive, cost-effective approach to assessing and mitigating pollution.

5.2.1 Environmental monitoring

Biosensors represent a class of analytical instruments that integrate a biological sensing element with a physicochemical transducer. These devices are engineered to quantitatively detect and measure the concentration of various chemical entities, facilitating precise analytical assessments. In environmental monitoring, biosensors are particularly valuable for detecting pollutants at low concentrations, offering a rapid and specific response to various environmental contaminants [74–81].

Metal/metalloids rank prominently among environmental pollutants, distinguished by their toxicity and enduring presence in ecosystems. Traditional methods for detecting metal/metalloids are often labor-intensive, time-consuming, and require sophisticated equipment. Biosensors, however, offer a promising alternative. For instance, microbial biosensors utilize microorganisms that exhibit specific responses to metal/metalloids, enabling the detection of pollutants like Hg, silver, Cu, and Cd [77]. These biosensors can provide real-time data on metal/metalloid concentrations, facilitating immediate decision-making and remediation actions [80].

5.2.2 Advancements and applications

Recent advancements in biosensor technology have significantly enhanced their application in environmental monitoring. Electrochemical biosensors, for example,

Geographic area	Pollutants	Remediation/assessment method	Key findings	Impact of using GIS	Reference
Fosu Lagoon, Ghana	Pb, Cu, Cd, Mn	Atomic Absorption Spectrometry	Seasonal variations in metal concentrations, sometimes exceeding WHO limits, highlight the need for dynamic monitoring and remediation strategies	Incorporating GIS could pinpoint contamination hotspots, optimize remediation efforts, and enhance phytoremediation strategies	[65]
Iran, with a focus on arid and semi-arid regions prone to dust storms	Dust particles are particularly interested in dust phenomena's metal/metalloid content. The study does not specify metal/metalloid concentrations but emphasizes the environmental and health risks associated with metal/metalloid-laden dust	The study leverages a combination of remote sensing data, geographic information systems (GIS), and statistical methods to monitor and evaluate dust phenomena, including the spatial and temporal distribution of dust storms	Remote sensing and GIS are effective tools for the spatial and temporal analysis of dust storms. The study highlights the increasing trend in dust storm occurrences and the effectiveness of remote sensing technologies, particularly MODIS imagery, in monitoring these events	GIS and remote sensing technologies have significantly contributed to understanding the distribution and intensity of dust storms. They offer valuable insights into identifying the sources and migration patterns of dust, facilitating targeted remediation and mitigation strategies	[66]
Northern Part of the Nile Delta, Egypt	Fe, Mn, Zn, Cu, Pb, Ni, Co, Cd	The study employed RS and GIS techniques to assess soil degradation and the environmental risks of metal/metalloids. It employed Landsat ETM+ images and digital elevation models (DEMs) for spatial analysis	The region experiences significant soil degradation due to salinization, alkalization, compaction, and waterlogging. Most soils exhibited moderate to high degrees of degradation, with a notable chemical and physical degradation risk. Metal/metalloid analysis revealed varied levels of pollution across different geomorphological units, indicating a need for targeted remediation	GIS and remote sensing proved instrumental in mapping soil degradation and assessing metal/metalloid risks, providing a valuable tool for environmental management and the development of sustainable agricultural practices in the Nile Delta	[67]

Geographic area	Pollutants	Remediation/assessment method	Key findings	Impact of using GIS	Reference
Northern Nile Delta, Egypt	As, Co, Cu, Ni, V, Zn	Assessment through GIS and multivariate analysis (PCA and cluster analysis) along with the use of geoaccumulation index, contamination factor (CF), Improved Nemerow's Pollution Index (Pn), and Potential Ecological Risk Index (PERI)	The study area exhibits varying degrees of soil metal/metalloid contamination, with certain regions displaying high levels of pollutants, indicating strong to extremely strong contamination in some samples. The application of GIS and multivariate statistical analysis highlighted the anthropogenic influence on metal/metalloid concentrations	GIS played a critical role in mapping the spatial distribution of metal/metalloid, aiding in the identification of contaminated zones, and facilitating the development of targeted remediation strategies	[68]
Northern slope of the Eastern Tianshan Mountains, Xinjiang, NW China	Zn, Cu, Cr, Pb, As, Hg	Geoaccumulation index, single-factor pollution index, potential ecological risk, and multivariate analysis were used to assess soil metal/metalloid pollution and health risks across commercial, industrial, and agricultural areas	Elevated Pb, As, and Hg levels were noted, surpassing Xinjiang soil background values by significant margins in various functional areas, indicating notable soil contamination. In contrast, average concentrations of Zn, Cu, and Cr were below these background values, suggesting lesser pollution concerns for these metals	GIS and remote sensing techniques provide a nuanced understanding of the spatial distribution of metal/metalloid contamination, pinpointing areas with heightened ecological risks. This spatial insight is crucial for targeted remediation efforts and policy planning	[69]
West side of Nile River, El-Minia governorate, Egypt	Cr, Co, Cu, Cd, Pb, Zn	The study utilized GIS and multivariate analysis, including Principal Component Analysis (PCA) and Contamination Factors (CF), to assess soil contamination levels. Soil samples were analyzed for metal/metalloid concentration using Inductively Coupled Plasma Mass Spectrometry (ICP-MS)	Significant contamination with metal/metalloid, notably Cd, Pb, and Zn, was observed. Approximately 70.7% of the area exhibited considerable metal/metalloid concentration, indicating moderate to high soil contamination. The assessment suggests that the misuse of low-quality water for irrigation and excessive use of mineral fertilizers are significant contributors to the contamination	GIS played a pivotal role in mapping the spatial distribution of metal/metalloid contamination, facilitating the identification of contaminated zones, and evaluating contamination levels across the study area. This spatial analysis aids in the formulation of targeted mitigation strategies to address soil contamination issues	[70]

Geographic area	Pollutants	Remediation/assessment method	Key findings	Impact of using GIS	Reference
Beni Amir Irrigated Perimeter, Tadla Plain, Morocco	Zn, Cr, Pb, Cu, Cd	Utilization of Geographic Information Systems (GIS), geoaccumulation index (Igeo), enrichment factor (EF), contamination factors (CF), and Pollution Load Index (PLI) for analyzing soil samples and assessing metal/metalloid contamination	The study area exhibits significant metal/metalloid contamination exceeding WHO and FAO limits, particularly around industrial areas and urban agglomerations. GIS was instrumental in revealing the spatial distribution and intensity of contamination	GIS technologies enabled precise mapping of contamination hotspots, illustrating the areas most affected by metal/metalloid pollution, which aids in targeted remediation efforts and environmental planning	[71]
Mollisol dairy farm in Dulbert Mongolian Autonomous County, Daqing City, Heilongjiang Province, China	As, Hg, Cd, Pb, Cr, Cu, Ni, Zn, Sb	GIS-based spatial analysis and pollution index method were used to evaluate soil, water, and manure contamination levels. The study applied inverse distance weighted interpolation for spatial distribution assessment	High contamination levels of As in soil indicate extreme contamination; moderate contamination levels of Ni in surface water and Sb in cow drinking water. The study underscores the significant environmental impact of metal/metalloid accumulation in soil and the potential risks to the food chain	GIS was crucial for visualizing the spatial distribution of contamination and optimizing sample collection. This technology facilitated a detailed understanding of the contamination patterns and helped prioritize areas for remediation efforts	[72]
Matehuala, San Luis Potosi, Mexico	As, Ca, Mg, Na, K, Sr., Mn, Fe	The study utilized the Positive Matrix Factorization (PMF) model and Geographic Information Systems (GIS) to assess contamination levels and identify the sources of metal pollutants in surface soil samples from agricultural and recreational areas	High As, Sr., Mn, and Fe levels were noted, indicating significant soil contamination, which was further analyzed for source apportionment. The study distinguished between natural mineral dissolution and anthropogenic contributions from past mining activities and current agricultural practices	GIS played a critical role in mapping the spatial distribution of contaminants, allowing for detailed visualization of contamination hotspots and facilitating the effective targeting of remediation efforts	[73]

Table 2.
Illustrative case studies on applying remote sensing and geographic information systems in monitoring and mitigating metal/metalloid contamination. This table presents a curated selection of research efforts that leverage RS and GIS for enhanced understanding and management of metal/metalloid stress in various geographical contexts. It details the pollutants analyzed, the assessment or remediation methods applied, and the critical outcomes of these studies, highlighting GIS's instrumental role in mapping contamination, identifying risk areas, and guiding remedial actions effectively.

have been developed to detect pesticides and metal/metalloids, leveraging the high sensitivity and specificity of electrochemical methods [82]. Integrating nanomaterials into biosensor designs has improved their performance, offering lower detection limits and faster response times [83].

Biosensors are particularly useful in monitoring soil and water quality. For instance, soil microbial fuel cell-based biosensors have emerged as a novel approach for real-time and rapid soil pollution monitoring [74, 84]. In water quality monitoring, biosensors can detect various pollutants, from metal/metalloids to organic compounds, ensuring drinking water's safety and aquatic ecosystems' health [78].

Despite their potential, biosensors' application in environmental monitoring faces challenges, including field validation, developing more robust and versatile sensors, and integrating biosensors into comprehensive monitoring networks [75, 76]. Future research is expected to address these challenges, enhance biosensors' sensitivity, selectivity, and portability, and expand their application to a broader range of environmental pollutants [81].

6. Crop improvement through breeding and biotechnological approaches for metal resistance

6.1 Traditional breeding for tolerance

Integrating traditional breeding techniques with molecular tools and recent biotechnological advancements offers a promising pathway toward developing crop varieties endowed with superior metal/metalloid stress tolerance. These innovative breeding strategies contribute to safeguarding food security and ensuring the safety of agricultural produce in regions plagued by metal/metalloid contamination. Notable works by the authors in [85–87] underscore the effectiveness of selecting and hybridizing plants bearing desirable traits, such as enhanced metal/metalloid tolerance. This approach has historically enhanced crop resilience to various environmental stresses, underscoring the enduring value of traditional breeding methodologies complemented by contemporary scientific advancements.

6.2 Molecular breeding and marker-assisted selection

Recent strides in molecular breeding, particularly marker-assisted selection (MAS), have significantly bolstered our ability to identify and select genes conferring metal tolerance. This approach, exemplified by the work of the authors in [86, 88], leverages high-throughput sequencing technologies for the rapid identification of Quantitative Trait Loci and candidate genes responsible for metal/metalloid tolerance. The precision of MAS in pinpointing genetic markers associated with desirable traits has revolutionized the breeding process, enabling the development of crop varieties tailored to thrive in contaminated soils.

6.3 Integration of biotechnological tools

The integration of CRISPR/Cas9 gene editing represents a pivotal advancement in crop improvement research. This technique allows for precise genetic modifications by directing the Cas9 enzyme to specific DNA sequences, facilitating targeted edits to enhance plant resilience to metal/metalloid stress.

The pioneering applications by the authors in [89, 90] demonstrate CRISPR/Cas9's capability to modify genetic pathways for improved abiotic stress tolerance, significantly enhancing crop yield and quality on contaminated soils. This technology complements traditional breeding and provides a transformative approach to understanding and manipulating the genetic bases of metal resistance in plants.

This strategy promises to advance food security in metal-contaminated regions and open new research avenues for sustainable agriculture by merging CRISPR/Cas9 with existing breeding methods.

6.4 Genetic modification for enhanced uptake or exclusion

Genetic modification is a transformative approach within the spectrum of strategies to enhance metal/metalloid resistance in crops. This section has detailed the application of genetic engineering tools to develop transgenic plants with enhanced capabilities for either the uptake or exclusion of metal/metalloids, thereby addressing environmental metal/metalloid stress more effectively. Pioneering efforts by researchers in [91] and the insights provided by the authors [4] illustrate the potential of genetic interventions in augmenting the phytoremediation capacity of plants.

This section has outlined several key genetic strategies, including enhancing metal accumulation through gene manipulation in metal homeostasis and improving plant tolerance to oxidative stress via bolstering antioxidant defense systems. These enhancements not only facilitate the direct management of metal/metalloids in contaminated environments but also support the development of crops with intrinsic resistance to such stresses.

Moreover, integrating genetic modifications that enable plants to volatilize metal/metalloid-explored by the authors in [92], represents a significant advancement in phytoremediation technology. These capabilities, particularly the enhanced mercury volatilization discussed by the authors in [93], leverage natural plant detoxification pathways augmented with engineered enzymatic functions to transform toxic metal forms into less harmful states.

However, deploying genetically modified plants in environmental remediation is not without challenges. It necessitates a comprehensive evaluation of potential ecological impacts to ensure that the benefits of such plants in phytoextraction substantially outweigh any adverse effects on biodiversity and ecosystem health, a concern highlighted by the authors in [5].

In summary, genetic modification, as part of a broader integrated strategy involving traditional breeding and cutting-edge biotechnological tools, offers promising prospects for developing robust agricultural systems that thrive in metal-contaminated environments. This comprehensive approach aims to enhance crop resilience and productivity and contributes to the sustainable management of environmental pollutants through advanced phytoremediation techniques.

7. Conclusion

This chapter underscores the imperative for a holistic and integrated approach to alleviating metal/metalloid stress in agricultural ecosystems, seamlessly integrating traditional agronomic practices with the forefront of technological innovation. It accentuates the necessity of a diverse arsenal of strategies, from applying phytoremediation and soil amendments to leveraging genetic engineering to effectively combat metal/metalloid toxicity in crops. This chapter not only underscores the vital role of interdisciplinary research in forging resilient and sustainable agricultural methodologies but also calls for an expanded focus on refining and global dissemination of these remediation technologies. As we navigate the complexities of metal/metalloid contamination, it is incumbent upon the global research and farming communities to foster collaborative, cross-disciplinary endeavors. Such concerted efforts are essential to enhance the efficacy, scalability, and accessibility of these technologies, thereby fortifying agricultural productivity and ensuring global food security amidst escalating environmental challenges.

Conflict of interest

The authors declare no conflict of interest.

Disclaimer

The views expressed in this paper are those of the author(s) and do not necessarily reflect those of the SFDA or its stakeholders. Guaranteeing the accuracy and validity of the data is the sole responsibility of the research author(s).

Author details

Hany Almotairy
Suadi Food and Drug Authority (SFDA), Riyadh, Saudi Arabia

*Address all correspondence to: hmmotairy@sfda.gov.sa

References

[1] Alengebawy A, Abdelkhalek ST, Qureshi SR, Wang MQ. Heavy metals and pesticides toxicity in agricultural soil and plants: Ecological risks and human health implications. Toxics. 2021;**9**(3):42. DOI: 10.3390/toxics9030042

[2] Vasilachi IC, Stoleru V, Gavrilescu M. Analysis of heavy metal impacts on cereal crop growth and development in contaminated soils. Agriculture. 2023;**13**(10):1983. DOI: 10.3390/agriculture13101983

[3] Angulo-Bejarano PI, Puente-Rivera J, Cruz-Ortega R. Metal and metalloid toxicity in plants: An overview on molecular aspects. Plants. 2021;**10**(4):635. DOI: 10.3390/plants10040635

[4] Yan A, Wang Y, Tan SN, Mohd Yusof ML, Ghosh S, Chen Z. Phytoremediation: A promising approach for revegetation of heavy metal-polluted land. Frontiers in Plant Science. 2020;**11**:359. DOI: 10.3389/fpls.2020.00359

[5] Suman J, Uhlik O, Viktorova J, Macek T. Phytoextraction of heavy metals: A promising tool for cleanup of polluted environment? Frontiers in Plant Science. 2018;**9**:1476. DOI: 10.3389/fpls.2018.01476

[6] Nedjimi B. Phytoremediation: A sustainable environmental technology for heavy metals decontamination. SN Applied Sciences. 2021;**3**(3):286. DOI: 10.1007/s42452-021-04301-4

[7] Islam MD, Hasan MM, Rahaman A, Haque P, Islam MS, Rahman MM. Translocation and bioaccumulation of trace metals from industrial effluent to locally grown vegetables and assessment of human health risk in Bangladesh. SN Applied Sciences. 2020;**2**:1-11. DOI: 10.1007/s42452-020-3123-3

[8] Priya AK, Muruganandam M, Ali SS, Kornaros M. Cleanup of heavy metals from contaminated soil by phytoremediation: A multidisciplinary and eco-friendly approach. Toxics. 2023;**11**(5):422. DOI: 10.3390/toxics11050422

[9] Riyazuddin R, Nisha N, Ejaz B, Khan MIR, Kumar M, Ramteke PW, et al. A comprehensive review on the heavy metal toxicity and sequestration in plants. Biomolecules. 2021;**12**(1):43. DOI: 10.3390/biom12010043

[10] Dutta S, Mitra M, Agarwal P, Mahapatra K, De S, Sett U, et al. Oxidative and genotoxic damages in plants in response to heavy metal stress and maintenance of genome stability. Plant Signaling & Behavior. 2018;**13**(8):e1460048. DOI: 10.1080/15592324.2018.1460048

[11] Pasricha S, Mathur V, Garg A, Lenka S, Verma K, Agarwal S. Molecular mechanisms underlying heavy metal uptake, translocation and tolerance in hyperaccumulators-an analysis: Heavy metal tolerance in hyperaccumulators. Environmental Challenges. 2021;**4**:100197. DOI: 10.1016/j.envc.2021.100197

[12] Vielee ST, Wise JP Jr. Among Gerontogens, heavy metals are a class of their own: A review of the evidence for cellular senescence. Brain Sciences. 2023;**13**(3):500. DOI: 10.3390/brainsci13030500

[13] Sychta K, Słomka A, Kuta E. Insights into plant programmed cell death

induced by heavy metals—Discovering a Terra incognita. Cell. 2021;**10**:65. DOI: 10.3390/cells10010065

[14] Sarker A, Kim JE, Islam ARMT, Bilal M, Rakib MRJ, Nandi R, et al. Heavy metals contamination and associated health risks in food webs—A review focuses on food safety and environmental sustainability in Bangladesh. Environmental Science and Pollution Research. 2022;**29**(3):3230-3245. DOI: 10.1007/s11356-021-17153-7

[15] Mawari G, Kumar N, Sarkar S, Daga MK, Singh MM, Joshi TK, et al. Heavy metal accumulation in fruits and vegetables and human health risk assessment: Findings from Maharashtra, India. India. Environmental Health Insights. 2022;**16**:11786302221119151. DOI: 10.1177/11786302221119151

[16] Rai PK, Lee SS, Zhang M, Tsang YF, Kim KH. Heavy metals in food crops: Health risks, fate, mechanisms, and management. Environment International. 2019;**125**:365-385. DOI: 10.1016/j.envint.2019.01.067

[17] Berni R, Luyckx M, Xu X, Legay S, Sergeant K, Hausman JF, et al. Reactive oxygen species and heavy metal stress in plants: Impact on the cell wall and secondary metabolism. Environmental and Experimental Botany. 2019;**161**:98-106. DOI: 10.1016/j.envexpbot.2018.10.017

[18] Sabreena, Hassan S, Bhat SA, Kumar V, Ganai BA, Ameen F. Phytoremediation of Heavy metals: An indispensable contrivance in green remediation technology. Plants. 2022;**11**(9):1255. DOI: 10.3390/plants11091255

[19] Siyar R, Doulati Ardejani F, Norouzi P, Maghsoudy S, Yavarzadeh M, Taherdangkoo R, et al. Phytoremediation potential of native hyperaccumulator plants growing on heavy metal-contaminated soil of Khatunabad copper smelter and refinery, Iran. Water. 2022;**14**(22):3597. DOI: 10.3390/w14223597

[20] Saeed M, Ilyas N, Bibi F, Shabir S, Mehmood S, Akhtar N, et al. Nanoremediation approaches for the mitigation of heavy metal contamination in vegetables: An overview. Nanotechnology Reviews. 2023;**12**(1):20230156. DOI: 10.1515/ntrev-2023-0156

[21] Nemati B, Baneshi MM, Akbari H, Dehghani R, Mostafaii G. Phytoremediation of pollutants in oil-contaminated soils by Alhagi camelorum: Evaluation and modeling. Scientific Reports. 2024;**14**(1):5502. DOI: 10.1038/s41598-024-56214-y

[22] Barathi S, Lee J, Venkatesan R, Vetcher AA. Current status of biotechnological approaches to enhance the phytoremediation of heavy metals in India—A review. Plants. 2023;**12**(22):3816. DOI: 10.3390/plants12223816

[23] Park JK, Oh K. Advancements in phytoremediation research for soil and water resources: Harnessing plant power for environmental cleanup. Sustainability. 2023;**15**(18):13901. DOI: 10.3390/su151813901

[24] Wang J, Delavar MA. Techno-economic analysis of phytoremediation: A strategic rethinking. Science of The Total Environment. 2023;**902**:165949. DOI: 10.1016/j.scitotenv.2023.165949

[25] Sharma JK, Kumar N, Singh NP, Santal AR. Phytoremediation technologies and their mechanism for removal of heavy metal from contaminated soil: An approach for a

sustainable environment. Frontiers in Plant Science. 2023;**14**:1076876. DOI: 10.3389/fpls.2023.1076876

[26] Kafle A, Timilsina A, Gautam A, Adhikari K, Bhattarai A, Aryal N. Phytoremediation: Mechanisms, plant selection and enhancement by natural and synthetic agents. Environmental Advances. 2022;**8**:100203. DOI: 10.1016/j.envadv.2022.100203

[27] Gao J, Han H, Gao C, Wang Y, Dong B, Xu Z. Organic amendments for in situ immobilization of heavy metals in soil: A review. Chemosphere. 2023;**335**:139088. DOI: 10.1016/j.chemosphere.2023.139088

[28] Shaffique S, Hussain S, Kang SM, Imran M, Kwon EH, Khan MA, et al. Recent progress on the microbial mitigation of heavy metal stress in soybean: Overview and implications. Frontiers in Plant Science. 2023;**14**:1188856. DOI: 10.3389/fpls.2023.1188856

[29] Lwin CS, Seo BH, Kim HU, Owens G, Kim KR. Application of soil amendments to contaminated soils for heavy metal immobilization and improved soil quality—A critical review. Soil Science and Plant Nutrition. 2018;**64**(2):156-167. DOI: 10.1080/00380768.2018.1440938

[30] Goldan E, Nedeff V, Barsan N, Culea M, Panainte-Lehadus M, Mosnegutu E, et al. Assessment of manure compost used as soil amendment—A review. Processes. 2023;**11**(4):1167. DOI: 10.3390/pr11041167

[31] Montreemuk J, Stewart TN, Prapagdee B. Bacterial-assisted phytoremediation of heavy metals: Concepts, current knowledge, and future directions. Environmental Technology & Innovation. 2023;**33**:103488. DOI: 10.1016/j.eti.2023.103488

[32] Sun Q, Zhang Y, Ming C, Wang J, Zhang Y. Amended compost alleviated the stress of heavy metals to pakchoi plants and affected the distribution of heavy metals in soil-plant system. Journal of Environmental Management. 2023;**336**:117674. DOI: 10.1016/j.jenvman.2023.117674

[33] Begum N, Qin C, Ahanger MA, Raza S, Khan MI, Ashraf M, et al. Role of arbuscular mycorrhizal fungi in plant growth regulation: Implications in abiotic stress tolerance. Frontiers in Plant Science. 2019;**10**:1068. DOI: 10.3389/fpls.2019.01068

[34] Irfan M, Mudassir M, Khan MJ, Dawar KM, Muhammad D, Mian IA, et al. Heavy Metals immobilization and improvement in maize (*Zea mays* L.) growth amended with biochar and compost. Scientific Reports. 2021;**11**(1):18416. DOI: 10.1038/s415

[35] Jócsák I, Knolmajer B, Szarvas M, Rabnecz G, Pál-Fám F. Literature review on the effects of heavy metal stress and alleviating possibilities through exogenously applied agents in Alfalfa (Medicago sativa L.). Plants. 2022;**11**(16):2161. DOI: 10.3390/plants11162161

[36] Guo M, Song W, Tian J. Biochar-facilitated soil remediation: Mechanisms and efficacy variations. Frontiers in Environmental Science. 2020;**8**:521512. DOI: 10.3389/fenvs.2020.521512

[37] Senthilkumar R, Prasad DMR. Sorption of heavy metals onto biochar. In: Applications of Biochar for Environmental Safety. Vol. 207. 2020. Available from: https://www.intechopen.com/chapters/72272

[38] Adamczyk-Szabela D, Wolf WM. The impact of soil pH on heavy metals uptake and photosynthesis efficiency in *Melissa*

officinalis, *Taraxacum officinalis, Ocimum basilicum*. Molecules. 2022;**27**(15):4671. DOI: 10.3390/molecules27154671

[39] Kicińska A, Pomykała R, Izquierdo-Diaz M. Changes in soil pH and mobility of heavy metals in contaminated soils. European Journal of Soil Science. 2022;**73**(1):e13203. DOI: 10.1111/ejss.13203

[40] Wen J, Wu C, Bi X, Zhang S, Ouyang H, Ye J, et al. Soil pH change induced by smelting activities affects secondary carbonate production and long-term Cd activity in subsoils. Applied Geochemistry. 2023;**152**:105663. DOI: 10.1016/j.apgeochem.2023.105663

[41] Xu D, Shen Z, Dou C, Dou Z, Li Y, Gao Y, et al. Effects of soil properties on heavy metal bioavailability and accumulation in crop grains under different farmland use patterns. Scientific Reports. 2022;**12**(1):9211. DOI: 10.1038/s41598-022-13140-1

[42] Hoeksema JD, Bever JD, Chakraborty S, Chaudhary VB, Gardes M, Gehring CA, et al. Evolutionary history of plant hosts and fungal symbionts predicts the strength of mycorrhizal mutualism. Communications Biology. 2018;**1**(1):116. DOI: 10.1038/s42003-018-0120-9

[43] Boorboori MR, Zhang HY. Arbuscular mycorrhizal fungi are an influential factor in improving the phytoremediation of arsenic, cadmium, lead, and chromium. Journal of Fungi. 2022;**8**(2):176. DOI: 10.3390/jof8020176

[44] Riaz M, Kamran M, Fang Y, Wang Q, Cao H, Yang G, et al. Arbuscular mycorrhizal fungi-induced mitigation of heavy metal phytotoxicity in metal contaminated soils: A critical review. Journal of Hazardous Materials. 2021;**402**:123919. DOI: 10.1016/j.jhazmat.2020.123919

[45] Adeyemi NO, Atayese MO, Sakariyawo OS, Azeez JO, Sobowale SPA, Olubode A, et al. Alleviation of heavy metal stress by arbuscular mycorrhizal symbiosis in *Glycine max* (L.) grown in copper, lead and pp contaminated soils. Rhizosphere. 2021;**18**:100325. DOI: 10.1016/j.rhisph.2021.100325

[46] Ma Y, Tiwari J, Bauddh K. Plant-mycorrhizal fungi interactions in phytoremediation of geogenic contaminated soils. Frontiers in Microbiology. 2022;**13**:843415. DOI: 10.3389/fmicb.2022.843415

[47] Zhang W, Yu L, Han B, Liu K, Shao X. Mycorrhizal inoculation enhances nutrient absorption and induces insect-resistant defense of Elymus nutans. Frontiers in Plant Science. 2022;**13**:898969. DOI: 10.3389/fpls.2022.898969

[48] Solanki MK, Kashyap PL, Kumari B, Ansari RA, Sumbul A, Rizvi R, et al. Mycorrhizal fungi and its importance in plant health amelioration. In: Microbiomes and Plant Health. United States: Academic Press; 2021. pp. 205-223. DOI: 10.1016/B978-0-12-819715-8.00006-9

[49] Frąc M, Hannula ES, Bełka M, Salles JF, Jedryczka M. Soil mycobiome in sustainable agriculture. Frontiers in Microbiology. 2022;**13**:1033824. DOI: 10.3389/fmicb.2022.1033824

[50] Wahab A, Muhammad M, Munir A, Abdi G, Zaman W, Ayaz A, et al. Role of arbuscular mycorrhizal fungi in regulating growth, enhancing productivity, and potentially influencing ecosystems under abiotic and biotic stresses. Plants. 2023;**12**(17):3102. DOI: 10.3390/plants12173102

[51] Suárez JP, Herrera P, Kalinhoff C, Vivanco-Galván O, Thangaswamy S. Generalist arbuscular mycorrhizal fungi

dominated heavy metal polluted soils at two artisanal and small– scale gold mining sites in southeastern Ecuador. BMC Microbiology. 2023;**23**(1):42. DOI: 10.1186/s12866-022-02748-y

[52] Shi J, Wang X, Wang E. Mycorrhizal symbiosis in plant growth and stress adaptation: From genes to ecosystems. Annual Review of Plant Biology. 2023;**74**:569-607. DOI: 10.1146/annurev-arplant-061722-090342

[53] Cai Z, Sun Y, Deng Y, Zheng X, Sun S, Sinkkonen A, et al. Enhanced electrokinetic remediation of cadmium (Cd)-contaminated soil with interval power breaking. International Journal of Environmental Research. 2022;**16**(3):31. DOI: 10.1007/s41742-022-00409-6

[54] Selvi A, Rajasekar A, Theerthagiri J, Ananthaselvam A, Sathishkumar K, Madhavan J, et al. Integrated remediation processes toward heavy metal removal/recovery from various environments-a review. Frontiers in Environmental Science. 2019;7:66. DOI: 10.3389/fenvs.2019.00066

[55] Federal Remediation Technologies Roundtable. Electrokinetic-enhanced remediation technologies. Available from: https://www.frtr.gov/matrix/Electrokinetic-Enhanced-Remediation/# [Accessed: April 10, 2023]

[56] Xu Q, Wu B, Chai X. In situ remediation technology for heavy metal contaminated sediment: A review. International Journal of Environmental Research and Public Health. 2022;**19**(24):16767. DOI: 10.3390/ijerph192416767

[57] Cameselle C, Gouveia S, Cabo A. Enhanced electrokinetic remediation for the removal of heavy metals from contaminated soils. Applied Sciences. 2021;**11**(4):1799. DOI: 10.3390/app11041799

[58] Garcia-Blas N, Jimenez-Relinque E, Castellote M. Surfactants in electrokinetic remediation of sediments to enhance the removal of metals. Journal of Soils and Sediments. 2022;**22**(11):2853-2864. DOI: 10.1007/s11368-022-03299-5

[59] Gidudu B, Chirwa EM. The role of pH, electrodes, surfactants, and electrolytes in electrokinetic remediation of contaminated soil. Molecules. 2022;**27**(21):7381. DOI: 10.3390/molecules27217381

[60] Dhanapal AR, Thiruvengadam M, Vairavanathan J, Venkidasamy B, Easwaran M, Ghorbanpour M. Nanotechnology approaches for the remediation of agricultural polluted soils. ACS Omega. 2024;**9**(12):13522-13533. DOI: 10.1021/acsomega.3c09776

[61] Kumar R, Rauwel P, Rauwel E. Nanoadsorbants for the removal of heavy metals from contaminated water: Current scenario and future directions. PRO. 2021;**9**(8):1379. DOI: 10.3390/pr9081379

[62] Kromah V, Zhang G. Aqueous adsorption of heavy metals on metal sulfide nanomaterials: Synthesis and application. Water. 2021;**13**(13):1843. DOI: 10.3390/w13131843

[63] Rajput VD, Minkina T, Upadhyay SK, Kumari A, Ranjan A, Mandzhieva S, et al. Nanotechnology in the restoration of polluted soil. Nanomaterials. 2022;**12**(5):769. DOI: 10.3390/nano12050769

[64] Anthony T. Assessment of heavy metal contamination in wetlands soils around an industrial area using combined GIS-based pollution indices and remote sensing techniques. Air, Soil and Water Research. 2023;**16**:11786221231214062. DOI: 10.1177/11786221231214062

[65] Adu-Boahen K, Boateng I. Mapping seasonal variation in the distribution and concentration of Heavy Metals using water quality index and geographic information system based applications. Journal of Geographical Research. 2021;**4**(2):31-42. DOI: 10.30564/jgr.v4i2.3100

[66] Abdollahi S, Madadi M, Ostad-Ali-Askari K. Monitoring and investigating dust phenomenon on using remote sensing science, geographical information system and statistical methods. Applied Water Science. 2021;**11**(7):111. DOI: 10.1007/s13201-021-01419-z

[67] Abowaly ME, Ali RA, Moghanm FS, Gharib MS, Moustapha ME, Elbagory M, et al. Assessment of soil degradation and hazards of some heavy metals, using remote sensing and GIS techniques in the Northern part of the Nile Delta, Egypt. Agriculture. 2022;**13**(1):76. DOI: 10.3390/agriculture13010076

[68] El Behairy RA, El Baroudy AA, Ibrahim MM, Mohamed ES, Rebouh NY, Shokr MS. Combination of GIS and multivariate analysis to assess the soil heavy metal contamination in some arid zones. Agronomy. 2022;**12**(11):2871. DOI: 10.3390/agronomy12112871

[69] Asaiduli H, Abliz A, Abulizi A, Sun X, Ye P. Assessment of soil heavy metal pollution and health risks in different functional areas on the northern slope of the eastern Tianshan Mountains in Xinjiang, NW China. International Journal of Environmental Research and Public Health. 2023;**20**(6):4843. DOI: 10.3390%2Fijerph20064843

[70] Hammam AA, Mohamed WS, Sayed SEE, Kucher DE, Mohamed ES. Assessment of soil contamination using gis and multi-variate analysis: A case study in El-Minia Governorate, Egypt. Agronomy. 2022;**12**(5):1197. DOI: 10.3390/agronomy12051197

[71] Oumenskou H, El Baghdadi M, Barakat A, Aquit M, Ennaji W, Karroum LA, et al. Assessment of the heavy metal contamination using GIS-based approach and pollution indices in agricultural soils from Beni Amir irrigated perimeter, Tadla plain, Morocco. Arabian Journal of Geosciences. 2018;**11**(22):692. DOI: 10.1007/s12517-018-4021-5

[72] Qi Z, Gao X, Qi Y, Li J. Spatial distribution of heavy metal contamination in mollisol dairy farm. Environmental Pollution. 2020;**263**:114621. DOI: 10.1016/j.envpol.2020.114621

[73] Saha A, Gupta BS, Patidar S, Martínez-Villegas N. Spatial distribution and source identification of metal contaminants in the surface soil of Matehuala, Mexico based on positive matrix factorization model and GIS techniques. Frontiers in Soil Science. 2022;**2**:1041377. DOI: 10.3389/fsoil.2022.1041377

[74] Wang SH, Wang JW, Zhao LT, Abbas SZ, Yang Z, Yong YC. Soil microbial fuel cell based self-powered cathodic biosensor for sensitive detection of heavy metals. Biosensors. 2023;**13**(1):145. DOI: 10.3390/bios13010145

[75] Huang CW, Lin C, Nguyen MK, Hussain A, Bui XT, Ngo HH. A review of biosensor for environmental monitoring: Principle, application, and corresponding achievement of sustainable development goals. Bioengineered. 2023;**14**(1):58-80. DOI: 10.1080/21655979.2022.2095089

[76] Gavrilaș S, Ursachi CȘ, Perța-Crișan S, Munteanu FD. Recent

trends in biosensors for environmental quality monitoring. Sensors. 2022;**22**(4):1513. DOI: 10.3390/s22041513

[77] Alfadaly RA, Elsayed A, Hassan RY, Noureldeen A, Darwish H, Gebreil AS. Microbial sensing and removal of heavy metals: Bioelectrochemical detection and removal of chromium (vi) and cadmium (ii). Molecules. 2021;**26**(9):2549. DOI: 10.3390/molecules26092549

[78] Hui Y, Huang Z, Alahi MEE, Nag A, Feng S, Mukhopadhyay SC. Recent advancements in electrochemical biosensors for monitoring the water quality. Biosensors. 2022;**12**(7):551. DOI: 10.3390/bios12070551

[79] Mathur S, Singh D, Ranjan R. Genetic circuits in microbial biosensors for heavy metal detection in soil and water. Biochemical and Biophysical Research Communications. 2023;**652**:131-137. DOI: 10.1016/j.bbrc.2023.02.031

[80] Velusamy K, Periyasamy S, Kumar PS, Rangasamy G, Pauline JMN, Ramaraju P, et al. Biosensor for Heavy Metals detection in wastewater: A review. Food and Chemical Toxicology. 2022;**168**:113307. DOI: 10.1016/j.fct.2022.113307

[81] Sharma K, Sharma M. Optical biosensors for environmental monitoring: Recent advances and future perspectives in bacterial detection. Environmental Research. 2023;**236**:116826. DOI: 10.1016/j.envres.2023.116826

[82] Hara TO, Singh B. Electrochemical biosensors for detection of pesticides and heavy metal toxicants in water: Recent trends and progress. ACS ES&T Water. 2021;**1**(3):462-478. DOI: 10.1021/acsestwater.0c00125

[83] Zare-Shehneh N, Mollarasouli F, Ghaedi M. Recent advances in carbon nanostructure-based electrochemical biosensors for environmental monitoring. Critical Reviews in Analytical Chemistry. 2023;**53**(3):520-536. DOI: 10.1080/10408347.2021.1967719

[84] Abbas SZ, Wang JY, Wang H, Wang JX, Wang YT, Yong YC. Recent advances in soil microbial fuel cells based self-powered biosensor. Chemosphere. 2022;**303**:135036. DOI: 10.1016/j.chemosphere.2022.135036

[85] Melotto M, Brandl MT, Jacob C, Jay-Russell MT, Micallef SA, Warburton ML, et al. Breeding crops for enhanced food safety. Frontiers in Plant Science. 2020;**11**:428. DOI: 10.3389/fpls.2020.00428

[86] Tiwari S, Lata C. Heavy metal stress, signaling, and tolerance due to plant-associated microbes: An overview. Frontiers in Plant Science. 2018;**9**:452. DOI: 10.3389/fpls.2018.00452

[87] Zeb A, Liu W, Lian Y, Zheng Z, Meng L, Chen C, et al. Selection and breeding of pollution-safe cultivars (PSCs)—An eco-friendly technology for safe utilization of heavy metal (loid) contaminated soils. Environmental Technology & Innovation. 2022;**25**:102142. DOI: 10.1016/j.eti.2021.102142

[88] Gallo-Franco JJ, Sosa CC, Ghneim-Herrera T, Quimbaya M. Epigenetic control of plant response to heavy metal stress: A new view on aluminum tolerance. Frontiers in Plant Science. 2020;**11**:602625. DOI: 10.3389/fpls.2020.602625

[89] Joshi A, Yang SY, Song HG, Min J, Lee JH. Genetic databases and gene editing tools for enhancing crop resistance against abiotic stress. Biology. 2023;**12**(11):1400. DOI: 10.3390/biology12111400

[90] Gajardo HA, Gómez-Espinoza O, Boscariol Ferreira P, Carrer H, Bravo LA. The potential of CRISPR/Cas technology to enhance crop performance on adverse soil conditions. Plants. 2023;**12**(9):1892. DOI: 10.3390/plants12091892

[91] Venegas-Rioseco J, Ginocchio R, Ortiz-Calderón C. Increase in phytoextraction potential by genome editing and transformation: A review. Plants. 2021;**11**(1):86. DOI: 10.3390/plants11010086

[92] Koźmińska A, Wiszniewska A, Hanus-Fajerska E, Muszyńska E. Recent strategies of increasing metal tolerance and phytoremediation potential using genetic transformation of plants. Plant Biotechnology Reports. 2018;**12**:1-14. DOI: 10.1007/s11816-017-0467-2

[93] Ugrina M, Jurić A. Current trends and future perspectives in the remediation of polluted water, soil and air—A review. Processes. 2023;**11**(12):3270. DOI: 10.3390/pr11123270

Chapter 8

Nanoparticles in Agriculture: Enhancing Crop Resilience and Productivity against Abiotic Stresses

Hafiza Fizzah Zulfiqar, Bakht Afroze, Sana Shakoor, Muhammad Saad Bhutta, Muhammad Ahmed, Sameera Hassan, Fatima Batool and Bushra Rashid

Abstract

The agricultural sector faces unprecedented challenges to ensure food security as the global population soars and climate change intensifies. Abiotic stresses are well-known for diminishing agricultural output and constraining crop yield generation worldwide. While conventional methods for managing crop stress fall short of meeting global demands, the integration of nanotechnology in agriculture offers a sustainable approach, providing a cornerstone for resilient and resource-efficient crop production in the face of evolving environmental challenges. Through targeted delivery systems and tailored formulations, nanoparticles exhibit the potential to enhance plant physiological processes, nutrient uptake efficiency, and stress tolerance mechanisms. This chapter describes the potential role of nanoparticles in abiotic stress management and activation of plant defence-related genes, improving the yield and quality of crops by combating nutrient deficiency and inducing stress tolerance. Moreover, it also discusses the potent molecular mechanisms upon application of nanoparticles for inducing tolerance to various abiotic stresses. However, while nanoparticle-based approaches hold great promise, their implementation also raises concerns regarding environmental impact, toxicity, regulatory frameworks, and socioeconomic implications.

Keywords: abiotic stresses, nanoparticles, drought, salinity, heavy metal, molecular pathway, sustainable agriculture, crop plant, food security

1. Introduction

Over the last few decades, crop production has decreased even as food demands have risen due to an increasing human population [1]. According to the reports, 2 billion people have micronutrient deficiencies, 800 million people experience chronic hunger, and 653 million people will still be undernourished in 2030 [2]. In 2011, the Food and Agriculture organization estimated a loss of 1.3 billion tons of food per year around the globe. As a result, food security will remain a major concern as the world's population is predicted to reach 10 billion in 2050 [3]. Climate change is considered a major limiting factor, having a detrimental effect on plant growth, development, crop yield, and productivity. Abiotic stresses such as drought, salinity, soil erosion, flooding, and nutrient depletion are the byproducts of climate change and are one of the key factors for stunted plant growth, resulting in reduced yield and quality of crop plants [4]. The development of stress-tolerant plants could be a promising and practical solution to this challenge. However, conventional breeding methods have not yielded the desired results in improving crop plant tolerance to stress through intergeneric or inter-specific hybridization [5]. Thus, novel and promising agricultural strategies capable of reforming modern agriculture are required to increase food productivity while limiting pesticide and fertilizer use to reduce abiotic stress. Nanotechnology is currently an emerging and rapidly developing field of study that has gained tremendous attention recently due to its broad-spectrum applications in various sectors, particularly in plant biotechnology and agriculture. Nanotechnology, in particular, can help provide effective solutions to agricultural problems and achieve a sustainable and secure future for agriculture [6]. Nanotechnology uses nanoparticles as a tool to mitigate abiotic stress factors and to improve crop yield and food quality.

Nanoparticles (NPs) are small materials ranging in size from 1 to 100 nm. In contrast to their larger-sized counterparts, NPs have distinct and diverse physicochemical properties including high adsorption efficacy, large surface area-to-volume ratio, and effective linking and work efficiency. These qualities make them a suitable candidate for disease management strategies as bactericides/fungicides/pesticides to boost plant health. Moreover, NPs can also help to alleviate nutrient deficiency symptoms and supplement essential elements when used as macro and micro-nano fertilizers in plants [7]. Titanium dioxide (TiO_2), iron oxide (Fe_3O_4), zinc oxide (ZnO), silicon oxide (SiO_2), copper (Cu-NPs), and selenium (Se-NPs) have recently caught significant attention due to their non-threatening use in agriculture [8–10]. Different biological, physical, and chemical techniques can be used for NPs synthesis. The impact of NPs on plants varies and primarily depends upon the size, origin, time of application, and concentration [11]. Recent studies reported that the application of NPs resulted in improved plant tolerance against abiotic stresses by increasing the antioxidant activities, decreasing hydrogen peroxide (H_2O_2) and malondialdehyde (MDA) accumulation, and maintaining the efficiency of the photosynthetic apparatus. Moreover, they also have the ability to reach the photo system-II (PS-II) reaction center and increase the transmission of electrons and light absorption by penetrating chloroplast, resulting in improved photosynthetic efficiency and plant growth [12, 13].

Despite their extensive and commercial use in different products, concerns about NP's toxicological and environmental impacts are circulating in the research community [14]. It has been reported that the excessive use of NPs can result in physiological abnormalities and oxidative stress in plants, leading to reduced antioxidant activities and gas exchange characteristics. Moreover, disruption of cell division, mitotic index, and root growth has also been reported [15, 16]. Most of the published literature emphasizes the positive effects of NPs on growth and crop productivity. However, they lack an in-depth understanding of NP-plant interaction at the biochemical, physiological, and molecular levels. Therefore, this chapter will focus on a detailed understanding of interaction of NPs with plants at the molecular level and help to fill in research gaps that will enable potential future use of NPs. This comprehensive review will be a valuable source of information to conduct more studies and develop new eco-friendly NPs of different metals to counter the devastating impact of abiotic stresses on plants.

2. Effects of abiotic stress on crop plant

Climate change has led to an increase in abiotic stress conditions, which are gradually affecting crop production in various respects. Different physiological and biochemical changes occur in plants during stress, resulting in disruptions in plant growth, development, and productivity. Nanobiotechnology could be the most efficient adaptation strategy to cope with climate and abiotic stress factors, thereby achieving sustainable production [17, 18]. These stressors encompass heat, drought, high soil salinity, and heavy metal (HM) toxicity—cadmium (Cd), lead (Pb), arsenic (As), nickel (Ni), chromium (Cr), cobalt (Co), and zinc (Zn) [17]. Drought and heat stress are chief contributors to reduced crop yield around the globe. Prolonged drought and heat increase reactive oxygen species (ROS) production, affecting oxidative processes and causing reductions in stomatal opening, leaf size, water potential, and root growth, resulting in a general decrease in crop yield (**Figure 1**) [18, 19].

Salinity and heavy metal stress are also considered among the environmental factors that limit crop yield in many countries. Soil salinity causes low osmotic potential around the seeds, preventing water uptake and thus resulting in inhibition of seed germination. Oxidative stress in legumes induced by sodium chloride (NaCl) decreases seed nutrient quality, lowers nodulation, and inhibits plant growth. Ascorbate and carotenoids are important non-enzymatic salinity defense mechanisms, whereas proline (Pro) is a known osmoregulatory stress-related compound [20, 21].

Various factors contribute to heavy metal accumulation in agricultural soils, such as chemical fertilizers, atmospheric deposition, sewage sludge, and rapid industrial growth [22]. Heavy metals significantly impede plant growth by disrupting essential functions like food intake, antioxidant enzymes, and photosynthetic machinery. The increased generation of reactive oxygen species (ROS) during heavy metal stress worsens the decline in agricultural productivity [20, 21]. The annual economic cost of

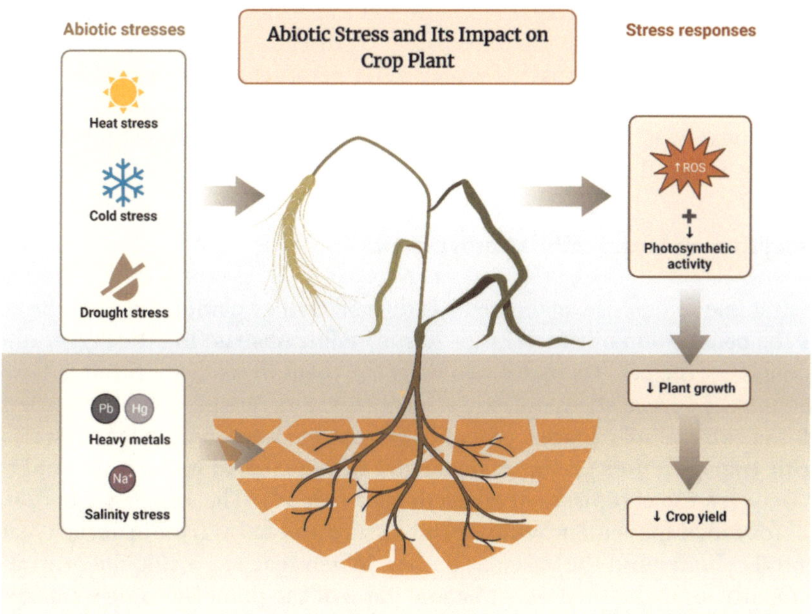

Figure 1.
Abiotic stresses and their response in crop plant. (The figure is created using BioRender https://biorender.com/).

heavy metal pollution exceeds $10 billion, primarily attributed to anthropogenic and industrial waste deposition. Mitigating heavy metal-induced stress requires a comprehensive approach, including transgenic crop engineering, marker-assisted breeding, and traditional breeding. Sustainable agricultural practices are vital for enhancing plant resilience in the face of heavy metal contamination [23].

Efforts to mitigate the impact of abiotic stressors on crop productivity require integrated strategies that address drought, salinity, cold, heavy metal contamination, and other environmental challenges. Sustainable agricultural practices, advanced breeding techniques, and a focus on resilience are essential components of a comprehensive approach to ensure food security in the face of a changing climate [24, 25].

3. Mechanisms of nanoparticle-plant interaction in crop plant

Plants and nanoparticles interact in a complicated and multidimensional way that is controlled by molecular processes, physiological reactions, barrier crossing, uptake, and defense [26]. Therefore, comprehending these dynamics is crucial in assessing the possible advantages and drawbacks of employing nanoparticles in agricultural and environmental settings [27]. In addition to that, optimizing the characteristics of nanoparticles, reducing their negative impacts, and maximizing their potential for environmental remediation and sustainable agriculture techniques are also essential [28]. At specified concentrations, distinct NPs have been shown to boost the growth and development of plants through standard mechanisms [29]. A detailed mechanism of action for nanoparticles within plants is given below (**Figure 2**).

3.1 Uptake mechanisms

Adsorbed nanoparticles (NPs) from the soil can adhere to the surface of the roots. Subsequently, they can enter the root system by direct absorption, root junctions, or wounds. Additionally, these particles may enter plants through their above-ground sections, primarily through cuticles (a non-polar channel) and stomata or trichomes (a polar pathway) [30, 31]. Artificial exposure might happen when they are sprayed on or directly injected into the leaf [31].

3.2 Absorption and intracellular movement

The first line of defense against NPs is the cell wall of plants, wherein the cellulose permits the penetration of tiny NPs (5–20 nm) while obstructing the bigger ones [32]. After navigating the cell, there are two ways for them to navigate through tissues: the apoplast and the symplast. Apoplastic transport occurs outside the plasma membrane through the extracellular spaces, cell walls of adjacent cells, and xylem vessels while symplastic transport uses plasmodesmata and sieve plates to move water and substances between the cytoplasm of adjacent cells [33, 34]. The apoplastic pathway is essential for radial movement within plant tissues and helps nanomaterials reach the root central cylinder and the vascular tissues. When reached inside the central cylinder, nanoparticles make their way to aerial parts of the plant by using a transpiration stream [35]. However, because Caspary strips function as a resistive barrier, nanoparticles can form aggregates and concentrate in the endodermis. The nanoparticles must then enter the symplast for effective translocation to shoots [36].

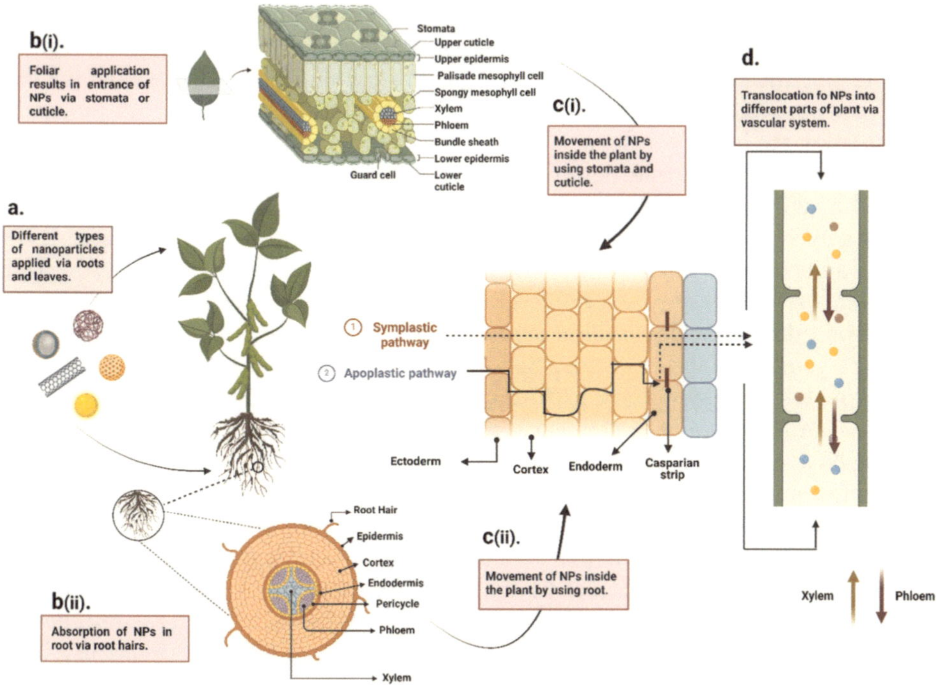

Figure 2.
Detailed pathway of NPs uptake, movement, and translocation inside plant. a. Application of different types of NPs via different routes, i.e., ground and foliar application. b(i). The small NPs (diameters range from 3 to 5 nm) are reported to penetrate plant roots along with osmotic pressure and capillary forces or pass directly through the root epidermal cells. The epidermal cells of the root cell wall are semipermeable, containing small pores that restrict the large NPs. Some NPs induced new pores in the epidermal cell wall, which facilitated its entrance. b(ii). The NPs applied by the leaves can enter the leaves through the stomata or cuticles. The cuticle acts as a primary leaf barrier, restricting the entry of NPs to a size of <5 nm. The NPs > 10 mm enter through stomata, and their cellular transport occurs through apoplastic and symplastic routes into the vascular system of the plant. c(i). Foliar application of NPs (between 10 and 50 nm) is favored through the adjacent cell's cytoplasm (symplastic route). Thus, larger NPs (between 50 and 200 nm) are translocated between the cells (apoplastic route). c(ii). After crossing the cell walls, NPs are apoplastically transported through extracellular spaces until they reach the central vascular cylinder. NPs, however, need to symplastically cross the Casparian strip barrier to enter the central vascular cylinder. This happens by binding to the endodermal cell membrane's carrier proteins through endocytosis, pore formation, and transport. NPs travel from one cell to another through the plasmodesmata as internalized in the cytoplasm. d. Internalized NPs entered via roots reach the central vascular cylinder, allowing the xylem to move unidirectionally upward. Meanwhile, NPs entered through leaves are transported along with the sugar flow through the phloem sieve tubes. As a result of vascular transport by phloem, NPs can travel bidirectionally and accumulate in roots, stems, fruits, grains, and young leaves to varying degrees because these organs serve as potent sinks for the sap. (The figure is created using BioRender https://biorender.com/).

Using plasmodesmata and retina structures, symplastic transport is the transfer of chemicals and water between the cytoplasm of adjacent cells [37]. Since the transfer of nanomaterials to the symplast allows for their efficient translocation to the shoots, this mechanism is thought to be more significant for the transportation of nanoparticles. The symplastic transit of nanoparticles across the membrane and their accumulation influences the charge inside the cell, which modifies the electrochemical potential of the membrane. These changes can have an impact on the movement of other materials over the membrane, so to sustain turgor pressure, water and nutrient intake, and plant development, the membrane's electrochemical potential must stay in equilibrium [38].

In the case of foliar applications, nanomaterials must cross the barrier the cuticle presents, following the lipophilic or the hydrophilic pathway [39]. Numerous other pathways also allow nanoparticles to penetrate the intracellular area [40]. They could get through by attaching to transport proteins, ion channels, aquaporins, endocytosis, or by starting the creation of new pores, some of which might be bigger than the ones that already exist [41]. The processes of nanoparticle adsorption, translocation, and accumulation are heavily influenced not only by the type of plant but also by the physical and chemical properties of the nanoparticles themselves [42].

The way NPs move inside the plant is a key indicator of their accumulation and storage in different parts of the plant. For example, if an NP took the apoplastic pathway and used the xylem for translocation, then it would most probably move from roots to shoots and aerial parts of plants but would not move downwards toward roots. Thus, application methods for these types of NPs should be carried out at the root level, ensuring a good distribution in the plant. On the other hand, if an NP translocates via phloem and takes the symplastic pathway, then it will accumulate in plant organs acting as sinks, such as fruits and grains. Thus, appropriate measures should be taken while applying NPs to plants to avoid further human or animal ingestion of nanomaterials. Moreover, translocation is not essentially limited to a specific cell type, and lateral movement of nanomaterials between xylem and phloem is also possible [43].

The crystallinity of the nanoparticles utilized is one of the most critical elements influencing accumulation [42], the hydrodynamic dimension of the NPs, the nature of the surrounding soil, such as whether it is "organic" or inorganic, and the influence of zeta potentials [44]. The plant's large concentration of nanoparticles results in a high amount of phytotoxicity. The phytotoxic impact of NPs is determined by their size, shape, exposure period, and concentration [45].

3.3 Nanoparticle-plant interaction

Plants' regular cellular activities can be disrupted by NPs, resulting in oxidative stress. This disruption is frequently caused by NPs interfering with the electron transport chain in mitochondria and chloroplasts, where these particles cause plants to produce ROS, resulting in an oxidative burst [46]. This increased ROS level can induce cellular damage such as lipid peroxidation, DNA damage, and protein alterations [47]. But despite its destructive nature, it is essential for various biological activities like stress tolerance. However, fortunately, plants are gifted with antioxidant defense systems to counterbalance the negative effects of high ROS levels. Enzymatic (e.g., superoxide dismutase, catalase, and guaiacol peroxidase) and non-enzymatic (e.g., ascorbate, glutathione, and carotenoids) molecules are produced in response to scavenge excess ROS [48]. NP-plant interactions can disturb hormonal homeostasis, influencing plant metabolism and stress responses. In response to NPs, many hormonal pathways may be activated or downregulated, regulating plant growth, development, and stress tolerance [49]. A study conducted on *Capsicum annuum* indicated increased cytokinin levels in response to AgNP stress while another study reported on Gossypium sp. showed decreased levels of ABA and auxin in response to CuO-NPs, suggesting the alteration of plant hormonal balance and metabolism upon plant-NP interaction [50].

Nanoparticles cannot work successfully unless they interfere with the molecular processes and gene expression of plants. The diverse gene expressions of micro-RNA were studied in nanoparticle-treated cells. Nanoparticles have an effect on miR-398 and miR-408, which are responsible for seed germination, root length, seedling development, and free radical scavengers [51]. The random drop in miR-164 expression, which is involved in auxin hormone signaling, is, nevertheless, linked to the nanoparticle-mediated rise in roots [52].

4. Synthesis of nanoparticles and their role in alleviating abiotic stress

As their name implies, nanoparticles have distinct physicochemical features and range in size from 1 to 100 nm. Usually, both top-down and bottom-up synthesis methods are used to synthesize nanoparticles [53]. Under the top-down method, bigger molecules break down to generate nanoparticles. Examples of top-down methods include electro-explosion, mechanical milling, chemical etching, laser ablation, and sputtering. Metal-oxide nanoparticles are mostly synthesized by [54]. In contrast, simpler materials mix and react to generate nanoparticles during bottom-up synthesis. Examples of bottom-up techniques include pyrolysis, condensation, the sol-gel process, and spinning. The bottom-up approach includes the green production of nanoparticles, typically from biological material [55]. Besides, NPs can also be synthesized by using the Green synthesis approach which involves reacting processed extracts of plants, fungi, and microorganisms with metal salt solutions, followed by reduction and capping to create metal nanoparticles coated with target biomolecules. Green-synthesized nanoparticles help improve results and facilitate efficient transportation to the target spot due to their small size [56].

The success of nanoparticle production depends on the following characteristics:

1. Characterizations of morphology by microscopy (scanning and transmission electron microscopy).

2. Characterization of structures by Zeta-size analyzers, Raman, XRD, and XPS.

3. UV-visible and photoluminescence optical characterizations [57].

4.1 Commonly used nanoparticles for abiotic stress

Several metals including silver (Ag), copper (Cu), gold (Au), iron (Fe), titanium (Ti), and zinc (Zn) and their oxides have been used for the green synthesis of NPs using plants and their extracts, microorganisms and membranes, and DNA of viruses including diatoms.

Recently, it has been reported that the cerium oxide (CeO) NPs application maintained the quantum yield of photosystem (PS) II and CO_2 assimilation through ROS scavenging, particularly hydrogen peroxide, induced by abiotic stress. The application of titania (TiO_2) NPs improved the activity of catalase (CAT), glutathione peroxidase (GPOX), and superoxide dismutase (SOD) and reduced oxidative stress in Duckweed (*Lemna minor*) plants [58, 59]. Several beneficial and stress-countering effects of NPs application in various crops have been reported, such as improved growth in *Solanum lycopersicum* L. [60] and *Allium cepa* L. [61]; *S. lycopersicum* L. [62], *Oryza sativa* [63], *Capsicum annuum* [64]. Therefore, it can be concluded that the stress ameliorative potential of NPs can be exploited to combat various negative effects caused by abiotic stresses in crop plants. Different types of NPs that are found to be effective in mitigating stress response are listed in **Table 1**.

4.2 Delivery methods of nanoparticles to crop plants

The delivery of nanoparticles to crop plants can be achieved through various methods to enhance plant growth, increase nutrient uptake, and mitigate environmental stresses. Different application methods are described in detail below (**Figure 3**).

Nanoparticles	Dosage (mg/L)	Duration (days)	Plant species	Abiotic stress	Application	References
Selenium dioxide	20–40 mg	30	*Phaseolus vulgaris*	Salinity	Enhanced the plant growth and yield	[65]
Calcium oxide	25 mg	30	*Hordeum vulgare*	Heavy metal	Increased plant growth, photosynthesis efficiency, and antioxidant enzymes	[66]
Silicon	2–50 mg	45	*Citrus x sinensis*	Salinity	Improved oxidative stress tolerance	[67]
Gold	10 mg	07	*Triticum aestivum*	Salinity	Improved the plant defense systems	[68]
Cerium oxide	50–100 mg	21	*Oryza sativa*	Salinity	Improved the crop yield by modulating the plant's physiological and biochemical mechanisms	[69]
Cerium oxide	50–100 mg	30	*Gossypium*	Salinity	Improved the plant growth by maintaining cytosolic K+/Na + ratio	[70]
Zinc oxide	0.5–1 mg	1–7	*Carthamus tinctorius*	Salinity	Enhanced plant germination and salinity tolerance by improving the activities of antioxidant enzymes	[71]
Iron oxide	10 mg	07	*Oryza sativa*	Cadmium and drought stress	Increased the tissue dry weight and reduced the Cd accumulation	[63]
Iron oxide	10 mg	05	*Triticum aestivum*	Salinity and heavy metal	Facilitates photosynthetic pigments and restricts cadmium uptake	[72]
Silver	50–75 mg	10	*Satureja hortensis*	Salinity	Improved plant growth and germination	[73]
Iron	5–20 mg	15	*Triticum aestivum*	Heavy metal and drought stress	Improved photosynthesis and alleviated oxidative stress	[12]
Zinc oxide	0.5–1 g	07	*Triticum aestivum*	Heavy metal and drought stress	Increased plant growth and reduction in chromium bioavailability	[74]
Selenium	100 mg	03	*Triticum aestivum*	Drought	Enhanced plant growth and development	[75]

Nanoparticles	Dosage (mg/L)	Duration (days)	Plant species	Abiotic stress	Application	References
Titanium oxide	2–5 ppm	01	*Dracocephalum moldavica*	Salinity	Promote plant growth and ameliorate salinity stress	[76]
Copper	50, 100, 200 mg	30	*Triticum aestivum*	Heavy metal	Increased biomass, antioxidant enzyme contents, and photosynthesis efficiency	[77]
Silica	2–50 mg	21	*Cucumis sativus*	Drought and salinity	Improved the growth and productivity of cucumber plants by balancing nutrient uptake	[78]
Titanium oxide	1.5–10 ppm	07	*Vicia faba*	Salinity	Improved growth and enhanced tolerance against salinity	[79]
Cerium oxide	50–100 mg	14	*Glycine max*	Salinity	Enhanced plant growth by regulating photosynthesis and water use efficiency	[80]
Chitosan	100 mg	01	*Zea mays*	Salinity	Mitigates the deleterious effects of salinity	[81]
Titanium oxide	2–5 ppm	03	*Cicer arietinum*	Cold	Increased plant growth and antioxidant activity	[82]
Titanium oxide	50–100 mg	30	*Solanum lycopersicum*	Heat	Enhanced plant growth and photosynthesis efficiency	[83]

Table 1.
List of the NPs applications in plant abiotic stress.

4.2.1 Seed coating

Nanoparticles are applied as coatings on the surface of seeds. Seeds are immersed or coated with a solution containing nanoparticles before planting. This method ensures direct contact between nanoparticles and emerging roots, promoting early interaction with the plant [84].

4.2.2 Soil drenching

Nanoparticles are applied as a solution directly to the soil around the plant root zone. Nanoparticle solutions are poured or sprayed onto the soil surface, allowing the roots to absorb the particles. Effective for delivering nanoparticles to the root system, promoting nutrient absorption and root development [85].

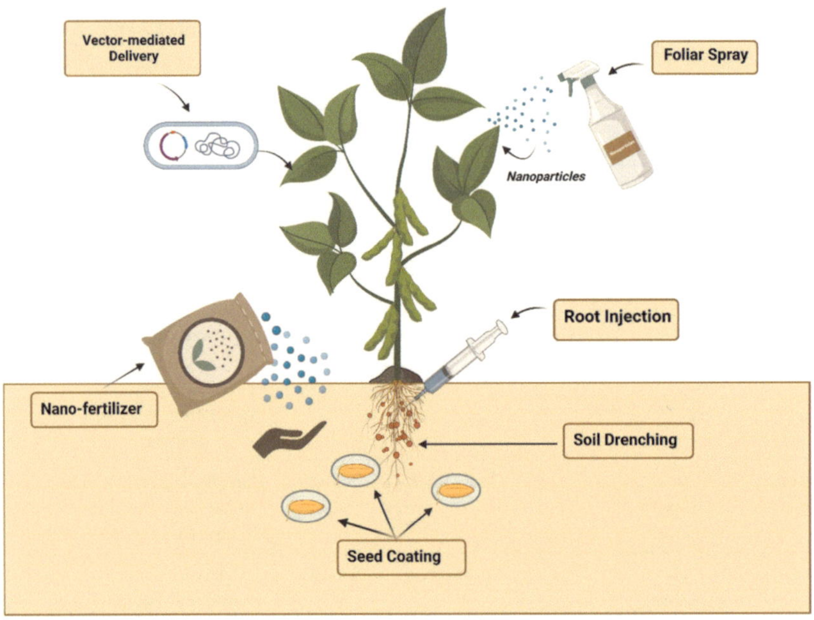

Figure 3.
Diagrammatic representation of different delivery methods used for nanoparticle application in crop plant. (The figure is created using BioRender https://biorender.com/).

4.2.3 Foliar spray

Nanoparticles are applied as a spray directly to the leaves and stems of plants. In this method, a nanoparticle solution is sprayed onto the plant's vegetative parts. Plants absorb foliar NPs usually through stomata, ion channels, protein carriers, cracks or water pores, endocytosis, stigma, wound and trichomes. This approach allows rapid uptake, controlled release of NPs at targeted locations and improved absorption and assimilation of foliar fertilizer [86].

4.2.4 Hydroponic systems

Nanoparticles are added to the nutrient solution in hydroponic systems. Nanoparticles are dissolved in the nutrient solution, and plants absorb them through their roots in a soilless growing medium, which provides precise control over nutrient delivery and uptake [87].

4.2.5 Root injection

Nanoparticles are injected directly into the plant root system. Nanoparticle solutions are injected into the soil around the root zone or directly into the roots, which offers a targeted and efficient delivery method [88].

4.2.6 Biological vector-mediated delivery

Nanoparticles are delivered into plants using biological vectors, such as bacteria or fungi. Engineered microorganisms carrying nanoparticles are introduced to the plant, facilitating uptake, which utilizes natural biological processes for nanoparticle delivery [89].

4.2.7 Nanoparticle-embedded fertilizers

Nanoparticles are incorporated into conventional fertilizers. Nanoparticles are blended with fertilizers, allowing for simultaneous delivery of nutrients and nanoparticles during standard fertilization practices, which integrates nanoparticle application with routine agricultural practices [90].

NPs were taken up by roots and/or leaves, and they moved through two different channels: apoplastic and symplastic [91]. After passing through the first barrier, the cell walls, NPs were distributed between the plasma membrane and the walls of the cells; osmotic pressure and capillary forces may have an impact on the NPs' subsequent migration [92]. In addition to ion channels and transport or carrier proteins like aquaporin, NPs can also enter cells by cell membrane barrier breaking or endocytosis [93]. The precise location and function of NPs in plant organelles are not exploited by any mechanistic method [94]. The translocation and uptake kinetics of NPs should be investigated further, as well as the biochemical, molecular, and physiological patterns of NPs uptake and transport kinetics, as these are crucial to comprehending the accumulation behavior of NPs in plants.

5. Molecular responses of plants to nanoparticles under abiotic stress

The complex and diverse array of molecular responses triggered in plants under abiotic stresses leads to the activation of stress phytohormones, accumulation of reactive oxygen species (ROS), kinase cascades, and stimulation of ion channels, contributing to electro-physiology and signaling [95–97]. Subsequently, a sophisticated species and stress-specific regulation of gene networks occur, targeting defense and repair mechanisms against stress-induced injuries [98]. Stress-induced proteins play roles in ROS neutralization, mitogen-activated protein (MAP) kinase, and salt overly sensitive (SOS) kinase signaling cascades [96, 99]. Recent studies reveal functional outcomes extending to ion and water transport, uptake, and transcription regulation for secondary response elements (**Figure 4**) [100, 101].

Nanoparticles (NPs) have proven to be useful in triggering these molecular responses when they are applied to plants facing abiotic stress conditions. Noteworthy NPs include aluminum oxide (Al_2O_3) NPs, silicon-NPs, zinc oxide (ZnO) NPs, multiwalled carbon nanotubes, and titanium dioxide (TiO_2) NPs. Research by Zhao et al. underscores the potential of nanomaterials like silver and copper oxide, inducing stress responses and defense mechanisms and enhancing crop stress resilience [102]. Proteomic and transcriptomic approaches have helped to investigate the molecular effects of NPs on various plant species, with observed morphological and physiological effects influenced by NP dosage, type, size, and shape.

5.1 NPs mitigate salinity stress in plants

Nanoparticles (NPs) play a crucial role in mitigating salinity stress in plants by influencing cellular mechanisms and gene expression [103]. Studies on micro-RNA expression in NP-treated cells revealed the impact of NPs on miR398

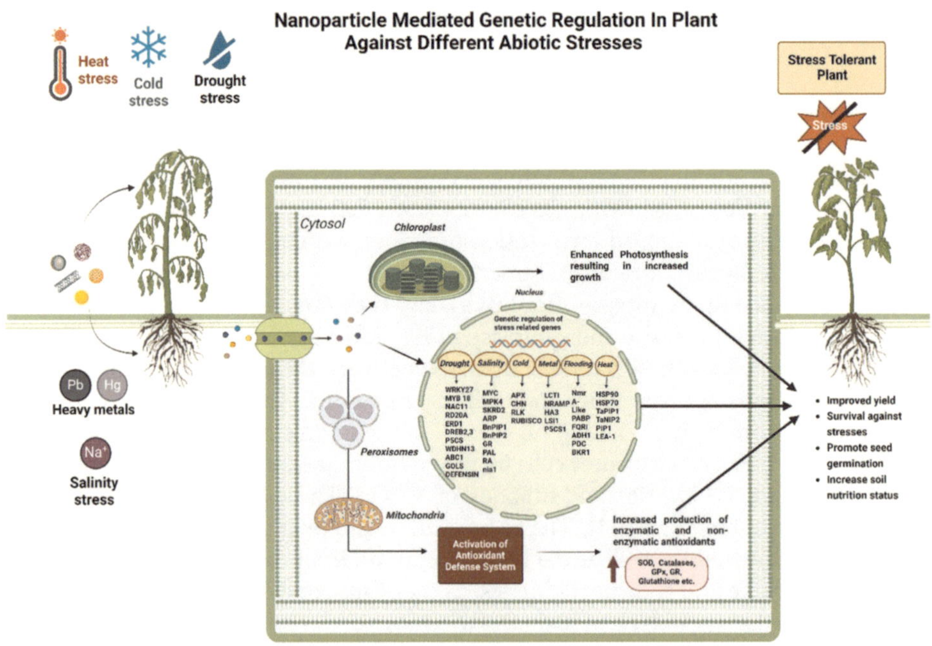

Figure 4.
Schematic pathway of genetic regulation in plants against different abiotic stresses upon application of different types of nanoparticles. (The figure is created using BioRender https://biorender.com/).

and miR408, crucial for regulating seed germination, root and seedling growth, antioxidants, and free radical scavengers [104, 105]. Foliar application of Zn-NPs on rapeseed plants under salinity stress altered gene expression, reducing some genes (e.g., *SKRD2, MYC,* and *MPK4*) and increasing others (e.g., *ARP* and *MPK*) related to physiological, hormonal, and developmental responses [106].

Carbon-based NPs, specifically multi-walled carbon nanotubes [107], affected genes involved in antioxidant and salt sensitivity systems in rapeseed plants under salinity stress [108]. According to Zhao et al., under salinity stress, multi-walled carbon nanotubes can change gene expression, affecting the antioxidant and salt overly sensitive 1 (SOS1) system in rapeseed plants [109]. Cerium oxide NPs decreased ROS levels and increased calcium content in plants, influencing ROS and Ca^{2+}-mediated signaling genes [110]. Silicon-NPs demonstrated positive effects on *Cannabis sativa* L., improving growth and inducing molecular changes [111]. Proteomic studies in tomato plants under salinity stress revealed that silicon affected cytochrome b6f (Cytb6f) light-harvesting complexes and ATP-synthase complex genes [112]. Copper-NPs enhanced tomato growth under salt stress by upregulating SOD and jasmonic acid genes, contributing to salinity tolerance [113]. The role of Cu-NPs in reducing the harmful impacts of salinity on plants was explored, showing potential in imparting salinity stress tolerance [114]. While AgNPs hold promise for crop improvement, their environmental and health impacts, particularly the release of toxic silver ions (Ag+), require careful consideration and further research [108, 115]. Overall, NPs offer a promising

Abiotic stress	Nanoparticles	Plant species	Genes	Gene regulation and plant response	References
Drought	Fe, Cu, Co, and ZnO	*Glycine L. max*	*WRKY27*	Increased expression, regulates ABA biosynthesis and signaling pathways involving ABA hormones	[116]
	Fe, Cu, Co, and ZnO	*Glycine L. max*	*MYB118*	improve expression in leaves, Fe-NP exhibits improved expression in both roots and leaves, ↑ drought tolerance by upregulating drought-related genes and, subsequently, ↓ ROS levels, regulating osmolytes, and increasing flavonoid biosynthesis.	[116]
	Fe, Cu, Co, and ZnO	*Glycine L. max*	*NAC11*	Increased expression in roots with all NPs, regulate gene expression by activating transcription factors like DREB2 or ERD1.	[116]
	Fe, Cu, Co, and ZnO	*Glycine L. max*	*RD20A*	↑ Expression in roots with Fe- or Co-NP, part of the ABA-dependent pathway, high expression in roots results in drought tolerance.	[116]
	Fe, Cu, Co, and ZnO	*Glycine L. max*	*ERD1*	↑ Expression in both root and leaves, functions in the ABA-independent pathway, ↑ Expression in the cascade of genes acting directly in response to abiotic stress.	[116]
	Fe, Cu, Co, and ZnO	*Glycine L. max*, *Cucumis melo L.*	*DREB2*, *DREB3*	Increased expression in roots with Fe, Cu, and ZnO-NPs, regulate ABA-independent pathways & improve drought tolerance.	[116, 117]
	ZnO	*Triticum aestivum*	*CAT1*	Upregulation of CAT1 in leaves, ↑ CAT activity, ↑ stress tolerance.	[118]
	ZnO	*Triticum aestivum*	*P5CS*	↑ Expression of proline biosynthesis in leaves, improves drought stress.	[118]
	ZnO	*Triticum aestivum*	*Wdhn13*	↑ Expression in leaves encodes LEA D-11 DHN (dehydrin), part of the ABA-dependent drought tolerance pathway.	[118, 119]
Drought	Silica	*Triticum aestivum*	*ABC1*	↑ Expression in leaves resulting in ↑ expression of plasma membrane-localized transporters	[119]

Abiotic stress	Nanoparticles	Plant species	Genes	Gene regulation and plant response	References
				→ ↑ movement of substances like abscisic acid → increased tolerance.	
	Silica	*Triticum aestivum*	*CHP zinc finger protein*	↑Regulation in leaves plays a role in mitigating the effects of drought tolerance.	[119]
	ZnO	*Cucumis melo L.*	*SOD, POD, CAT, APX*	↑ Expression of antioxidant genes→ inhibit H_2O_2, resulting in improved drought tolerance.	[117]
	Chitosan-GSNO	*Glycine L. max*	*DEFENSIN*	↑ Expression in drought stress, a multi-member family of cysteine-rich proteins having strong antimicrobial activity and play a key role in protection against drought.	[120]
	Chitosan-GSNO	*Glycine L. max*	*GOLS*	Increased expression, early response gene of drought stress	[120]
Salinity	Fe_2O_3	*Dracocephalum moldavica*	*PAL, TAT* genes	Over expression of PAL and TAT genes, ↑production of RA that showed a positive correlation with MDA and scavenging of DPPH radicals.	[121]
	Chitosan	*Catharanthus roseus*	*MAPK3, GS*, and *ORCA3*	High expression, improve the plant growth and development under salinity by increasing the alkaloid contents in the leaf and roots.	[122]
	BioSeNPs	*Brassica napus*	*BnPIP1-1* and *BnPIP2-1*	Over expression of BnPIP1-1 and BnPIP2-1, ↑ROS scavenging attributes (enzymatic and non-enzymatic antioxidants) and enhanced water potential.	[123]
	PEC	*Gossypium hirsutum*	*DEGs*	Upregulation of DEGs, 46% decreased ROS accumulation in roots, improved plant germination and morphological traits	[110]
	Fe_3O_4 and ZnO	*Solanum lycopersicum*	*APX, SOD,* and *GR*	↑expression of APX and SOD genes. ↓ Regulation of the GR gene which activated the antioxidants against salinity stress.	[124]
	Se	*Melissa officinalis*	*PAL* and *RA*	Overexpression of PAL and RA genes, increased activity of SOD, CAT, and POD.	[125]
Salinity	PNC	*Oryza sativa*	*nia1*	↑Expression of nia 1, increase the production of nitric oxide up to 34%, ion of nia1 gene.	[126]

Abiotic stress	Nanoparticles	Plant species	Genes	Gene regulation and plant response	References
	ZnO	*Brassica napus*	*MYC, MPK4, SKRD2,* and *ARP*	Downregulation of MYC, MPK4, and SKRD2 genes. 44% upregulation of ARP mRNA expression, which resulted in improved plant physiology and morphology.	[106]

Table 2.
Genetic regulation in plant upon application of nanoparticles in drought and salinity stress.

avenue for enhancing plant resilience to salinity stress, but comprehensive understanding and caution are essential for their safe and effective use in agriculture (**Table 2**).

5.2 NPs alleviate the impact of drought stress in plants

Sessile organisms such as plants are constantly exposed to a range of abiotic elements such as drought, resulting in a substantial decrease in yield. Drought stress impairs the photosynthesis, nutrient uptake, osmotic, and antioxidant activities of plants. Photorespiration can lead to overproduction of reactive oxygen species (ROS) in drought-stressed plants, resulting in denaturation of proteins, DNA damage, and lipid peroxidation, which hinders cell growth and elongation, resulting in poor plant growth and productivity [127].

Recent studies have reported the role of metal-based and carbon-based NPs in mitigating drought stress by inducing tolerance. Metal-based NPs, such as ZnO, TiO_2, Fe, and Cu-NPs, and carbon-based NMs such as graphene, fullerene, fullerol, and carbon NTs, have been extensively used to control drought stress by increasing water and nutrient uptake through stress tolerance and upregulation of genes involved in cell growth [128, 129].

The expression of the *P5CS* gene enhances a plant's ability to withstand diverse environmental stress, including both biotic and abiotic challenges, due to its role in encoding proline biosynthesis [130]. In response to various stress conditions, *MAPK2*, a member of the MAP kinase gene family, plays a pivotal role in regulating phytohormones and antioxidant defense mechanisms, working in tandem with Ca21 and ROS [131]. Transcriptional regulators *AREB/ABF* are crucial for overseeing the *AREB* gene, which controls abscisic acid and is vital for fostering resilience against demanding environments like drought and salt stress [131, 132]. The reduction in *ZFHD* gene expression mitigates the adverse impacts of salt and drought stress, governed by the abscisic acid biosynthesis pathway. Conversely, lowering the expression of the *TAS14* gene alleviates osmotic pressure and enhances solute aggregation, including K1 and sugars, thereby strengthening a plant's resistance to drought and salt stress [131]. Applying AgNPs (5 and 10 mg/L) to rapeseed plants brings about changes in the metabolic pathways of glucosinolate and phenolic-related genes, linked to both biotic and abiotic stresses, while suppressing carotenoid genes [133]. The use of Ag and Ag1-NPs on Arabidopsis plants induces the overexpression of genes related to oxidative stress and metal response, alongside the downregulation of ethylene and auxin-related genes [131]. Notably, three of the genes overexpressed by AgNPs are involved in thalianol biosynthesis, contributing to the plant's antioxidant defense mechanism (**Table 2**).

5.3 NPs mitigate extreme temperature effects in plants

The tolerance of plants to heat and cold stress involves the regulation of numerous genes, transcription factors, and proteins. The expression levels of these components, whether upregulated or downregulated, directly impact the plant's ability to survive under challenging conditions. For instance, in rice, the foliar application of ZnO-NPs induced the expression of genes related to the antioxidative system and chilling-response transcription factors in chilling-treated seedlings [134]. In soybeans, ZnO-NPs led to the upregulation of various genes, including *EREB, R2R3MYB, HSF-34, WRKY1, MAPK1, HDA3,* and *CAT*, resulting in increased photosynthetic pigments, proline concentration, antioxidant enzyme activity, and plant yields [135]. Cu-based NPs of 50 nm size were found to modulate genes associated with oxidative stress, brassinosteroid biosynthesis, and root formation [136]. Additionally, 40 nm-sized Cu nanoparticles in wheat were observed to accumulate secondary metabolites involved in cell signaling and defense responses [137]. The application of silicon in wheat induced the overexpression of aquaporin genes under heat stress, leading to increased relative water content [138]. In Arabidopsis thaliana, ZnO-NPs enhanced the alleviation of heat stress-induced genes, while nano-anatase in maize increased the concentration and activity of Rubisco activase, improving photosynthetic carbon production rates [139]. In maize seedlings, the application of lanthanum oxide (La_2O_3) affected the expression of aquaporin genes at root tips [140]. The varied effects of different NPs involve the regulation of numerous cold and heat stress-inducing/regulatory transcription factors and genes, ultimately enhancing plant stress tolerance through physiological, molecular, and biochemical modifications (**Table 3**) [130].

Stress	NPs	Plant	Genes	Major regulated genes	References
Cold Stress	TiO_2	*Cicer arietinum L*	*Rubisco*	Upregulation of LRubisco; SRubisco; chlorophyll a/b-binding protein; PEPC.	[82]
Cold Stress	TiO_2	*Cicer arietinum L*	*APX, CHN, RLK*	Upregulation of genes, Differential expression of genes related to cellular defense, cell signaling transcriptional regulation, and chromatin modification.	[141]
Heat Stress	CeO_2	*Zea mays*	*HSP70*	Upregulation of Heat Shock Proteins (HSP70) and excessive production of hydrogen peroxide H_2O_2.	[142]
Heat Stress	MWCNTs	*Solanum lycopersicum*	*HSP90*	Upregulation of HSP90	[143]
Flooding	AL_2O_3	*Glycine L. max*	*Nmr A-Like, PABP, FQRI.*	Differential expression of proteins related to protein synthesis/degradation; glycolysis; lipid metabolism. Upregulation of *Nmr A-Like, PABP2.* Downregulation of *FQRI.*	[144]
Flooding	Ag	*Glycine L. max*	*ADH1, PDC*	Differential expression of root proteins for cell signaling and cell metabolism.	[145]
Flooding	Ag	*Glycine L. max*	*BKR1*	Upregulation of proteins related to protein metabolism; cell division/ organization; amino acid metabolism; *BKR1*	[146]

Stress	NPs	Plant	Genes	Major regulated genes	References
Cd Stress	Si	*Oryza sativa*	*LCTI, NRAMP, HA3 and LSI1*	Downregulation of *LCTI* and *NRAMP5* Upregulation of *HA3* and *LSI1*	[147]
Fluoride stress	Ag	*Cajanus cajan*	*P5CS1*	Downregulation of NADPH oxidase and *P5CSI*	[148]

Table 3.
Genetic regulation in plant upon application of nanoparticles in different abiotic stresses.

5.4 NP intervention in alleviating the adverse effects of heavy metals in plants

Cong et al. outlined the impact of silicon nanoparticles (Si-NPs) in diminishing the absorption and toxicity of cadmium (Cd) in rice [149]. Si-NPs inhibit genes linked to the transportation and uptake of Cd from the root to the shoot, specifically targeting low-affinity cation transporter (*LCT1*) and natural resistance-associated macrophage protein 5 (*NRAMP5*). Upregulation is observed in genes related to the transport of Cd into vacuoles, such as heavy metal ATPase 3 (*HMA3*) and the silicon uptake gene, low silicon rice 1 (LSI1). By using Si-NPs, silicon is enhanced from the roots, concurrently suppressing Cd uptake. In the study by Ahmed et al., the Cd transporter genes, including *OsHMA2, OsHMA3,* and *OsLCT1*, were identified [13]. The application of iron oxide nanoparticles (FeO-NPs) and hydrogel NPs substantially decreased the expression of these genes in rice. The discussion also delves into the natural resistance-associated macrophage protein (*NRAMP*) gene family, which is responsible for heavy metal transport in various plant species like rice, potato, pepper, tomato, Arabidopsis, and soybean [150]. Si-NPs treatment leads to the downregulation of Cd uptake and transport genes, fostering wheat growth and mitigating heavy metal stress. Additionally, nanoscale zero-valent iron (nZVI) has been recognized for its ability to alleviate the accumulation of heavy metals in plants such as rice seedlings and facilitate growth by downregulating genes (*IRT1, IRT2, YSL2,* and *YSL15*) responsible for iron and cadmium uptake [151]. Venkatachalam et al. observed notable genomic alterations in plants treated with zinc oxide nanoparticles (ZnONPs), which manifested as the emergence of new DNA bands and/or the absence of standard bands in the random amplified polymorphic DNA (RAPD) pattern under cadmium (Cd) and lead (Pb) stress [152]. Detailed information about gene activation is listed in (**Table 3**).

5.5 Flood stress mitigated by NP in crop plants

Flooding induces severe oxygen depletion in plant roots, leading to significant physiological changes [153]. In a study using *Glycine max* L. (soybeans) exposed to Ag-NPs and Al_2O_3-NPs during flooding stress, alterations in the proteome and transcriptome were observed. The analysis revealed changes in 172 proteins, with 107 being root proteins. NPs influenced the expression of genes related to stress response, including downregulation of *FQR1, ADH,* and *PDC*, and upregulation of *NmrA-Like* [144]. Al_2O_3-NPs showed better alleviation of flood stress compared to Al-NPs. NPs affected energy metabolism, glycolysis, protein, and lipid metabolism during flooding. Notably, glyoxalase II downregulation and reduced *PDC* expression suggested a potential role of NPs in reducing cytotoxicity and alleviating anaerobic stress. NPs also influenced lipid metabolism, with *BKR1* expression increased during flooding and decreased during the recovery period [146]. The upregulation of translation initiation by *PAB2* and the antioxidant activity of *NmrA-like* family proteins contributed to stress alleviation. These effects were observed specifically with NPs of 15 nm size, highlighting the size-dependent impact on differentially expressed protein classes (**Table 3**) [146].

6. Nanotechnology for sustainable agriculture in crop plants

The integration of nanotechnology in agriculture offers a sustainable approach, providing a cornerstone for resilient and resource-efficient crop production in the face of evolving environmental challenges. Nano-enabled technologies, ranging from nanofertilizers to genetic engineering, present a paradigm shift in promoting plant growth and resilience [154]. The superior attributes of nanoparticles (NPs), including a high surface-area-to-volume ratio and efficient nutrient supply, distinguish them from conventional fertilizers [155]. Studies demonstrate the potential of NPs, such as iron oxide (Fe_3O_4) NPs and nano-silicon (Si), in enhancing crop growth under diverse stress conditions, encompassing heavy metal-contaminated soil and drought stress [156, 157]. Notably, green-synthesized silver nanoparticles (AgNPs) enhance antimicrobial activity in *Tephrosia apollinea* under drought stress [158], mitigating membrane damage and elevating hydrogen peroxide (H_2O_2) content in plant roots, thereby exemplifying the multifaceted benefits of NPs in sustainable agriculture [159].

Moreover, examples such as cerium oxide (CeO) NPs preserving photosystem yield [58] and titanium dioxide (TiO_2) NPs boosting antioxidant enzyme activity underscore the diverse avenues through which NPs positively influence plant health [160, 161]. Leveraging the stress ameliorative potential of NPs, particularly through green synthesis methods, emerges as a promising tool to counteract negative effects induced by abiotic stresses in various crops [162].

6.1 Applications of nanotechnology for crop protection

In the context of plant abiotic stress management, nanotechnology can play a role in enhancing plant tolerance to environmental stressors. Here are some applications of this emerging technology that have been studied for their potential to mitigate abiotic stress in plants:

6.1.1 Nanofertilizers

Nanofertilizers are nutrients encapsulated or coated within nanomaterial to enable controlled release, and its subsequent slow diffusion into the soil. These fertilizers help to restore degraded soils, manage soil health, improve seed germination, enhance plant growth, and increase crop yield like nitrogen nanoparticles designed to improve the efficiency of nitrogen delivery to plants. Phosphorus nanoparticles enhance the availability and uptake of phosphorus by plants. Potassium nanoparticles improve the delivery of potassium to plants for better growth and stress resistance [155, 163].

6.1.2 Nanomediators

They act as efficient carriers for targeted delivery of genetic material, regulating gene expression to improve stress response in plants. These nanoparticles can also aid in nutrient management by facilitating the controlled release and mitigation of persistent organic pollutants, Pesticides, and HM. Facilitate the targeted delivery of genetic material for gene therapy [164].

6.1.3 Nanopesticides

Nanopesticides revolutionize agricultural practices by offering targeted and controlled release of pesticides. This reduces environmental impact by minimizing pesticide usage while effectively managing both target and non-target plant pests. Additionally, nanopesticides contribute to plant disease management, enhancing crop health and ultimately improving overall yields. As a result, these innovative formulations showcase potential advancements in sustainable and efficient pest control strategies for modern agriculture [165, 166].

6.1.4 Nanosensors

Nanosensors, incorporating nanoparticles, play a pivotal role in detecting biological molecules and assessing environmental conditions. In agriculture, they help monitor heavy metal levels in soil, ensuring optimal plant growth and preventing contamination. Nanosensors also enhance the sensitivity and selectivity of gas sensors, allowing for precise detection of gases related to environmental health. Their versatility extends to identifying pollutants and contaminants in the environment, offering a promising avenue for real-time, high-precision environmental monitoring and management [167].

6.1.5 Nanomaterials

Nanomaterials such as carbon nanotubes (CNTs), graphene nanoparticles, and metal nanoparticles are employed to enhance plant resilience to various environmental stresses. In the context of salinity stress, these nanomaterials have shown promise in increasing plant tolerance by regulating ion balance and minimizing salt-induced damage. Additionally, they play a crucial role in reducing the bioaccumulation of heavy metals in plants and mitigating environmental contamination [168, 169].

7. Challenges and risks associated with nanobiotechnology

Although nanobiotechnology harbors immense potential for future agricultural technology revolutions, particularly considering climate change and growing populations, it also presents numerous challenges and associated risks. This requires more study and research on nanomaterials and their properties, as they behave very differently than in bulk form. Some materials are suspected of being toxic at the nanoscale due to significant, yet unknown, hazardous properties related to their unique physiochemical features. This can put manufacturers, formulators, handlers, applicators, and consumers at greater risk. Thus, nanoparticles can cause heterogeneous effects as they can leave toxic effects on non-target organisms upon contact, i.e., nanoparticles can come into direct contact with humans and cause toxic effects on them. The hazardous effects of nanoparticles on non-target organisms include dermal absorption of nanoparticles and translocation to go deep into the lungs and brain through inhalation and crossing the brain barriers, respectively.

Nanoparticles can enter various human body parts and cause harm by disrupting cellular pathways, enzymatic actions, and organ functions. Nanomaterial disposal could result in the formation of a new class of non-biodegradable pollutants in the environment [170, 171]. The potential risks associated with nanomaterials lead to the challenges of basic and applied nanobiotechnology in different fields including crop production. The major possible challenges are the following [170, 171]:

1. High-quality mass production of nano-based products at an economical cost.

2. The development of a personalized nanomaterial production system to fulfill local needs.

3. Environmental and human health safety and protection during the use and disposal of nanomaterials.

4. The challenge is to overcome the gap between basic and applied nano-based research.

5. The production cost, severe risks, and practical knowledge gaps are also considered as chief concerns/challenges in nanobiotechnology applications.

Hence, a collaborative effort, based on comprehensive scientific research and technological advancements, between regulatory bodies and workforce research should be made to cater to the above-mentioned challenges and associated risks. This would help to provide the necessary information to devise appropriate guidelines for comprehensive risk management and applications of nanobiotechnology for sustainable crop production.

8. Conclusion and future prospects

The impact of climatic variations in recent years has contributed significantly to reduce crop productivity and gave birth to various environmental stress factors. Abiotic stresses are the key player responsible for reduced crop yield, quality and stunted plant growth around the globe. Efforts to cater to challenges associated with abiotic stress led to the development of nanobiotechnology, which encompasses small-size nanoparticles that aid in enhancing the farming sector and managing global food demand. Several NPs, such as TiO_2, SiO_2, and AgNPs, can reduce the negative effects of abiotic stress by activating the genetic makeup of plant defense mechanisms via the induction of ROS production and phytotoxicity. Their small size allows them to penetrate easily into plant tissues, after which they positively influence plant morphological, physiological, and biochemical processes, promote plant development, and improve crop productivity in plants under various abiotic stresses. Furthermore, NPs have a large surface area that improves the absorption and delivery of various targeted nutrients. Despite the positive effects of nanoparticles, there are some risks and challenges associated with them that need to be addressed. These risks will aid in exploring their dimensions and help to improve crop productivity by understanding the underlying interaction pathways at the molecular level.

It is important to conduct future research exploring different uses of nanoparticles to enhance crop plant productivity. Nanoenzymes including Mn_3O_4-NPs gain much attention due to their potential role in mitigating stress conditions in plants. Mechanisms responsible for nanopriming-induced seed germination, breaking seed dormancy, and their interactions with seeds need to be investigated. A better understanding of plant–NP interaction will enable researchers to design tailor-made nanomaterials targeting agricultural challenges. Moreover, targeted delivery approaches using nanomaterials will prove to be another emerging way to enhance plant productivity. Recently, a study conducted with this approach used nanomaterials to convert chloroplasts into a "chloroplast factory," resulting in improved plant photosynthesis under low light conditions [172]. In addition to that, the use of nanomaterial CRISPR-Cas genome editing in cargo delivery will also improve the efficiency of genetic engineering to enhance plant stress tolerance [173]. However, the management of biosafety hazards associated with the use of nanoparticles demands the development of strict policies and procedures so that we can utilize and benefit from their potential role in enhancing crop tolerance to abiotic stresses.

Abbreviations

HM	heavy metal
Cd	cadmium
Pb	lead
As	arsenic
Ni	nickel
Cr	chromium
Co	cobalt
Zn	zinc
ROS	reactive oxygen species
NPs	nanoparticles
CuO	copper oxide
DNA	deoxyribose nucleic acid
RNA	ribose nucleic acid
UV	ultraviolet
PS	photosystem
TiO2	titania
CAT	catalase
GPOX	glutathione peroxidase
SOD	superoxide dismutase
MAP	mitogen-activated protein
SOS	salt overly sensitive
ZnO	zinc oxide
MAPK	mitogen-activated protein kinase
Myc	myelocytomatosis oncogene
ARP	auxin-repressed protein
EREB	ethylene responsive element binding protein
HSF	heat shock transcription factor
LCT1	low-affinity cation transporter
NRAMP5	natural resistance-associated macrophage protein 5
nZVI	nanoscale zero-valent iron
RAPD	random amplified polymorphic DNA
CeO	cerium oxide

Authors contribution

All authors have contributed equally to writing and drafting this article. Hafiza Fizzah Zulfiqar also contributed to designing the images in this article.

Conflict of interest

There are no conflicts of interest among the authors.

Author details

Hafiza Fizzah Zulfiqar[1*], Bakht Afroze[2], Sana Shakoor[1], Muhammad Saad Bhutta[1], Muhammad Ahmed[1], Sameera Hassan[1], Fatima Batool[1] and Bushra Rashid[1]

1 Centre of Excellence in Molecular Biology, University of the Punjab, Lahore, Pakistan

2 University of Melbourne, Melbourne, Australia

*Address all correspondence to: fizzah.zulfiqar@cemb.edu.pk

References

[1] Khan M, Khan AU, Hasan MA, Yadav KK, Pinto MM, Malik N, et al. Agro-nanotechnology as an emerging field: A novel sustainable approach for improving plant growth by reducing biotic stress. Applied Sciences. 2021;**11**(5):2282

[2] Ashraf J, Javed A. Food security and environmental degradation: Do institutional quality and human capital make a difference? Journal of Environmental Management. 2023;**331**: 117330

[3] Gustavsson J, Cederberg C, Sonesson U, Van Otterdijk R, Meybeck A. Global Food Losses and Food Waste. Rome: FAO; 2011

[4] Khalid MF, Hussain S, Ahmad S, Ejaz S, Zakir I, Ali MA, et al. Impacts of abiotic stresses on growth and development of plants. In: Plant Tolerance to Environmental Stress. Florida: CRC Press; 2019. pp. 1-8

[5] Bedair H, Ghosh S, Abdelsalam IM, Keerio AA, AlKafaas SS. Potential implementation of trees to remediate contaminated soil in Egypt. Environmental Science and Pollution Research. 2022;**29**(52):78132-78151

[6] Salem HM, Ismael E, Shaalan M. Evaluation of the effects of silver nanoparticles against experimentally induced necrotic enteritis in broiler chickens. International Journal of Nanomedicine. 2021;**16**: 6783-6796

[7] Singhal P, Jan AT, Azam M, Haq QMR. Plant abiotic stress: A prospective strategy of exploiting promoters as alternative to overcome the escalating burden. Frontiers in Life Science. 2016;**9**(1):52-63

[8] Hafez Y, Attia K, Alamery S, Ghazy A, Al-Doss A, Ibrahim E, et al. Beneficial effects of biochar and chitosan on antioxidative capacity, osmolytes accumulation, and anatomical characters of water-stressed barley plants. Agronomy. 2020;**10**(5):630

[9] Alabdallah NM, Hasan MM. Plant-based green synthesis of silver nanoparticles and its effective role in abiotic stress tolerance in crop plants. Saudi Journal of Biological Sciences. 2021;**28**(10):5631-5639

[10] Hashem AH, Abdelaziz AM, Askar AA, Fouda HM, Khalil AM, Abd-Elsalam KA, et al. Bacillus megaterium-mediated synthesis of selenium nanoparticles and their antifungal activity against *Rhizoctonia solani* in faba bean plants. Journal of Fungi. 2021;**7**(3): 195

[11] Rubilar O, Rai M, Tortella G, Diez MC, Seabra AB, Durán N. Biogenic nanoparticles: Copper, copper oxides, copper sulphides, complex copper nanostructures and their applications. Biotechnology Letters. 2013;**35**: 1365-1375

[12] Adrees M, Khan ZS, Ali S, Hafeez M, Khalid S, et al. Simultaneous mitigation of cadmium and drought stress in wheat by soil application of iron nanoparticles. Chemosphere. 2020;**238**:124681

[13] Ahmed T, Noman M, Manzoor N, Shahid M, Abdullah M, Ali L, et al. Nanoparticle-based amelioration of drought stress and cadmium toxicity in rice via triggering the stress responsive genetic mechanisms and nutrient acquisition. Ecotoxicology and Environmental Safety. 2021;**209**: 111829

[14] Seabra AB, Pasquôto T, Ferrarini ACF, Santos MC, Haddad PS, de Lima R. Preparation, characterization, cytotoxicity, and genotoxicity evaluations of thiolated-and S-nitrosated superparamagnetic iron oxide nanoparticles: Implications for cancer treatment. Chemical Research in Toxicology. 2014;**27**(7):1207-1218

[15] Wang X, Li Q, Pei Z, Wang S. Effects of zinc oxide nanoparticles on the growth, photosynthetic traits, and antioxidative enzymes in tomato plants. Biologia Plantarum. 2018;**62**:801-808

[16] Kumari M, Khan SS, Pakrashi S, Mukherjee A, Chandrasekaran N. Cytogenetic and genotoxic effects of zinc oxide nanoparticles on root cells of *Allium cepa*. Journal of Hazardous Materials. 2011;**190**(1–3):613-621

[17] Gull A, Lone AA, Wani NUI. Abiotic and Biotic Stress in Plants. London: IntechOpen; 2019. p. 174. DOI: 10.5772/intechopen.77845

[18] Fahad S, Bajwa AA, Nazir U, Anjum SA, Farooq A, Zohaib A, et al. Crop production under drought and heat stress: Plant responses and management options. Frontiers in Plant Science. 2017; **8**:1147

[19] Yadav S, Modi P, Dave A, Vijapura A, Patel D, Patel M. Effect of abiotic stress on crops. Sustainable Crop Production. 2020;**3**:1-22

[20] Berni R, Luyckx M, Xu X, Legay S, Sergeant K, Hausman J-F, et al. Reactive oxygen species and heavy metal stress in plants: Impact on the cell wall and secondary metabolism. Environmental and Experimental Botany. 2019;**161**: 98-106

[21] Huihui Z, Xin L, Zisong X, Yue W, Zhiyuan T, Meijun A, et al. Toxic effects of heavy metals Pb and Cd on mulberry (*Morus alba* L.) seedling leaves: Photosynthetic function and reactive oxygen species (ROS) metabolism responses. Ecotoxicology and Environmental Safety. 2020;**195**: 110469

[22] Jayakumar M, Surendran U, Raja P, Kumar A, Senapathi V. A review of heavy metals accumulation pathways, sources and management in soils. Arabian Journal of Geosciences. 2021;**14**: 1-19

[23] Li C, Zhou K, Qin W, Tian C, Qi M, Yan X, et al. A review on heavy metals contamination in soil: Effects, sources, and remediation techniques. Soil and Sediment Contamination: An International Journal. 2019;**28**(4): 380-394

[24] Selvaraju R, Gommes R, Bernardi M. Climate science in support of sustainable agriculture and food security. Climate Research. 2011;**47**(1–2):95-110

[25] Albahri G, Alyamani AA, Badran A, Hijazi A, Nasser M, Maresca M, et al. Enhancing essential grains yield for sustainable food security and bio-safe agriculture through latest innovative approaches. Agronomy. 2023;**13**(7):1709

[26] Yang J, Cao W, Rui Y. Interactions between nanoparticles and plants: Phytotoxicity and defense mechanisms. Journal of Plant Interactions. 2017;**12**(1): 158-169

[27] Younis SA, Kim K-H, Shaheen SM, Antoniadis V, Tsang YF, Rinklebe J, et al. Advancements of nanotechnologies in crop promotion and soil fertility: Benefits, life cycle assessment, and legislation policies. Renewable and Sustainable Energy Reviews. 2021;**152**: 111686

[28] Singh H, Sharma A, Bhardwaj SK, Arya SK, Bhardwaj N, Khatri M. Recent advances in the applications of nano-agrochemicals for sustainable agricultural development. Environmental Science: Processes & Impacts. 2021;**23**(2):213-239

[29] Shi R, Liu W, Lian Y, Wang X, Men S, Zeb A, et al. Toxicity mechanisms of nanoplastics on crop growth, interference of phyllosphere microbes, and evidence for foliar penetration and translocation. Environmental Science & Technology. 2023;**58**(2):1010-1021

[30] Khan MR, Adam V, Rizvi TF, Zhang B, Ahamad F, Jośko I, et al. Nanoparticle–plant interactions: Two-way traffic. Small. 2019;**15**(37):1901794

[31] Anjum SA, Tanveer M, Ashraf U, Hussain S, Shahzad B, Khan I, et al. Effect of progressive drought stress on growth, leaf gas exchange, and antioxidant production in two maize cultivars. Environmental Science and Pollution Research. 2016;**23**:17132-17141

[32] Molnár Á, Rónavári A, Bélteky P, Szőllősi R, Valyon E, Oláh D, et al. ZnO nanoparticles induce cell wall remodeling and modify ROS/RNS signalling in roots of Brassica seedlings. Ecotoxicology and Environmental Safety. 2020;**206**:111158

[33] Sattelmacher B. The apoplast and its significance for plant mineral nutrition. New Phytologist. 2001;**149**(2):167-192

[34] Roberts A. Plasmodesmata and the control of symplastic transport. Plant, Cell and Environment. 2003;**26**:103-124

[35] Sun D, Hussain HI, Yi Z, Siegele R, Cresswell T, Kong L, et al. Uptake and cellular distribution, in four plant species, of fluorescently labeled mesoporous silica nanoparticles. Plant Cell Reports. 2014;**33**:1389-1402

[36] Ramírez Anguiano AC, Velasco Ramírez AP, Zamudio Ojeda A, Jiménez Torres HD, Velázquez Juárez G, Velázquez López JM, et al. Effects of Nanomaterials/Nanocomposites on Trace Element Uptake and Phytotoxicity. Nanomaterials and Nanocomposites Exposures to Plants: Response, Interaction, Phytotoxicity and Defense Mechanisms. Singapore: Springer; 2023. pp. 127-156

[37] Pérez-de-Luque A. Interaction of nanomaterials with plants: What do we need for real applications in agriculture? Frontiers in Environmental Science. 2017;**5**:12

[38] Servin A, Elmer W, Mukherjee A, De la Torre-Roche R, Hamdi H, White JC, et al. A review of the use of engineered nanomaterials to suppress plant disease and enhance crop yield. Journal of Nanoparticle Research. 2015;**17**:1-21

[39] Schönherr J. A mechanistic analysis of penetration of glyphosate salts across astomatous cuticular membranes. Pest Management Science: Formerly Pesticide Science. 2002;**58**(4):343-351

[40] Tripathi DK, Tripathi A, Shweta SS, Singh Y, Vishwakarma K, Yadav G, et al. Uptake, accumulation and toxicity of silver nanoparticle in autotrophic plants, and heterotrophic microbes: A concentric review. Frontiers in Microbiology. 2017;**8**:07

[41] Avellan A, Yun J, Zhang Y, Spielman-Sun E, Unrine JM, Thieme J, et al. Nanoparticle size and coating chemistry control foliar uptake pathways, translocation, and leaf-to-rhizosphere transport in wheat. ACS Nano. 2019;**13**(5):5291-5305

[42] Ma X, Yan J. Plant uptake and accumulation of engineered metallic nanoparticles from lab to field conditions. Current Opinion in Environmental Science & Health. 2018; **6**:16-20

[43] Pate JS. Exchange of solutes between phloem and xylem and circulation in the whole plant. In: Zimmermann MH, Milburn JA, editors. Transport in Plants I: Phloem Transport. Berlin, Heidelberg: Springer; 1975. pp. 451-473

[44] Peng C, Tong H, Shen C, Sun L, Yuan P, He M, et al. Bioavailability and translocation of metal oxide nanoparticles in the soil-rice plant system. Science of the Total Environment. 2020;**713**:136662

[45] Arruda SCC, Silva ALD, Galazzi RM, Azevedo RA, Arruda MAZ. Nanoparticles applied to plant science: A review. Talanta. 2015;**131**:693-705

[46] Pagano L, Maestri E, Caldara M, White JC, Marmiroli N, Marmiroli M. Engineered nanomaterial activity at the organelle level: Impacts on the chloroplasts and mitochondria. ACS Sustainable Chemistry & Engineering. 2018;**6**(10):12562-12579

[47] Wakeel A, Xu M, Gan Y. Chromium-induced reactive oxygen species accumulation by altering the enzymatic antioxidant system and associated cytotoxic, genotoxic, ultrastructural, and photosynthetic changes in plants. International Journal of Molecular Sciences. 2020;**21**(3):728

[48] Singh A, Kumar A, Yadav S, Singh IK. Reactive oxygen species-mediated signaling during abiotic stress. Plant Gene. 2019;**18**:100173

[49] Hasanuzzaman M, Bhuyan M, Zulfiqar F, Raza A, Mohsin SM, Mahmud JA, et al. Reactive oxygen species and antioxidant defense in plants under abiotic stress: Revisiting the crucial role of a universal defense regulator. Antioxidants (Basel). 2020; **9**(8):1-52

[50] Sham A, Al-Ashram H, Whitley K, Iratni R, El-Tarabily KA, AbuQamar SF. Metatranscriptomic analysis of multiple environmental stresses identifies RAP2. 4 gene associated with Arabidopsis immunity to *Botrytis cinerea*. Scientific Reports. 2019;**9**(1):17010

[51] Zhang B. MicroRNA: A new target for improving plant tolerance to abiotic stress. Journal of Experimental Botany. 2015;**66**(7):1749-1761

[52] Wahab A, Munir A, Saleem MH, AbdulRaheem MI, Aziz H, Mfarrej MFB, et al. Interactions of metal-based engineered nanoparticles with plants: An overview of the state of current knowledge, research progress, and prospects. Journal of Plant Growth Regulation. 2023;**42**:1-21

[53] Abid N, Khan AM, Shujait S, Chaudhary K, Ikram M, Imran M, et al. Synthesis of nanomaterials using various top-down and bottom-up approaches, influencing factors, advantages, and disadvantages: A review. Advances in Colloid and Interface Science. 2022;**300**: 102597

[54] Baig N, Kammakakam I, Falath W. Nanomaterials: A review of synthesis methods, properties, recent progress, and challenges. Materials Advances. 2021;**2**(6):1821-1871

[55] Lin HK, Yan T-H, Bashir S, Liu JL. Synthesis of nanomaterials using bottom-up methods. In: Advanced Nanomaterials and Their Applications Renewable Energy. 2022. pp. 61-110

[56] Salem SS, Fouda A. Green synthesis of metallic nanoparticles and their prospective biotechnological applications: An overview. Biological Trace Element Research. 2021;**199**: 344-370

[57] Zafar S. Synthesis, properties, and characterization of biological nanomaterial. In: Nanomaterials: The Building Blocks of Modern Technology: Synthesis, Properties and Applications. Singapore: Springer; 2023. pp. 91-114

[58] Djanaguiraman M, Nair R, Giraldo JP, Prasad PVV. Cerium oxide nanoparticles decrease drought-induced oxidative damage in sorghum leading to higher photosynthesis and grain yield. ACS Omega. 2018;**3**(10): 14406-14416

[59] Kumar V, Sharma M, Khare T, Wani SH. Impact of nanoparticles on oxidative stress and responsive antioxidative defense in plants. In: Nanomaterials in Plants, Algae, and Microorganisms. Massachusetts: Elsevier; 2018. pp. 393-406

[60] Shankramma K, Yallappa S, Shivanna M, Manjanna J. Fe_2O_3 magnetic nanoparticles to enhance *S. lycopersicum* (tomato) plant growth and their biomineralization. Applied Nano. 2016; **6**:983-990

[61] Anandaraj K, Natarajan N. Effect of nanoparticles for seed quality enhancement in onion [*Allium cepa* (Linn) cv. CO (On)] 5. International Journal of Current Microbiology and Applied Sciences. 2017;**6**: 3714-3724

[62] Sancho-Gil JM, Hernández-Hernández F. La profesión docente en la era del exceso de información y la falta de sentido. Revista de Educación a Distancia (RED). 2018;**56**:1-20

[63] Ahmed T, Noman M, Manzoor N, Shahid M, Hussaini KM, Rizwan M, et al. Green magnesium oxide nanoparticles-based modulation of cellular oxidative repair mechanisms to reduce arsenic uptake and translocation in rice (*Oryza sativa* L.) plants. Environmental Pollution. 2021;**288**:117785

[64] Salama HE, Aziz MSA. Development of active edible coating of alginate and aloe vera enriched with frankincense oil for retarding the senescence of green capsicums. Lwt. 2021; **145**:111341

[65] Rady MM, Elrys AS, Selem E, Mohsen AA, Arnaout SM, El-Sappah AH, et al. Spirulina platensis extract improves the production and defenses of the common bean grown in a heavy metals-contaminated saline soil. Journal of Environmental Sciences. 2023;**129**: 240-257

[66] Nazir MM, Noman M, Ahmed T, Ali S, Ulhassan Z, Zeng F, et al. Exogenous calcium oxide nanoparticles alleviate cadmium toxicity by reducing Cd uptake and enhancing antioxidative capacity in barley seedlings. Journal of Hazardous Materials. 2022;**438**:129498

[67] Mahmoud LM, Shalan AM, El-Boray MS, Vincent CI, El-Kady ME, Grosser JW, et al. Application of silicon nanoparticles enhances oxidative stress tolerance in salt stressed 'Valencia' sweet orange plants. Scientia Horticulturae. 2022;**295**:110856

[68] Wahid I, Rani P, Kumari S, Ahmad R, Hussain SJ, Alamri S, et al. Biosynthesized gold nanoparticles maintained nitrogen metabolism, nitric oxide synthesis, ions balance, and stabilizes the defense systems to improve salt stress tolerance in wheat. Chemosphere. 2022;**287**:132142

[69] Zhou C-Q, Lu C-H, Mai L, Bao L-J, Liu L-Y, Zeng EY. Response of rice (*Oryza sativa* L.) roots to nanoplastic treatment at seedling stage. Journal of Hazardous Materials. 2021;**401**:123412

[70] Liu J, Li G, Chen L, Gu J, Wu H, Li Z. Cerium oxide nanoparticles improve cotton salt tolerance by enabling better ability to maintain cytosolic K+/Na+ ratio. Journal of Nanobiotechnology. 2021;**19**(1):153

[71] Yasmin H, Mazher J, Azmat A, Nosheen A, Naz R, Hassan MN, et al. Combined application of zinc oxide nanoparticles and biofertilizer to induce salt resistance in safflower by regulating ion homeostasis and antioxidant defence responses. Ecotoxicology and Environmental Safety. 2021;**218**:112262

[72] Manzoor N, Ahmed T, Noman M, Shahid M, Nazir MM, Ali L, et al. Iron oxide nanoparticles ameliorated the cadmium and salinity stresses in wheat plants, facilitating photosynthetic pigments and restricting cadmium uptake. Science of the Total Environment. 2021;**769**:145221

[73] Nejatzadeh F. Effect of silver nanoparticles on salt tolerance of *Satureja hortensis* l. during in vitro and in vivo germination tests. Heliyon. 2021; 7(2):1-11

[74] Noman M, Shahid M, Ahmed T, Tahir M, Naqqash T, Muhammad S, et al. Green copper nanoparticles from a native Klebsiella pneumoniae strain alleviated oxidative stress impairment of wheat plants by reducing the chromium bioavailability and increasing the growth. Ecotoxicology and Environmental Safety. 2020;**192**: 110303

[75] Ikram M, Raja NI, Javed B, Z-u-R M, Hussain M, Hussain M, et al. Foliar applications of bio-fabricated selenium nanoparticles to improve the growth of wheat plants under drought stress. Green Processing and Synthesis. 2020; **9**(1):706-714

[76] Gohari G, Mohammadi A, Akbari A, Panahirad S, Dadpour MR, Fotopoulos V, et al. Titanium dioxide nanoparticles (TiO_2 NPs) promote growth and ameliorate salinity stress effects on essential oil profile and biochemical attributes of *Dracocephalum moldavica*. Scientific Reports. 2020; **10**(1):912

[77] Khan ZS, Rizwan M, Hafeez M, Ali S, Javed MR, Adrees M. The accumulation of cadmium in wheat (*Triticum aestivum*) as influenced by zinc oxide nanoparticles and soil moisture conditions. Environmental Science and Pollution Research. 2019;**26**: 19859-19870

[78] Alsaeedi A, El-Ramady H, Alshaal T, El-Garawany M, Elhawat N, Al-Otaibi A. Silica nanoparticles boost growth and productivity of cucumber under water deficit and salinity stresses by balancing nutrients uptake. Plant Physiology and Biochemistry. 2019;**139**:1-10

[79] Abdel Latef AAH, Srivastava AK, El-sadek MSA, Kordrostami M, Tran LSP. Titanium dioxide nanoparticles improve growth and enhance tolerance of broad bean plants under saline soil conditions. Land Degradation & Development. 2018; **29**(4):1065-1073

[80] Cao Z, Stowers C, Rossi L, Zhang W, Lombardini L, Ma X. Physiological effects of cerium oxide nanoparticles on the photosynthesis and water use efficiency of soybean (*Glycine max* (L.) Merr.). Environmental Science Nano. 2017;**4**(5):1086-1094

[81] Oliveira HC, Gomes BC, Pelegrino MT, Seabra AB. Nitric oxide-releasing chitosan nanoparticles alleviate the effects of salt stress in maize plants. Nitric Oxide. 2016;**61**:10-19

[82] Hasanpour H, Maali-Amir R, Zeinali H. Effect of TiO_2 nanoparticles on metabolic limitations to photosynthesis under cold in chickpea. Russian Journal of Plant Physiology. 2015;**62**:779-787

[83] Qi M, Liu Y, Li T. Nano-TiO_2 improve the photosynthesis of tomato leaves under mild heat stress. Biological Trace Element Research. 2013;**156**: 323-328

[84] Javed T, Afzal I, Shabbir R, Ikram K, Zaheer MS, Faheem M, et al. Seed coating technology: An innovative and sustainable approach for improving seed quality and crop performance. Journal of the Saudi Society of Agricultural Sciences. 2022;**21**(8):536-545

[85] Naseer I, Javad S, Singh A, Maqsood S, Iqbal S, Jabeen K. Alleviation Mechanism of Drought Stress in Plants Using Metal Nanoparticles–A Perspective Analysis. Sustainable Agriculture Reviews 53: Nanoparticles: A New Tool to Enhance Stress Tolerance. Cham: Springer; 2022. pp. 115-149

[86] Hong J, Wang C, Wagner DC, Gardea-Torresdey JL, He F, Rico CM. Foliar application of nanoparticles: Mechanisms of absorption, transfer, and multiple impacts. Environmental Science: Nano. 2021;**8**(5):1196-1210

[87] Sambo P, Nicoletto C, Giro A, Pii Y, Valentinuzzi F, Mimmo T, et al. Hydroponic solutions for soilless production systems: Issues and opportunities in a smart agriculture perspective. Frontiers in Plant Science. 2019;**10**:923

[88] Su Y, Ashworth V, Kim C, Adeleye AS, Rolshausen P, Roper C, et al. Delivery, uptake, fate, and transport of engineered nanoparticles in plants: A critical review and data analysis. Environmental Science: Nano. 2019; **6**(8):2311-2331

[89] Banerjee A, Sarkar A, Acharya K, Chakraborty N. Nanotechnology: An emerging hope in crop improvement. Letters in Applied NanoBioScience. 2021;**10**(4):2784-2803

[90] Guo J, Mattos BD, Tardy BL, Moody VM, Xiao G, Ejima H, et al. Porous inorganic and hybrid systems for drug delivery: Future promise in combatting drug resistance and translation to botanical applications. Current Medicinal Chemistry. 2019; **26**(33):6107-6131

[91] Rani S, Kumari N, Sharma V. Uptake, translocation, transformation and physiological effects of nanoparticles in plants. Archives of Agronomy and Soil Science. 2023;**69**(9): 1579-1599

[92] Kurczyńska E, Godel-Jędrychowska K, Sala K, Milewska-Hendel A. Nanoparticles—Plant interaction: What we know, where we are? Applied Sciences. 2021;**11**(12):5473

[93] de la Rosa G, Vázquez-Núñez E, Molina-Guerrero C, Serafín-Muñoz AH, Vera-Reyes I. Interactions of nanomaterials and plants at the cellular level: Current knowledge and relevant gaps. Nanotechnology for Environmental Engineering. 2021;**6**:1-19

[94] Jat SK, Bhattacharya J, Sharma MK. Nanomaterial based gene delivery: A promising method for plant genome engineering. Journal of Materials Chemistry B. 2020;**8**(19):4165-4175

[95] Al-Khayri JM, Rashmi R, Ulhas RS, Sudheer WN, Banadka A, Nagella P, et al. The role of nanoparticles in response of plants to abiotic stress at physiological, biochemical, and molecular levels. Plants-Basel. 2023; **12**(2):292

[96] Fraire-Velázquez S, Rodríguez-Guerra RI, Sánchez-Calderón L. Abiotic and Biotic Stress Response Crosstalk in Plants. London: IntechOpen. 2011. pp. 1-27

[97] Bari R, Jones JD. Role of plant hormones in plant defence responses. Plant Molecular Biology. 2009;**69**: 473-488

[98] Ben Rejeb I, Pastor V, Mauch-Mani B. Plant responses to simultaneous biotic and abiotic stress: Molecular mechanisms. Plants. 2014;**3**(4):458-475

[99] Scandalios J. Oxidative stress: Molecular perception and transduction of signals triggering antioxidant gene defenses. Brazilian Journal of Medical and Biological Research. 2005;**38**: 995-1014

[100] Saibo NJ, Lourenço T, Oliveira MM. Transcription factors and regulation of photosynthetic and related metabolism under environmental stresses. Annals of Botany. 2009;**103**(4):609-623

[101] Blumwald E. Sodium transport and salt tolerance in plants. Current Opinion in Cell Biology. 2000;**12**(4):431-434

[102] Zhao L, Bai T, Wei H, Gardea-Torresdey JL, Keller A, White JC. Nanobiotechnology-based strategies for enhanced crop stress resilience. Nature Food. 2022;**3**(10):829-836

[103] Etesami H, Fatemi H, Rizwan M. Interactions of nanoparticles and salinity stress at physiological, biochemical and molecular levels in plants: A review. Ecotoxicology and Environmental Safety. 2021;**225**:112769

[104] Kumar V, Guleria P, Kumar V, Yadav SK. Gold nanoparticle exposure induces growth and yield enhancement in *Arabidopsis thaliana*. Science of the Total Environment. 2013;**461**:462-468

[105] Tolaymat T, Genaidy A, Abdelraheem W, Dionysiou D, Andersen C. The effects of metallic engineered nanoparticles upon plant systems: An analytic examination of scientific evidence. Science of the Total Environment. 2017;**579**:93-106

[106] Hezaveh TA, Pourakbar L, Rahmani F, Alipour H. Interactive effects of salinity and ZnO nanoparticles on physiological and molecular parameters of rapeseed (*Brassica napus* L.). Communications in Soil Science and Plant Analysis. 2019;**50**(6):698-715

[107] Khan MN, Mobin M, Abbas ZK, AlMutairi KA, Siddiqui ZH. Role of nanomaterials in plants under challenging environments. Plant Physiology and Biochemistry. 2017;**110**: 194-209

[108] Zulfiqar F, Ashraf M. Nanoparticles potentially mediate salt stress tolerance in plants. Plant Physiology and Biochemistry. 2021;**160**:257-268

[109] Zhao G, Zhao Y, Lou W, Su J, Wei S, Yang X, et al. Nitrate reductase-dependent nitric oxide is crucial for multi-walled carbon nanotube-induced plant tolerance against salinity. Nanoscale. 2019;**11**(21):10511-10523

[110] An J, Hu P, Li F, Wu H, Shen Y, White JC, et al. Emerging investigator series: Molecular mechanisms of plant salinity stress tolerance improvement by seed priming with cerium oxide

nanoparticles. Environmental Science: Nano. 2020;**7**(8):2214-2228

[111] Guerriero G, Sutera FM, Torabi-Pour N, Renaut J, Hausman J-F, Berni R, et al. Phyto-courier, a silicon particle-based nano-biostimulant: Evidence from Cannabis sativa exposed to salinity. ACS Nano. 2021;**15**(2):3061-3069

[112] Muneer S, Park YG, Manivannan A, Soundararajan P, Jeong BR. Physiological and proteomic analysis in chloroplasts of *Solanum lycopersicum* L. under silicon efficiency and salinity stress. International Journal of Molecular Sciences. 2014;**15**(12):21803-21824

[113] Iqbal MN, Rasheed R, Ashraf MY, Ashraf MA, Hussain I. Exogenously applied zinc and copper mitigate salinity effect in maize (*Zea mays* L.) by improving key physiological and biochemical attributes. Environmental Science and Pollution Research. 2018;**25**: 23883-23896

[114] Hernández-Hernández H, Juárez-Maldonado A, Benavides-Mendoza A, Ortega-Ortiz H, Cadenas-Pliego G, Sánchez-Aspeytia D, et al. Chitosan-PVA and copper nanoparticles improve growth and overexpress the SOD and JA genes in tomato plants under salt stress. Agronomy. 2018;**8**(9):175

[115] Tortella G, Rubilar O, Durán N, Diez M, Martínez M, Parada J, et al. Silver nanoparticles: Toxicity in model organisms as an overview of its hazard for human health and the environment. Journal of Hazardous Materials. 2020; **390**:121974

[116] Linh TM, Mai NC, Hoe PT, Lien LQ, Ban NK, Hien LTT, et al. Metal-based nanoparticles enhance drought tolerance in soybean. Journal of Nanomaterials. 2020;**2020**:4056563

[117] Rehman A, Weng J, Li P, Shah IH, Rahman S, Khalid M, et al. Green synthesized zinc oxide nanoparticles confer drought tolerance in melon (*Cucumis melo* L.). Environmental and Experimental Botany. 2023;**212**:105384

[118] Sadati SYR, Godehkahriz SJ, Ebadi A, Sedghi M. Zinc oxide nanoparticles enhance drought tolerance in wheat via physio-biochemical changes and stress genes expression. Iranian Journal of Biotechnology. 2022;**20**(1): e3027

[119] Boora R, Rani N, Kumari S, Yashveer S, Kumari N, Grewal S. Efficacious role of silica nanoparticles in improving growth and yield of wheat under drought stress through stress-gene upregulation. Plant Nano Biology. 2023; **6**:100051

[120] Methela NJ, Pande A, Islam MS, Rahim W, Hussain A, Lee D-S, et al. Chitosan-GSNO nanoparticles: A positive modulator of drought stress tolerance in soybean. BMC Plant Biology. 2023;**23**(1):639

[121] Zia-ur-Rehman M, Anayatullah S, Irfan E, Hussain SM, Rizwan M, Sohail MI, et al. Nanoparticles assisted regulation of oxidative stress and antioxidant enzyme system in plants under salt stress: A review. Chemosphere. 2023;**314**:137649

[122] Hassan F, Ali E, Gaber A, Fetouh M, Mazrou R. Chitosan nanoparticles effectively combat salinity stress by enhancing antioxidant activity and alkaloid biosynthesis in *Catharanthus roseus* (L.) G. Don. Plant Physiology and Biochemistry. 2021;**162**:291-300

[123] El-Badri AM, Batool M, Mohamed IA, Wang Z, Wang C, Tabl KM, et al. Mitigation of the salinity stress in rapeseed (Brassica napus L.)

productivity by exogenous applications of bio-selenium nanoparticles during the early seedling stage. Environmental Pollution. 2022;**310**:119815

[124] Aazami MA, Rasouli F, Ebrahimzadeh A. Oxidative damage, antioxidant mechanism and gene expression in tomato responding to salinity stress under in vitro conditions and application of iron and zinc oxide nanoparticles on callus induction and plant regeneration. BMC Plant Biology. 2021;**21**(1):597

[125] Ghasemian S, Masoudian N, Saeid Nematpour F, Safipour AA. Selenium nanoparticles stimulate growth, physiology, and gene expression to alleviate salt stress in *Melissa officinalis*. Biologia. 2021;**76**(10):2879-2888

[126] Zhou H, Wu H, Zhang F, Su Y, Guan W, Xie Y, et al. Molecular basis of cerium oxide nanoparticle enhancement of rice salt tolerance and yield. Environmental Science: Nano. 2021; **8**(11):3294-3311

[127] Waqas Mazhar M, Ishtiaq M, Hussain I, Parveen A, Hayat Bhatti K, Azeem M, et al. Seed nano-priming with zinc oxide nanoparticles in rice mitigates drought and enhances agronomic profile. PLoS One. 2022;**17**(3):e0264967

[128] Linh TM, Mai NC, Hoe PT, Lien LQ, Ban NK, Hien LTT, et al. Metal-based nanoparticles enhance drought tolerance in soybean. Journal of Nanomaterials. 2020;**2020**:1-13

[129] Shekhawat GS, Mahawar L, Rajput P, Rajput VD, Minkina T, Singh RK. Role of engineered carbon nanoparticles (CNPs) in promoting growth and metabolism of *Vigna radiata* (L.) Wilczek: Insights into the biochemical and physiological responses. Plants. 2021;**10**(7):1317

[130] Khalid MF, Iqbal Khan R, Jawaid MZ, Shafqat W, Hussain S, Ahmed T, et al. Nanoparticles: The plant saviour under abiotic stresses. Nanomaterials. 2022;**12**(21):3915

[131] Pérez-Labrada F, Hernández-Hernández H, López-Pérez MC, González-Morales S, Benavides-Mendoza A, Juárez-Maldonado A. Nanoparticles in plants: Morphophysiological, biochemical, and molecular responses. In: Plant Life under Changing Environment. Massachusetts: Elsevier; 2020. pp. 289-322

[132] Yoshida T, Fujita Y, Maruyama K, Mogami J, Todaka D, Shinozaki K, et al. Four *A. rabidopsis* AREB/ABF transcription factors function predominantly in gene expression downstream of SnRK2 kinases in abscisic acid signalling in response to osmotic stress. Plant, Cell & Environment. 2015; **38**(1):35-49

[133] Thiruvengadam M, Gurunathan S, Chung I-M. Physiological, metabolic, and transcriptional effects of biologically-synthesized silver nanoparticles in turnip (*Brassica rapa* ssp. rapa L.). Protoplasma. 2015;**252**: 1031-1046

[134] Song Y, Jiang M, Zhang H, Li R. Zinc oxide nanoparticles alleviate chilling stress in rice (*Oryza sativa* L.) by regulating antioxidative system and chilling response transcription factors. Molecules. 2021;**26**(8):2196

[135] Mirakhorli T, Ardebili ZO, Ladan-Moghadam A, Danaee E. Bulk and nanoparticles of zinc oxide exerted their beneficial effects by conferring modifications in transcription factors, histone deacetylase, carbon and nitrogen assimilation, antioxidant biomarkers, and secondary metabolism in soybean. PLoS One. 2021;**16**(9):e0256905

[136] Ali S, Rizwan M, Hussain A, et al. Silicon nanoparticles enhanced the growth and reduced the cadmium accumulation in grains of wheat (*Triticum aestivum* L.). Plant Physiology and Biochemistry. 2019;**140**:1-8

[137] Zhao L, Huang Y, Hu J, Zhou H, Adeleye AS, Keller AA. 1H NMR and GC-MS based metabolomics reveal defense and detoxification mechanism of cucumber plant under nano-Cu stress. Environmental Science & Technology. 2016;**50**(4):2000-2010

[138] Younis A, Khattab H, Emam M. Impacts of silicon and silicon nanoparticles on leaf ultrastructure and TaPIP1 and TaNIP2 gene expressions in heat stressed wheat seedlings. Biologia Plantarum. 2020;**64**(1):1-10

[139] Wu J, Wang T. Synergistic effect of zinc oxide nanoparticles and heat stress on the alleviation of transcriptional gene silencing in *Arabidopsis thaliana*. Bulletin of Environmental Contamination and Toxicology. 2020;**104**:49-56

[140] Yue L, Ma C, Zhan X, White JC, Xing B. Molecular mechanisms of maize seedling response to La 2 O_3 NP exposure: Water uptake, aquaporin gene expression and signal transduction. Environmental Science: Nano. 2017; **4**(4):843-855

[141] Amini S, Maali-Amiri R, Mohammadi R, Kazemi-Shahandashti S-S. cDNA-AFLP analysis of transcripts induced in chickpea plants by TiO_2 nanoparticles during cold stress. Plant Physiology and Biochemistry. 2017;**111**: 39-49

[142] Zhao L, Peng B, Hernandez-Viezcas JA, Rico C, Sun Y, Peralta-Videa JR, et al. Stress response and tolerance of *Zea mays* to CeO_2 nanoparticles: Cross talk among H_2O_2, heat shock protein, and lipid peroxidation. ACS Nano. 2012;**6**(11): 9615-9622

[143] Khodakovskaya MV, de Silva K, Nedosekin DA, Dervishi E, Biris AS, Shashkov EV, et al. Complex genetic, photothermal, and photoacoustic analysis of nanoparticle-plant interactions. National Academy of Sciences of the United States of America. 2011;**108**(3):1028-1033

[144] Mustafa G, Sakata K, Komatsu S. Proteomic analysis of flooded soybean root exposed to aluminum oxide nanoparticles. Journal of Proteomics. 2015;**128**:280-297

[145] Mustafa G, Sakata K, Hossain Z, Komatsu S. Proteomic study on the effects of silver nanoparticles on soybean under flooding stress. Journal of Proteomics. 2015;**122**:100-118

[146] Mustafa G, Sakata K, Komatsu S. Proteomic analysis of soybean root exposed to varying sizes of silver nanoparticles under flooding stress. Journal of Proteomics. 2016;**148**:113-125

[147] Cui J, Liu T, Li F, Yi J, Liu C, Yu H. Silica nanoparticles alleviate cadmium toxicity in rice cells: Mechanisms and size effects. Environmental Pollution. 2017;**228**:363-369

[148] Yadu B, Chandrakar V, Korram J, Satnami ML, Kumar M, Keshavkant S. Silver nanoparticle modulates gene expressions, glyoxalase system and oxidative stress markers in fluoride stressed *Cajanus cajan* L. Journal of Hazardous Materials. 2018;**353**:44-52

[149] Cong W, Miao Y, Xu L, Zhang Y, Yuan C, Wang J, et al. Transgenerational memory of gene expression changes induced by heavy metal stress in rice

(*Oryza sativa* L.). BMC Plant Biology. 2019;**19**(1):1-14

[150] Tian W, He G, Qin L, Li D, Meng L, Huang Y, et al. Genome-wide analysis of the NRAMP gene family in potato (*Solanum tuberosum*): Identification, expression analysis and response to five heavy metals stress. Ecotoxicology and Environmental Safety. 2021;**208**:111661

[151] Guha T, Barman S, Mukherjee A, Kundu R. Nano-scale zero valent iron modulates Fe/Cd transporters and immobilizes soil Cd for production of Cd free rice. Chemosphere. 2020;**260**:127533

[152] Venkatachalam P, Jayaraj M, Manikandan R, Geetha N, Rene ER, Sharma N, et al. Zinc oxide nanoparticles (ZnONPs) alleviate heavy metal-induced toxicity in Leucaena leucocephala seedlings: A physiochemical analysis. Plant Physiology and Biochemistry. 2017;**110**: 59-69

[153] Kozlowski T. Responses of woody plants to flooding and salinity. Tree Physiology. 1997;**17**(7):490

[154] Sharma B, Tiwari S, Kumawat KC, Cardinale M. Nano-biofertilizers as bio-emerging strategies for sustainable agriculture development: Potentiality and their limitations. Science of The Total Environment. 2023;**860**:160476

[155] Iqbal M, Umar S, Mahmooduzzafar. Nano-fertilization to enhance nutrient use efficiency and productivity of crop plants. Nanomaterials and Plant Potential. Cham: Springer. 2019:473-505

[156] Giglou MT, Giglou RH, Esmaeilpour B, Azarmi R, Padash A, Falakian M, et al. A new method in mitigation of drought stress by chitosan-coated iron oxide nanoparticles and growth stimulant in peppermint. Industrial Crops and Products. 2022;**187**: 115286

[157] Koleva L, Umar A, Yasin NA, Shah AA, Siddiqui MH, Alamri S, et al. Iron oxide and silicon nanoparticles modulate mineral nutrient homeostasis and metabolism in cadmium-stressed *Phaseolus vulgaris*. Frontiers in Plant Science. 2022;**13**:806781

[158] Ali MA, Mosa KA, El-Keblawy A, Alawadhi H. Exogenous production of silver nanoparticles by *Tephrosia apollinea* living plants under drought stress and their antimicrobial activities. Nanomaterials. 2019;**9**(12):1716

[159] Raza A, Charagh S, Salehi H, Abbas S, Saeed F, Poinern GE, et al. Nano-enabled stress-smart agriculture: Can nanotechnology deliver drought and salinity-smart crops? Journal of Sustainable Agriculture and Environment. 2023;**2**(3):189-214

[160] Ebrahimi A, Galavi M, Ramroudi M, Moaveni P. Effect of TiO_2 nanoparticles on antioxidant enzymes activity and biochemical biomarkers in pinto bean (*Phaseolus vulgaris* L.). Journal of Molecular Biology Research. 2016;**6**(1):58

[161] Sallam MF, Ahmed HM, Diab KA, El-Nekeety AA, Abdel-Aziem SH, Sharaf HA, et al. Improvement of the antioxidant activity of thyme essential oil against biosynthesized titanium dioxide nanoparticles-induced oxidative stress, DNA damage, and disturbances in gene expression in vivo. Journal of Trace Elements in Medicine and Biology. 2022; **73**:127024

[162] Sangeetha J, Thangadurai D, Hospet R, Purushotham P, Karekalammanavar G, Mundaragi AC, et al. Agricultural nanotechnology: Concepts, benefits, and risks.

Nanotechnology: An Agricultural Paradigm. Singapore: Springer; 2017:1-17

[163] Jakhar AM, Aziz I, Kaleri AR, Hasnain M, Haider G, Ma J, et al. Nano-fertilizers: A sustainable technology for improving crop nutrition and food security. NanoImpact. 2022;**27**: 100411

[164] Jadhav S, Jadhav EB, Sonone SS, Sankhla MS, Kumar R. Carbon-based nanomaterials: Nanobiosensors detecting environmental pollution. In: Nanotechnology for Environmental Pollution Decontamination: Tools, Methods, and Approaches for Detection and Remediation. London: Taylor & Francis; 2022. p. 263

[165] Pradhan S, Mailapalli DR. Nanopesticides for pest control. Sustainable Agriculture Reviews. 2020; **40**:43-74

[166] Hajji-Hedfi L, Chhipa H. Nano-based pesticides: Challenges for pest and disease management. Euro-Mediterranean Journal for Environmental Integration. 2021; **6**(3):69

[167] Potadar RS, Moholkar DN, Pawar KD. Nanosensors for the detection of heavy trace metals in soil. In: Nanosensors for Smart Agriculture. Netherlands: Elsevier; 2022. pp. 329-353

[168] El-Saadony MT, Saad AM, Soliman SM, Salem HM, Desoky E-SM, Babalghith AO, et al. Role of nanoparticles in enhancing crop tolerance to abiotic stress: A comprehensive review. Frontiers in Plant Science. 2022;**13**:946717

[169] Gopinath KP, Rajagopal M, Krishnan A, Sreerama SK. A review on recent trends in nanomaterials and nanocomposites for environmental applications. Current Analytical Chemistry. 2021;**17**(2):202-243

[170] Nawaz A, Rehman H, Usman M, Wakeel A, Shahid MS, Alam S, et al. Nanobiotechnology in crop stress management: An overview of novel applications. Discover Nano. 2023; **18**(1):74

[171] Upadhyay SK, Kumar S, Rout C, Vashistha G, Aggarwal D. Risks and concerns of use of nanoparticles in agriculture. In: Rajput VD, Verma KK, Sharma N, Minkina T, editors. The Role of Nanoparticles in Plant Nutrition under Soil Pollution: Nanoscience in Nutrient Use Efficiency. Cham: Springer International Publishing; 2022. pp. 371-394

[172] Santana I, Wu H, Hu P, Giraldo JP. Targeted delivery of nanomaterials with chemical cargoes in plants enabled by a biorecognition motif. Nature Communications. 2020;**11**(1): 2045

[173] Demirer GS, Silva TN, Jackson CT, Thomas JB, et al. Nanotechnology to advance CRISPR–Cas genetic engineering of plants. Nature Nanotechnology. 2021;**16**(3):243-250

Chapter 9

Plant Proteome in Response to Abiotic Stresses

Suvarna, R. Yashaswini, S.P. Prem Sagar, Prakash H. Kuchanur, V.C. Raghavendra, B.K. Prasad, A. Amaregouda and Ayyanagouda Patil

Abstract

Abiotic stresses, including drought, heat, salinity, waterlogging, and toxic metal stress, can negatively impact plant growth, leading to reduced crop yield and quality. Plants employ two key strategies, avoidance and tolerance, to adapt to such stress, making cellular and metabolic adjustments to withstand adverse conditions. Acclimatization results in significant changes in a plant's proteome, crucial for its stress response. Proteins encoded by a plant's genome play a vital role in environmental adaptation, supporting biotechnological advancements in plant breeding, utilizing data from proteomic investigations. Proteomics provides unique insights into post-translational modifications and their impact on crop yield. Subcellular proteomics aids in understanding cellular responses and interactions during growth and responses to biotic and abiotic stresses. Proteomic tools, like mass spectrometry, liquid chromatography, protein microarrays, and antibody-based assays, are integral to proteomic studies, contributing to our understanding of protein functions and interactions. Developing stress-tolerant crops is crucial for enhancing crop productivity and growth.

Keywords: drought, proteome, stress response, salinity, stress-tolerant crops

1. Introduction

Abiotic stress encompasses various environmental factors such as drought, salt, temperature fluctuations, waterlogging, and nutritional imbalances, all of which significantly impede plant growth and yield. These stresses create unfavourable conditions for plants. Predictions indicate that approximately half of all crop losses are attributed to abiotic stresses. Profound changes in the physiological, molecular, and biochemical processes within plants limits their distribution, alter growth patterns, and diminishes yield. Drought, a major abiotic stress, triggers oxidative stress through the overproduction of reactive oxygen species (ROS), leading to cell membrane damage and activating various stress-signalling pathways. Key responses to drought stress include root development, stomatal closure, photosynthesis, hormone production, and ROS scavenging. Salt stress negatively impacts plant growth and reproduction through nutritional and hormonal imbalances, ion toxicity, oxidative and osmotic stress, and increased susceptibility to diseases. The three major ways plants are damaged by salt stress involve altered soil properties causing water stress, destabilization of cell membranes, and protein degradation due to toxic ion effects. Physiological and metabolic changes occur, affecting seed germination, photosynthesis, biosynthetic processes, and overall growth reduction. Different crops respond differently to salinity, with glycophytes experiencing growth and yield reduction, while halophytes thrive. Plants cope by accumulating compatible solutes, redistributing ions, and increasing endogenous abscisic acid content, resulting in changes in genetic expression. Elevated levels of ions, such as Na^+ and Cl^-, induce ionic toxicity, disrupting ion homeostasis and limiting essential nutrient availability for plant growth and metabolism [1].

High temperatures exert a detrimental impact on plant yield, particularly during the vulnerable reproductive stage, causing a substantial reduction in seed set and overall crop yield. This vulnerability is especially pronounced in food crops like rice, wheat, soybean, maize, cotton, sorghum, and tomato. Intense high temperatures contribute to pollen abortion, leading to incomplete pollination and hindering successful reproduction. The delicate balance between pollen and pistil interactions is disrupted under these conditions, further compromising the reproductive process [2]. During waterlogging reduced light intensity, restricted gas diffusion, nutrient leaching, and hypoxia are pronounced effects. Plants exhibit active responses to stress, with the nature of their reaction contingent upon the intensity and duration of the stress.

Kosova et al. (2018) delineated five phases characterizing plant responses to stress. The initial alarm phase occurs when plants first encounter stress, followed by the acclimation phase during which biological processes adapt to stress, leading to increased tolerance. Subsequently, the maintenance phase ensues, where plants sustain their tolerance levels. The exhaustion phase follows, marked by a decline in tolerance due to prolonged stress and the failure to maintain homeostasis achieved during acclimation. The final stage is the recovery phase, wherein plants, upon stress withdrawal, regain normal homeostasis [3]. Each phase may be distinguished by a unique protein composition. Acclimation is an energy-intensive, dynamic process, with metabolic adjustments reflected at both the metabolic and proteomic levels in response to stress [4, 5].

In recent decades, research employing diverse *Omics* approaches has put forth numerous abiotic stress-related mechanisms as potential contributors to the creation of resilient plant varieties. The comprehensive utilization of functional *Omics* plays a crucial role in elucidating the intricate connection between an organism's genome and its observable characteristics in varying environmental conditions. Among all the *Omics* approaches, proteomics offers a comprehensive exploration of the structural, functional, abundance and interactions of proteins at a specific point of time. This technique holds an advantage over other "omics" tools, given that proteins play a pivotal role in the majority of cellular processes. Beyond complementing changes observed at the genomic and transcriptome level, proteomics has the capacity to identify translational and post-translational regulations, providing additional insights into intricate biological phenomena like abiotic stress responses in plants. This chapter delves in to the proteomics studies focused on plant responses to drought, salinity, flooding, and waterlogging.

1.1 Proteome

Marc Wilkins coined the term "proteome" in 1994 to refer to the complete set of proteins in a given organism during a specific time period, essentially representing the protein complement of the genome. In contrast to the genome, a static structure inherited from parents that defines the plant genotype, alterations in the plant epigenome, transcriptome, proteome, and metabolome collectively influence the plant phenotype.

Proteins play a direct role in the plant stress response, serving as both structural components and regulators of the plant epigenome, transcriptome, and metabolome. Additionally, the functionality of a protein is not solely determined by its molecular structure; it is also influenced by factors such as cellular localization, post-translational modifications, and interactions with other proteins. These aspects collectively contribute to the nuanced and multifaceted role that proteins play in the intricate mechanisms of plant stress adaptation and response. Proteomics, a state-of-the-art molecular technique, provides distinct advantages over genome-based technologies by directly focusing on functional molecules rather than the genetic code or mRNA abundance [6].

1.2 Importance of proteomics for abiotic stress

The field of plant stress proteomics is a dynamic discipline focused on investigating the plant proteome and the biological functions of proteins in plants that are subjected to various stress conditions. This area of study aims to unravel the complex and dynamic changes occurring at the protein level in response to stress, providing insights into the molecular mechanisms underlying a plant's adaptive responses to environmental challenges [7].

In plant abiotic stress studies, it is common to analyse proteomes by contrasting stressed plants with controls to correlate changes in protein accumulation with phenotypic responses. Additionally, comparing genotypes with varying stress sensitivity (sensitive *vs* tolerant) is crucial for understanding the influence of differentially abundant proteins in tolerant genotypes.

2. Proteomic tools to address abiotic stresses in plants

In response to abiotic stress, plants activate specific cellular processes leading to the accumulation of protective proteins and adjustments in metabolism and development. This involves substantial alterations in their gene expression patterns, which are regulated by various transcription factor (TF) families and signal-sensing networks, including phosphorylation/dephosphorylation events and other post-translational modifications (PTMs). These PTMs can either stabilize TFs or spot them for degradation. However, detecting and quantifying these elements through proteomics is tricky due to the typically low abundance of TFs and the rapid, transient nature of protein PTMs like phosphorylation and cysteine oxidation/reduction [8].

Quantitative proteomic methods can be categorized into gel-based or gel-free approaches, as well as "label-free" or "label-based" methods. The latter can be further divided into various types of labelling approaches, including chemical and metabolic labelling.

2.1 Low throughput-based approaches

2.1.1 Antibody-based methods

Techniques like ELISA (enzyme-linked immunosorbent assay) and western blotting rely on the presence of antibodies specifically directed at particular proteins or epitopes. These methods are utilized to identify proteins and quantify their expression levels.

2.1.2 Gel based approaches

2.1.2.1 Two-dimensional gel electrophoresis (2D-DIGE)

Since 1975, 2-DE has undergone various advancements, emerging as a primary method for protein separation and the preferred choice for analysing differential protein expression. In this technique, proteins undergo isoelectric focusing (IEF) based on their net charge at different pH values, followed by a second-dimensional separation based on molecular weight (MW) in an orthogonal direction.

Two types of reagents are integral to 2-DE buffers for protein solubility and denaturation. Chaotropes like urea and thiourea, employed at high concentrations, unfold proteins by weakening noncovalent bonds. The second type involves ionic detergents, exemplified by SDS (sodium dodecyl sulfate), which, with its long hydrocarbon chain, binds to hydrophobic amino acids, promoting denaturation.

2D-DIGE has proven effective in examining symbiosis and pathogenesis-related proteins in *Medicago truncatula*. Additionally, it has been employed to investigate the effects of abiotic stresses, including drought in oak, frost in Arabidopsis, as well as ozone and heavy metals in poplar. In context with the field crops, Natarajan et al., 2013 [9] analysed proteome of common bean and depicted that, 2D-PAGEfacilitated the identification of a significant number of low abundant proteins, demonstrating its efficacy in dissecting the common bean proteome with high resolution and sensitivity.

Limitations: less reproducibility, poor representation of less abundant proteins, hydrophobicity and problems in automation of the gel-based techniques.

2.1.2.2 Electrophoretic separations of innate proteins

This method facilitates the separation of protein complexes under native conditions, followed by the separation of individual proteins under denaturing conditions. This dual approach provides valuable insights into the stoichiometry of the complexes.

Recently, this approach was effectively employed to analyse the proteome of wheat chloroplast protein complexes.

Limitations: less resolution and difficult to optimize

2.1.2.3 One dimensional gel electrophoresis (1-DE)

One-dimensional gel electrophoresis (1-DE) has become the most widely adopted method, primarily serving two purposes: rapid determination of protein molecular weight (MW) and assessment of protein purity. This widely used technique finds applications in various areas, including comparing protein compositions of different samples, analysing the number and size of polypeptide subunits, conducting western blotting coupled with immunodetection, and, of course, serving as a second dimension in 2-DE maps.

The 1-DE approach has found extensive application in plant studies, as exemplified by research on *M. truncatula* plasma membrane alterations in response to arbuscular mycorrhizal symbiosis and investigations into the *Arabidopsis thaliana* chloroplast envelope. More recently, this method has been employed to construct a protein expression map of *Arabidopsis* roots, offering insight into the identity and cell type-specific localization of nearly 2,000 proteins.

2.1.2.4 Chromatography-based methods

Chromatography-based approaches are effective for the separation and purification of proteins from intricate biological mixtures like cell lysates. Ion-exchange chromatography, for instance, segregates proteins according to their charge, while size exclusion chromatography distinguishes proteins based on their molecular size. Affinity chromatography utilizes reversible interactions between specific affinity ligands and target proteins, such as the use of lectins to purify IgM and IgA molecules. These methods serve the dual purpose of purifying and identifying proteins of interest and preparing them for subsequent analysis, such as downstream mass spectrometry (MS).

2.2 High-throughput methods:

2.2.1 Mass spectrometry (MS)

This is a technique used to determine the molecular weight of proteins by measuring their mass-to-charge ratio (m/z). The process involves three steps. First, molecules need to be transformed into the gas phase. In the second step, ions are separated based on their m/z values in the presence of a mass analyzer. Finally, the separated ions and the quantity of each species with a specific m/z value are measured. Common ionization methods include matrix-assisted laser desorption ionization (MALDI), surface-enhanced laser desorption/ionization (SELDI), and electrospray ionization (ESI).

2.2.1.1 Mass spectrometry-based proteomics

Various "gel-free" methods exist for protein separation, such as isotope-coded affinity tag (ICAT), stable isotope labeling with amino acids in cell culture (SILAC), and isobaric tags for relative and absolute quantitation (iTRAQ). These approaches facilitate both quantitation and comparative/differential proteomics. Less quantitative techniques like multidimensional protein identification technology (MudPIT) are faster and simpler. Additionally, gel-free chromatographic methods for protein separation encompass gas chromatography (GC) and liquid chromatography (LC).

Tandem Mass Tag (TMT) is a sensitive and quantitative proteomic technique that enables the simultaneous identification and determination of the relative abundances of peptides from multiple samples. This method utilizes isobaric labeling in conjunction with tandem mass spectrometry (MS/MS) for analysis [10].

2.2.1.2 Isobaric tags for relative and absolute quantification (iTRAQ)

It is a multiplex protein labeling technique used for protein quantification, relying on tandem mass spectrometry. This method involves labeling proteins with isobaric tags, available in 8-plex and 4-plex configurations, for quantification purposes. The N-terminus and side chain amine groups in proteins are labeled and subsequently fractionated using liquid chromatography. The final analysis is conducted through Mass Spectrometry. iTRAQ is a suitable method that allows for the simultaneous identification and quantification of proteins [11]. Liu et al. (2020) assessed maize seed proteome, they explored the differential protein expression profiles between the two groups comparing transgenic and natural variation shedding light on the impact of genetic modification on the maize seed proteome. Their findings provide valuable insights into the molecular mechanisms underlying transgenic traits in maize seeds, offering potential implications for agricultural biotechnology and crop improvement strategies [12].

2.2.1.3 Reversed-phase chromatography (RP)

This is the most commonly used liquid chromatography method in proteomics, enabling the separation of neutral peptides based on their hydrophobicity. The process relies on the analyte's partition coefficient between the polar mobile phase and the hydrophobic (non-polar) stationary phase. Yu et al. (2013) effectively utilized RP-UPLC to swiftly separate and characterize water-soluble proteins in wheat grains through optimized conditions. This method ensures fast, high-resolution, and reproducible separations with minimal resource consumption, showcasing its potential for rapid analysis in wheat breeding and fundamental genetic studies. RP-UPLC emerges as a robust tool for exploring wheat grain proteins, promising valuable insights for agricultural research and germplasm screening [13].

2.2.1.4 Two-dimensional liquid chromatography (2D-LC)

To mitigate sample complexity and enhance proteome coverage, multidimensional analytical methods with orthogonal separation capabilities are essential. The separation of peptide mixtures in 2D-LC has been achieved through various orthogonal combinations, including anion exchange coupled to reversed-phase (AX/RP), size exclusion chromatography coupled to reversed-phase (SEC/RP), and affinity chromatography coupled to reversed-phase (AFC/RP).

2.2.1.5 OFFGEL electrophoresis (OGE)

Orthogonal Gel Electrophoresis (OGE) boasts high loading capacity and resolution power. In contrast to LC fractionation, OGE offers additional physicochemical information, including peptide isoelectric point (pI). This information proves highly valuable for corroborating MS results, sorting false positive rates, and enhancing the reliability of the identification procedure.

2.2.1.6 Stable isotope labeling by amino acids in cell culture (SILAC)

This is a quantitative proteomic method that employs labeled isotopically heavy amino acids, such as 13C6,15N2-lysine and 13C6,15N4-arginine. These labeled amino acids are metabolically incorporated into the entire proteome, enabling multiplexing for quantitative analysis.

2.2.1.7 Nuclear magnetic resonance (NMR) spectroscopy

NMR spectroscopy is a discipline within structural biology that utilizes NMR spectroscopy to gather information about the structure and dynamics of proteins, nucleic acids, and their complexes. The phases encompass sample preparation, measurements, the application of interpretive approaches, and the calculation and validation of the protein structure. Studies by Mazzei and Piccolo (2017) NMR technique offers high-resolution spectroscopy for semi-solid samples, bridging solid- and liquid-state NMR advantages [14]. Its rapid spinning and sample orientation minimize anisotropic processes, ensuring enhanced spectral quality. The method employs edited pulse sequences, providing simultaneous information on polar and non-polar components without sample extraction. With applications expanding into agricultural chemistry, HRMAS NMR proves valuable in characterizing soil components, plant tissues, and agrofood products.

2.2.2 Protein microarrays

Microarray technology involves the miniaturization of numerous assays on a single small plate. This concept originated from the earlier ambient analyte immunoassay introduced by Roger Ekins in 1989. Over the subsequent decade, this concept evolved into the DNA microarray, a technology capable of simultaneously determining the mRNA expression levels of thousands of genes.

2.2.2.1 Analytical protein microarrays

These are also known as antibody arrays, come in different formats. The "analyte-labelled" format detects proteins after antibody capture through direct labelling, revealing alterations in protein expression in various cell types. However, it has limitations in specificity and sensitivity. Another model, the "sandwich" assay, uses two antibodies for higher sensitivity, successfully distinguishing between samples like blood plasma and serum with minimal volume. While resembling a multiplexed ELISA, these arrays can detect dozens of analytes simultaneously due to potential cross-reactivity between antibodies.

2.2.2.2 Functional protein microarrays

These play a crucial role in understanding protein functionality at the proteome level. These microarrays, made with individually purified proteins, allow the exploration of various biochemical properties, including interactions like protein-protein, protein-DNA, and enzyme-substrate relationships. Zhu et al. (2001) pioneered their use in determining the substrate specificity of protein kinases, and since then, these microarrays have found applications in basic and clinical research [15]. They have been instrumental in providing the entire proteome of organisms on arrays, leading to significant biological discoveries. Additionally, protein microarrays facilitate the study of post-translational modifications, contributing vital insights into cellular protein synthesis and function.

2.2.2.3 Reverse-Phase Protein Microarrays

This technique employs an opposite format to traditional protein microarrays, offer expanded applications in analyzing samples obtained at different states by directly spotting tissue, cell lysates, or fractionated cell lysates on a glass slide. Poetz et al., 2005 successfully detected microscopic transition stages of pro-survival checkpoint proteins in different prostate cancer stages, correlating phosphorylation status with cancer progression [16].

2.2.3 X-ray protein crystallography

It is a technique that enables the determination of the three-dimensional positions of each atom within a protein. Protein crystallography involves multiple steps, starting with the production of protein using Escherichia coli and cloning the gene of

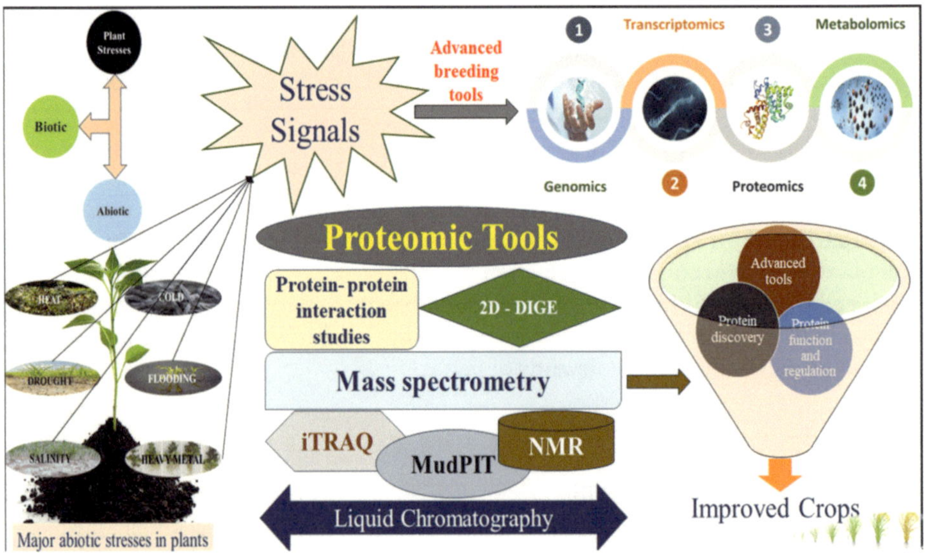

Figure 1.
A brief and schematic depiction of proteomic tools for different abiotic stresses.

interest into an expression plasmid. After inducing protein expression, cells are lysed, and the protein is purified. Crystallization is then attempted using various conditions, and a drop diffusion technique is commonly used. Once suitable crystals are obtained, X-ray diffraction data is collected by bombarding the crystals with X-rays. This data is analyzed computationally to infer the three-dimensional positions of atoms in the protein, producing an electron density map through a process called phasing. Brief workflow for crop improvement using proteomic tools (**Figure 1**).

3. Studies exploring changes in the plant proteome in response to various abiotic stress factors

In this chapter, we offer an update on the contribution of proteomic studies in elucidating the mechanisms that underlie stress responses in plants. The detailed view of plant proteome responses to various stresses are briefly discussed below:

3.1 Drought or water deficit stress

Water is essential for plant life, serving as the primary compound in cytosol for biochemical reactions and influencing cell expansion. However, water stress, caused by factors like drought or waterlogging, poses a significant challenge to global crop production. Plants vary in their optimal environments and responses to water availability extremes.

Proteins play a crucial role in conferring resistance to drought stress by participating in stress signalling, transcription regulation, cellular detoxification, and protection of macromolecules. Proteomic methods, studying the dynamically translated part of the genome, offer insights into molecular events not revealed by DNA or mRNA analysis (**Table 1**). The application of proteomic technologies has been instrumental in identifying mechanisms associated with drought stress and tolerance in various plant species. Some of the illustrations are provided below (**Table 2**).

Stress	Specific response	Aspecific response	
		Antioxidant enzymes	**HSPs**
Drought	Osmotin, Dehydrin, Aquaporin, LEA proteins	Aldose/Aldehyde reductase, Met synthase	smHSPs, HSP70, HSC70
Heat	Galactinol synthase, Choline kinase, Peptidyl prolyl isomerase, Glutaredoxin, Thaumatin	APX2, GST, Catalase, SOD	HSP110, HSP100, HSP90, HSP70, HSP60, smHSPs, CPN60, BiP
Salt	Osmotin, Dehydrin, Remorin1, HIR protein, GF14a, GF14b, ABP	COX6b-1, Triosephosphate isomerase, Enolase, UGPase, GST, GPX, Methyltransferase	HSP70, sti1, HSP90, smHSPs
Cold	Osmotin, Dehydrin, Glycine-rich protein, RNA binding protein CP29	APX, CA, Met synthase, GST, Thx	smHSPs, HSP70, HSP90, CPN20, CPN60, GRP78
Heavy metals	MT and PC proteins, Phytochelatins	GSH-derived peptides	HSP70

Table 1.
List of the major specific and aspecific proteins expressed upon several abiotic stresses in plant [17].

Crop	Stress treatment	Proteomics approach	Differentially expressed protein spots (identified proteins)	References
Drought				
Soybean cv Taegwang	Root – 4 d and recovery	2-DE MALDI-TOF	45 (28)	[18]
Arachis hypogea	Leaf proteins	1 and 2-DE/ MALDI-TOF-MS	49	[19]
Pea cv Green Feast	7 d of water withdrawal	2-DE LC–MS/MS	33 (32)	[20]
Maize (line FR697	primary root elongation zone – cell wall proteome ψw −1.6 MPa (48 h)	2-DE LC–ESI-MS TOF	152	[21]
Chickpea gen JG-62	nuclear fraction 0, 24, 48, 72, 96, 120, 144 h of water withdrawal	2-DE LC–MS/MS	205 (147)	[22]
Populus euphratica	35, 24, 8% rel. soil water and recovery (10 d after reirrigation)	2D-DIGE MALDI-TOF/TOF	375 (39)	[23]
Oryza sativa L.	cv.CT9993, cv. IR2266 leaves	2DE/ MALDI-TOF-MS	42	[24]
Triticum aestivum L.	Ningchun 4, Chinese spring, grain	2DE/ MALDI-TOF-MS	28	[25]
Hordeum vulgare Golden promise	Basrah roots and leaves	DIGE/ MALDI-TOF-MS	24 (L)	[26]
Arabidopsis thaliana	Plastid	2DE-DIGE/ MALDI-TOF-MS	18	[27]

Crop	Stress treatment	Proteomics approach	Differentially expressed protein spots (identified proteins)	References
Sugarcane	Stop watering	1-DE, 2-DE	30	[28]
Grapevine (*Vitis sylvestris*)	water deficit stress for 16 days	2D-PAGE	18	[29]
Maize (tolerant ND476 and sensitive ZX978)	water deficit (drought) treatment conditions for 12 days at the grain filling stage	iTRAQ analysis	165	[30]
Pandanus amaryllifolius	withholding water for 4, 7, 10, and 14 days	TMT-labelled LCMS/MS	74	[31]
Tomato (*Solanum lycopersicum*; cultivar M82)	Five-leaf stage withhold water	HPLC fractionation and LC-MS/MS analysis	294	[32]
Heat				
Common wheat (*T. aestivum*) cvs Fang	grain 40/25°C (72 h)	2-DE MALDI-TOF	48 (7)	[33]
Populus euphratica	leaf 42/37°C (3 d)	2-DE MALDI-TOF/TOF	62 (51)	[34]
Rice cv Dongjin	seedling leaf 42°C (12, 24 h)	2-DE MALDI-TOF	73 (48)	[34]
Carissa spinarum	leaf 42/35°C (5 d)	2-DE MALDI-TOF/TOF	49 (30)	[35]
Spinach (*Spinacia oleracea* L.)	temperature regime of 22/18°C, 10/14 h day/night cycle	Trypsin Digestion and iTRAQ Labeling	45	[36]
Chick pea	27/18°C Day/night temperatures till flowering	LCMS analysis, fractionation, MS parameters	82	[37]
Sorghum (ICSB338)	40°C for 72 h	iTRAQ Labelling	31	[38]
*Clematis forida*Thun (CfT)	38°C for 3 h; air humidity 65%	LC-MS/MS	17	[39]
Pepper		iTRAQ	91	[40]
Salinity				
Rice	150 mM NaCl	2-DE	63	[41]
Wheat	Mitochondria	2-DE/LC MS/MS	68	[42]
Tobacco (*Nicotiana tabacum*) cv Petit Havana	SR1 – leaf apoplast 100 mM NaCl (20 d)	2-DE LC–MS/MS	30 (20)	[43]
Durum wheat (*T. durum*)	leaf 100 mM NaCl (2 d)	2-DE MALDI-TOF	(38)	[44]
Soybean cv Enrei	root, hypocotyl, leaf 40 mM NaCl (7 d)	2-DE Edman sequencing MALDI-TOF	14 – root; 22 – hypocotyl; 19 – leaf	[45]

Crop	Stress treatment	Proteomics approach	Differentially expressed protein spots (identified proteins)	References
Rice cv Nipponbare	seedling 3 rd leaf 130 mM NaCl (4 d)	2-DE nanoESI-LC–MS/MS	55 (33)	[46]
Rapeseed (*Brassica napus* L.)	0, 150, and 300 mM NaCl	2-DE	44	[47]
Waterlogging				
Grapevine rootstock (*V. berlandieri* × *V. riparia*) plants	waterlogging at the sixth-leaf stage for 0 days (T0d), 10 days (T10d), and 20 days (T20d)	LC–MS/MS	1057	[48]
Peanut (*Arachis hypogaea* L.)	flooding depth of 1 cm for different duration time (3, 6, 9, 12, and 15 days)	HPLC Fractionation and LC MS/MS	6285	[49]
Wheat *(Triticum aestivum L.)*	Flowering stage for 15 days	iTRAQ and HPLC-MS techniques	710	[50]
Cold temperature				
Rapeseed (*Brassica napus* L.)	four-leaf stage were subjected to cold treatments (8°C, 16 h of light/4°C, 8 h of dark) for 7 days	iTRAQ and LC-MS techniques	241	[51]
Arabidopsis	4–5 expanded leaf stage, plants were transferred to – 4°C for 24 h)	SDS-PAGE	(21)	[52]
Citrus junos fruits	−7°C as a cold stress condition	TMT labelling and LC MS/MS	413	[53]
Coconut (*Cocos nucifera* L.)	8°C for 2 and 5 days	iTRAQ and LC-ESI-MS/MS	104	[54]
Saffron crocus (*Crocus sativus*)	16°C, non-flowering phenotype) for 30 days	iTRAQ	201	[55]
Toxic metals				
Arsenic *B. napus*	200 μmol L^{-1}	MS/MS and iTRAQ labeling	117	[56]
Cadmium Wheat	50 μM for 24 h	TMT and LC-MS/MS	532	[57]

Table 2.
Summary of proteomic analysis of plant responses to abiotic stress.

In rice, upregulation of three proteins namely S-like RNase homologue, actin depolymerizing factor, and Rubisco activase and downregulation of isoflavone reductase-like protein was observed under drought stress. In barley, greater regulation of ROS-homeostasis has been observed. In soyabean, the enzyme caffeoyl-CoA

O-methyltransferase, crucial in lignin synthesis, exhibited significant upregulation. This upregulation is likely linked to the increased accumulation of lignin, thereby associated with growth inhibition. In tomato plants subjected to osmotic stresses, there was an increase in levels of ABA (abscisic acid), phospholipid signalling, and mitogen-activated protein kinases (MAPKs) [58].

Drought stress triggers the activation of Ca^{2+}-binding proteins, such as calmodulin, in plants, serving as crucial transducers for Ca^{2+} signals. This involvement of Ca^{2+-}binding proteins, alongside Ca^{2+-}dependent protein kinases, calcineurin B–like proteins, and SOS_3, highlights their significant role in mediating ABA-dependent stress responses in plants. The functions of numerous proteins induced in higher plants during stress remain unknown, requiring inference based on available information from model plants [59].

3.2 Flooding or waterlogging stress

Waterlogging, caused by excessive soil water, is a major abiotic stress affecting about half of global crops annually, leading to a 25% yield reduction. This stress restricts oxygen availability to plant roots, triggering anaerobic metabolism and processes like glycolysis and pyruvate fermentation. In higher plants, the cell wall is the initial organelle to respond to flood stress signals, transmitting them to the cell's interior and influencing the cell signalling cascade, stress tolerance, or intolerance [60].

Proteins within the cell walls play essential roles in signal transduction, maintaining structural integrity, managing metabolism, and facilitating cell enlargement in response to waterlogging stress. Upregulation of several proteins *viz.*, fructose-bisphosphate aldolase, alcohol dehydrogenase, UDP-glucose pyro phosphorylase and GAPDH involved in the fermentation and glycolysis pathways. Under waterlogging stress in tomato leaves, a comparative proteomic analysis showed increased ion leakage, lipid peroxidation, and H_2O_2 content, coupled with a decrease in chlorophyll levels. Proteome profiling in different crops has demonstrated that prolonged waterlogging stress led to an elevation in glycolytic flux, meeting increased ATP demands [59].

3.3 Imbalances in mineral nutrition

Plants necessitate a minimum of 17 essential nutrients to successfully complete their life cycle. The absorption of mineral elements from the soil solution by plant roots and their subsequent distribution within the plant has been extensively studied for decades. Deficiencies in any of these mineral elements significantly limit crop growth and yield, while higher soil concentrations can adversely affect plant growth and development.

Plants employ various adaptations, including proton exudation, enzyme release, and structural changes in roots, to enhance phosphorus (P) uptake efficiency from the soil. Under P deficiency, specific proteins associated with root cell cycle and division, such as GTP-binding nuclear protein, mini-chromosome maintenance protein, and glycogen synthase kinase-3 homolog MsK-3, are up-regulated. Phosphorus (P) deficiency induces oxidative stress in plants, leading to an increase in reactive oxygen species (ROS) [59].

Proteomic analysis revealed the up-regulation of P deficiency-responsive proteins involved in oxidative stress defense, including superoxide dismutase (SOD), 2-Cys peroxiredoxin, ascorbate peroxidase (APX), and 1,4-benzoquinone reductase. Nitrogen (N) deficiency in plants leads to down-regulation of proteins involved in photosynthesis, such as Rubisco activase, Rubisco LS, and SS, particularly in leaves. N deficiency triggers complex responses in carbon (C) metabolism, affecting various

C metabolism-related proteins. Additionally, it accelerates leaf senescence, leading to the breakdown of proteins like RuBisCO, enabling the remobilization of N to active tissues. This condition also induces an increase in reactive oxygen species (ROS), with up-regulation of antioxidant proteins like 2-Cys peroxiredoxin, superoxide dismutase (SOD), and ascorbate peroxidase (APX) to mitigate oxidative stress [61].

3.4 Toxic metal stress

Metals play vital roles in plant metabolism, serving as essential micronutrients for various protein functions. However, an excess of these essential metals can lead to detrimental effects on plant health. Additionally, nonessential metals such as cadmium, chromium, lead, mercury, and arsenic, even in trace amounts, can negatively impact plants. The influence of these nonessential metals on plants is complex, and their adverse effects are determined by biological availability rather than the absolute concentration in the soil. This highlights the delicate balance required for optimal metal levels to ensure proper plant growth and development.

Metal toxicity occurs when ionic forms of metals form complexes or ligands with biomolecules. The presence of redox-active metals can lead to the production of reactive oxygen species (ROS) through Haber-Weiss and Fenton reactions, contributing to metal toxicity [59]. NADP(H)-oxido-reductase plays a crucial role in protecting cells from the harmful effects of toxic metals.

3.5 Salinity stress

Salinity adversely affects plant growth through osmotic water-deficit and ionic stress/toxicity. The osmotic effect results from a decrease in soil water osmotic potential due to dissolved salt ions, limiting water uptake and causing physiological drought or cellular dehydration. Ionic stress involves the entry of salts, particularly Na and Cl_2, into plant cells, leading to cell damage, growth impairment, and secondary stresses like oxidative damage [62]. Salinity-induced consequences include slow growth, developmental delays, reduced yields, economic losses, soil erosion, ecological imbalances, and potential harm to human health through the introduction of toxic elements into the food chain.

Proteins responsive to salt stress are involved in key processes such as stress signal transduction, transcription, protein metabolism, osmotic homeostasis, ion homeostasis, ROS homeostasis, photosynthesis, carbohydrate and energy metabolism, and cytoskeleton and cell wall dynamics. Salt-responsive proteins are identified in various plant organs, including vegetative organs (seeds, seedlings, radicles, hypocotyls, roots, shoots, leaves), reproductive organs (panicles, anthers, grains), and other tissues (callus and cell suspension cultures). Proteome studies also target subcellular organelles and membrane systems, including the plasma membrane, microsome, apoplast, mitochondria, and chloroplast [59].

3.6 Heat stress

High temperature stress is a prominent abiotic stress affecting plant growth, leading to enzyme denaturation and membrane fluidity issues. Protein synthesis is generally suppressed, while the induction of Heat Shock Proteins (HSPs) occurs. Proteomic studies on heat stress, including details on tissues, varieties, temperature duration, and analysis methods, have been extensively studied.

Plants have an optimal temperature range for optimal growth and development. Temperature extremes, both high and low, can impact plant sensitivity and growth. High-temperature treatment increases the abundance of low-molecular-weight sHSPs (small heat shock proteins). The expression of high-molecular-weight HSPs is minimally affected by high-temperature treatment [8]. The expression level of glyceraldehyde-3-phosphate dehydrogenase, an enzyme in the glycolytic pathway, is influenced by high temperatures.

3.7 Cold stress

Chilling or cold stress is a significant environmental factor affecting crop growth and productivity. Cold stress poses a serious limitation to crop development, particularly in various regions around the world.

Chilling temperatures directly impact plants by slowing metabolic processes and causing the loss of membrane functions, leading to cold stress. Plants are categorized into chilling-sensitive, chilling-tolerant/resistant, and freezing-tolerant types based on their ability to survive different temperature conditions. Chilling-sensitive plants experience metabolic dysfunction at temperatures slightly below the optimum, chilling-tolerant plants endure lower non-freezing temperatures, and freezing-tolerant plants can survive below freezing conditions [59].

UDP-glucose pyrophosphorylase and sucrose synthase are proteins linked to UDP-glucose formation. These proteins exhibit up-regulation in the leaf blade of Japonica rice seedlings in response to short-term cold treatment. Four antioxidant enzymes – APX, glutathionine S-transferase, thioredoxin h, and thioredoxin peroxidase – are induced by oxidative stress caused by ROS. A RING zinc finger protein-like, identified as a new cold-induced protein, exhibits marked responsiveness only to extreme cold stress.

Plant proteome analysis requires various bioinformatics tools as discussed in **Table 3**.

Crop	Proteome database	References
Rice	http://oryzapg.iab.keio.ac.jp/	[63]
	http://gene64.dna.affrc.go.jp/RPD/	[64]
Soybean	http://proteome.dc.affrc.go.jp/Soybean	[65]
	http://www.oilseedproteomics.missouri.edu/soybean.php	[66]
Maize	http://ppdb.tc.cornell.edu	[67]
Wheat	http://www.proteome.ir	[68]
	http://www.gramene.org/protein	[69]
Chickpea	http://bioinfo.noble.org/manuscript-support/legumedb	[70]
Pea	http://bioinfo.noble.org/manuscript-support/legumedb	[70]
Alfalfa	http://bioinfo.noble.org/manuscript-support/legumedb	[70]
Barley	http://www.gramene.org/protein	[69]
Sorghum	http://www.gramene.org/protein	[69]

Table 3.
Proteome databases helpful in the protein identification of different crops.

4. Conclusion

Proteomics serves as a fundamental frontier in the realm of biological research, offering a holistic perspective on the proteins within living systems. By systematically analysing the entire proteome of a cell, tissue, or organism, proteomics enables a comprehensive exploration of protein structure, function, and regulation. One of its key contributions lies in unravelling the functional roles of proteins, shedding light on their involvement in diverse biological processes, signalling pathways, and cellular functions. Proteomics is particularly adept at uncovering post-translational modifications (PTMs), providing crucial insights into the dynamic regulatory mechanisms that influence protein activity and stability. Furthermore, it plays a crucial role in translational research and clinical applications. In the broader context of environmental responses, such as the plant's reaction to abiotic stress, proteomics becomes indispensable in delineating the specific proteins involved in stress adaptation and understanding the molecular intricacies underlying these adaptive processes. As an integral component of systems biology, proteomics collaborates with genomics, transcriptomics, and metabolomics to provide a holistic understanding of biological systems, paving the way for advanced breeding of crop plants for abiotic stress and biotic stress at the molecular level.

Author details

Suvarna[1*], R. Yashaswini[1], S.P. Prem Sagar[1], Prakash H. Kuchanur[2], V.C. Raghavendra[1], B.K. Prasad[1], A. Amaregouda[3] and Ayyanagouda Patil[4]

1 Department of Genetics and Plant Breeding, College of Agriculture, Raichur, University of Agricultural Sciences, Raichur, Karnataka, India

2 Department of Genetics and Plant Breeding, College of Agriculture, Bheemarayanaudi, University of Agricultural Sciences, Raichur, Karnataka, India

3 Department of Crop Physiology, College of Agriculture, Raichur, UAS, Raichur, Karnataka, India

4 Department of Molecular Biology and Biotechnology, College of Agriculture, Raichur, UAS, Raichur, Karnataka, India

*Address all correspondence to: suvarna.gpb@gmail.com

References

[1] Sarraf M, Vishwakarma K, Kumar V, Arif N, Das S, Johnson R, et al. Metal/metalloid-based nanomaterials for plant abiotic stress tolerance: An overview of the mechanisms. Plants. 2022;**11**(3):316

[2] Parthasarathi T, Firdous S, David EM, Lesharadevi K, Djanaguiraman M. Effects of high temperature on crops. In: Advances in Plant Defense Mechanisms. London, UK: IntechOpen; 2022

[3] Kosova K, Vitamvas P, Urban MO, Prasil IT, Renaut J. Plant abiotic stress proteomics: The major factors determining alterations in cellular proteome. Frontiers in Plant Science. 2018;**8**(9):122

[4] Miller MA, O'Cualain R, Selley J, Knight D, Karim MF, Hubbard SJ, et al. Dynamic acclimation to high light in *Arabidopsis thaliana* involves widespread reengineering of the leaf proteome. Frontiers in Plant Science. 2017;**8**:1239

[5] Hu G, Wang K, Song J, Uversky VN, Kurgan L. Taxonomic landscape of the dark proteomes: Whole-proteome scale interplay between structural darkness, intrinsic disorder, and crystallization propensity. Proteomics. 2018;**18**(21-22):1800243

[6] Komatsu S, Hossain Z. Organ-specific proteome analysis for identification of abiotic stress response mechanism in crop. Frontiers in Plant Science. 2013;**4**:71

[7] Jogaiah S, Govind SR, Tran LS. Systems biology-based approaches toward understanding drought tolerance in food crops. Critical Reviews in Biotechnology. 2013;**33**(1):23-39

[8] Hakeem KR, Chandna R, Ahmad P, Iqbal M, Ozturk M. Relevance of proteomic investigations in plant abiotic stress physiology. OMICS: A Journal of Integrative Biology. 2012;**16**(11):621-635

[9] Natarajan SS, Pastor-Corrales MA, Khan FH, Garrett WM. Proteomic analysis of common bean (*Phaseolus vulgaris* L.) by two-dimensional gel electrophoresis and mass spectrometry. Journal of Basic & Applied Sciences. 2013;**9**:424

[10] Thompson A, Schäfer J, Kuhn K, Kienle S, Schwarz J, Schmidt G, et al. Tandem mass tags: A novel quantification strategy for comparative analysis of complex protein mixtures by MS/MS. Analytical Chemistry. 2003;**75**(8):1895-1904

[11] Wiese S, Reidegeld KA, Meyer HE, Warscheid B. Protein labeling by iTRAQ: A new tool for quantitative mass spectrometry in proteome research. Proteomics. 2007;**7**(3):340-350

[12] Liu S, Ma Y, Zheng Y, Zhao W, Zhao X, Luo T, et al. Cold-stress response of probiotic *Lactobacillus* plantarum K25 by iTRAQ proteomic analysis. Journal of Microbiology and Biotechnology. 2020;**30**(2):187

[13] Yu Z, Han C, Wang S, Lv D, Chen G, Li X, et al. Fast separation and characterization of water-soluble proteins in wheat grains by reversed-phase ultra performance liquid chromatography (RP-UPLC). Journal of Cereal Science. 2013;**57**(3):288-294

[14] Mazzei P, Piccolo A. HRMAS NMR spectroscopy applications in agriculture. Chemical and Biological Technologies in Agriculture. 2017;**4**(1):1-3

[15] Zhu H, Bilgin M, Bangham R, Hall D, Casamayor A, Bertone P, et al. Global analysis of protein activities using proteome chips. Science. 2001;**293**(5537):2101-2105

[16] Poetz O, Schwenk JM, Kramer S, Stoll D, Templin MF, Joos TO. Protein microarrays: Catching the proteome. Mechanisms of Ageing and Development. 2005;**126**(1):161-170

[17] Timperio AM, Egidi MG, Zolla L. Proteomics applied on plant abiotic stresses: Role of heat shock proteins (HSP). Journal of Proteomics. 2008;**71**(4):391-411

[18] Alam I, Lee DG, Kim KH, Park CH, Sharmin SA, Lee H, et al. Proteome analysis of soybean roots under waterlogging stress at an early vegetative stage. Journal of Biosciences. 2010;**35**:49-62

[19] Tamburino R, Vitale M, Ruggiero A, Sassi M, Sannino L, Arena S, et al. Chloroplast proteome response to drought stress and recovery in tomato (*Solanum lycopersicum* L.). BMC Plant Biology. 2017;**17**:1-4

[20] Kundratova K, Bartas M, Pecinka P, Hejna O, Rychla A, Curn V, et al. Transcriptomic and proteomic analysis of drought stress response in opium poppy plants during the first week of germination. Plants. 2021;**10**(9):1878

[21] Yahoueian SH, Bihamta MR, Babaei HR, Bazargani MM. Proteomic analysis of drought stress response mechanism in soybean (*Glycine max* L.) leaves. Food Science & Nutrition. 2021;**9**(4):2010-2020

[22] Li H, Yang M, Zhao C, Wang Y, Zhang R. Physiological and proteomic analyses revealed the response mechanisms of two different drought-resistant maize varieties. BMC Plant Biology. 2021;**21**:1-5

[23] Zhou Y, Li H, Chen H, Yang X, Yu T, Wang Y, et al. Proteomic investigation of molecular mechanisms in response to PEG-induced drought stress in soybean roots. Plants. 2022;**11**(9):1173

[24] Pei Y, Bai J, Guo X, Zhao M, Ma Q, Song X. Comparative proteome analysis of drought-sensitive and drought-tolerant maize leaves under osmotic stress. Canadian Journal of Plant Science. 2019;**99**(4):467-479

[25] Ge P, Ma C, Wang S, Gao L, Li X, Guo G, et al. Comparative proteomic analysis of grain development in two spring wheat varieties under drought stress. Analytical and Bioanalytical Chemistry. 2012;**402**:97-113

[26] Wendelboe-Nelson C, Morris PC. Proteins linked to drought tolerance revealed by DIGE analysis of drought resistant and susceptible barley varieties. Proteomics. 2012;**12**(22):74-85

[27] Horn R, Chudobova I, Hänsel U, Herwartz D, Koskull-Döring PV, Schillberg S. Simultaneous treatment with tebuconazole and abscisic acid induces drought and salinity stress tolerance in *Arabidopsis thaliana* by maintaining key plastid protein levels. Journal of Proteome Research. 2013;**12**(3):1266-1281

[28] Jangpromma N, Kitthaisong S, Lomthaisong K, Daduang S, Jaisil P, Thammasirirak S. A proteomics analysis of drought stress-responsive proteins as biomarker for drought-tolerant sugarcane cultivars. American Journal of Biochemistry and Biotechnology. 2010;**6**(2):89-102

[29] Azri W, Cosette P, Guillou C, Rabhi M, Nasr Z, Mliki A. Physiological

and proteomic responses to drought stress in leaves of two wild grapevines (*Vitis sylvestris*): A comparative study. Plant Growth Regulation. 2020;**91**:37-52

[30] Dong X, Guan L, Zhang P, Liu X, Li S, Fu Z, et al. Responses of maize with different growth periods to heat stress around flowering and early grain filling. Agricultural and Forest Meteorology. 2021;**303**:108378

[31] Amnan MA, Aizat WM, Khaidizar FD, Tan BC. Drought stress induces morpho-physiological and proteome changes of *Pandanus amaryllifolius*. Plants. 2022;**11**(2):221

[32] Liu B, Song L, Deng X, Lu Y, Lieberman-Lazarovich M, Shabala S, et al. Tomato heat tolerance: Progress and prospects. Scientia Horticulturae. 2023;**322**:112435

[33] Makonya GM, Ogola JB, Gabier H, Rafudeen MS, Muasya AM, Crespo O, et al. Proteome changes and associated physiological roles in chickpea (*Cicer arietinum*) tolerance to heat stress under field conditions. Functional Plant Biology. 2021;**49**(1):13-24

[34] Jegadeesan S, Chaturvedi P, Ghatak A, Pressman E, Meir S, Faigenboim A, et al. Proteomics of heat-stress and ethylene-mediated thermotolerance mechanisms in tomato pollen grains. Frontiers in Plant Science. 2018;**9**:1558

[35] Echevarria-Zomeno S, Fernandez-Calvino L, Castro-Sanz AB, Lopez JA, Vazquez J, Castellano MM. Dissecting the proteome dynamics of the early heat stress response leading to plant survival or death in *Arabidopsis*. Plant, Cell & Environment. 2016;**39**(6):1264-1278

[36] Zhao Q, Chen W, Bian J, Xie H, Li Y, Xu C, et al. Proteomics and phosphoproteomics of heat stress-responsive mechanisms in spinach. Frontiers in Plant Science. 2018;**26**(9):800

[37] Parankusam S, Bhatnagar-Mathur P, Sharma KK. Heat responsive proteome changes reveal molecular mechanisms underlying heat tolerance in chickpea. Environmental and Experimental Botany. 2017;**141**:132-144

[38] Ngcala MG, Goche T, Brown AP, Chivasa S, Ngara R. Heat stress triggers differential protein accumulation in the extracellular matrix of sorghum cell suspension cultures. PRO. 2020;**8**(4):29

[39] Jiang C, Bi Y, Mo J, Zhang R, Qu M, Feng S, et al. Proteome and transcriptome reveal the involvement of heat shock proteins and antioxidant system in thermotolerance of *Clematis florida*. Scientific Reports. 2020;**10**(1):8883

[40] Wang J, Liang C, Yang S, Song J, Li X, Dai X, et al. iTRAQ-based quantitative proteomic analysis of heat stress-induced mechanisms in pepper seedlings. PeerJ. 2021;**3**(9):11509

[41] Kholghi M, Toorchi M, Bandehagh A, Ostendorp A, Ostendorp S, Hanhart P, et al. Comparative proteomic analysis of salt-responsive proteins in canola roots by 2-DE and MALDI-TOF MS. Biochimica et Biophysica Acta (BBA)-Proteins and Proteomics. 2019;**1867**(3):227-236

[42] Song C, Zhang Y, Chen R, Zhu F, Wei P, Pan H, et al. Label-free quantitative proteomics unravel the impacts of salt stress on *Dendrobium huoshanense*. Frontiers in Plant Science. 2022;**13**:874579

[43] Rodriguez MC, Mehta D, Tan M, Uhrig RG. Quantitative proteome and PTMome analysis of Arabidopsis thaliana

root responses to persistent osmotic and salinity stress. Plant & Cell Physiology. 2021;**62**(6):1012-1029

[44] Yildiz M, Celik M, Terzi H. Proteomic analysis of the protective effect of sodium nitroprusside on leaves of barley stressed by salinity. European Journal of Biology. 2020;**79**(2):89-97

[45] Tammam A, El-Aggan W, Helaly A, Badr G, El-Dakak R. Proteomics and photosynthetic apparatus response to vermicompost attenuation of salinity stress *Vicia faba* leaves. Acta Physiologiae Plantarum. 2023;**45**(1):17

[46] Lakra N, Kaur C, Singla-Pareek SL, Pareek A. Mapping the 'early salinity response' triggered proteome adaptation in contrasting rice genotypes using iTRAQ approach. Rice. 2019;**12**(1):1-22

[47] Dolatabadi N, Toorchi M, Valizadeh M, Bandehagh A. The proteome response of salt-sensitive rapeseed (*Brassica napus* L.) genotype to salt stress. Notulae Botanicae Horti Agrobotanici Cluj-Napoca. 2019;**47**(1):17-23

[48] Wang X, Yan L, Wang B, Qian Y, Wang Z, Wu W. Comparative proteomic analysis of grapevine rootstock in response to waterlogging stress. Frontiers in Plant Science. 2021;**12**:749184

[49] Liu D, Zhan J, Luo Z, Zeng N, Zhang W, Zhang H, et al. Quantitative proteomics and relative enzymatic activities reveal different mechanisms in two peanut cultivars (*Arachis hypogaea* L.) under waterlogging conditions. Frontiers. Plant Science. 2021;**12**:716114

[50] Yang R, Li M, Harrison MT, Fahad S, Wei M, Li X, et al. iTRAQ proteomic analysis of Wheat (*Triticum aestivum* L.) genotypes differing in waterlogging tolerance. Frontiers. Plant Science. 2022;**13**:890083

[51] Mehmood SS, Lu G, Luo D, Hussain MA, Raza A, Zafar Z, et al. Integrated analysis of transcriptomics and proteomics provides insights into the molecular regulation of cold response in Brassica napus. Environmental and Experimental Botany. 2021;**187**:104480

[52] Dong X, Liu Z, Mi W, Xu C, Xu M, Zhou Y, et al. Overexpression of BrAFP1 gene from winter rapeseed (*Brassica rapa*) confers cold tolerance in Arabidopsis. Plant Physiology and Biochemistry. 2020;**155**:338-345

[53] Jiang J, Hou R, Yang N, Li L, Deng J, Qin G, et al. Physiological and TMT-labeled proteomic analyses reveal important roles of sugar and secondary metabolism in *Citrus junos* under cold stress. Journal of Proteomics. 2021;**237**:104145

[54] Yang Y, Saand MA, Abdelaal WB, Zhang J, Wu Y, Li J, et al. iTRAQ-based comparative proteomic analysis of two coconut varieties reveals aromatic coconut cold-sensitive in response to low temperature. Journal of Proteomics. 2020;**220**:103766

[55] Chen J, Zhou G, Dong Y, Qian X, Li J, Xu X, et al. Screening of key proteins affecting floral initiation of saffron under cold stress using iTRAQ-based proteomics. Frontiers in Plant Science. 2021;**12**:644934

[56] Farooq MA, Hong Z, Islam F, Noor Y, Hannan F, Zhang Y, et al. Comprehensive proteomic analysis of arsenic induced toxicity reveals the mechanism of multilevel coordination of efficient defense and energy metabolism in two Brassica napus cultivars. Ecotoxicology and Environmental Safety. 2021;**208**:111744

[57] Jian M, Zhang D, Wang X, Wei S, Zhao Y, Ding Q, et al. Differential

expression pattern of the proteome in response to cadmium stress based on proteomics analysis of wheat roots. BMC Genomics. 2020;**21**(1):1-3

[58] Komatsu S, Kuji R, Nanjo Y, Hiraga S, Furukawa K. Comprehensive analysis of endoplasmic reticulum-enriched fraction in root tips of soybean under flooding stress using proteomics techniques. Journal of Proteomics. 2012;**77**:531-560

[59] Khan PSV, Nagamallaiah GV, Rao MD, Sergeant K, Hausman JF. Abiotic stress tolerance in plants: Insights from proteomics. In: Ahmad P, Rasool S, editors. Emerging Technologies and Management of Crop Stress Tolerance, Volume 2: A Sustainable Approach. Academic Press; 2014. pp. 23-68

[60] Yanagawa Y, Komatsu S. Ubiquitin/proteasome-mediated proteolysis is involved in the response to flooding stress in soybean roots, independent of oxygen limitation. Plant Science. 2012;**185**:250-258

[61] White PJ, Brown P. Plant nutrition for sustainable development and global health. Annals of Botany. 2010;**105**(7):1073-1080

[62] Isarankura-Na-Ayudhya P, Isarankura-Na-Ayudhya C, Treeratanapaiboon L, Kasikun K, Thipkeaw K, Prachayasittikul V. Proteomic profiling of *Escherichia coli* in response to heavy metals stress. European Journal of Scientific Research. 2009;**25**:679-688

[63] Helmy M, Tomita M, Ishihama Y. OryzaPG-DB: Rice proteome database based on shotgun proteogenomics. BMC Plant Biology 2011;**11**(1):1-9

[64] Komatsu S, Kojima K, Suzuki K, Ozaki K, Higo K. Rice proteome database based on two-dimensional polyacrylamide gel electrophoresis: Its status in 2003. Nucleic Acid Research. 2004;**32**:388-392

[65] Sakata K, Ohyanagi H, Nobori H, Nakamura T, Hashiguchi A, Nanjo Y, et al. Soybean proteome database: A data resource for plant differential omics. Journal of Proteome Research. 2009;**8**(7):3539-3548

[66] Hajduch M, Ganapathy A, Stein JW, Thelen JJ. A systematic proteomic study of seed filling in soybean. Establishment of high-resolution two-dimensional reference maps, expression profiles, and an interactive proteome database. Plant Physiology. 2005;**137**(4):1397-1419

[67] Sun Q, Zybailov B, Majeran W, Friso G, Olinares PD, van Wijk KJ. PPDB, the plant proteomics database at Cornell. Nucleic Acids Research. 2009;**37**(9):69-74

[68] Hajheidari M, Abdollahian-Noghabi M, Askari H, Heidari M, Sadeghian SY, Ober ES, et al. Proteome analysis of sugar beet leaves under drought stress. Proteomics. 2005;**5**(4):950-960

[69] Ware DH, Jaiswal P, Ni J, Yap IV, Pan X, Clark KY, et al. Gramene, a tool for grass genomics. Plant Physiology. 2002;**130**(4):1606-1613

[70] Lei Z, Dai X, Watson BS, Zhao PX, Sumner LW. A legume specific protein database (LegProt) improves the number of identified peptides, confidence scores and overall protein identification success rates for legume proteomics. Phytochemistry. 2011;**72**(10):1020-1027

Chapter 10

Integrating Omics Approaches for Abiotic Stress Tolerance in Plants

Amal Morsy and Nahla El-Sherif

Abstract

Plants are exposed to a variety of challenging abiotic stress pressures such as salt, drought, waterlogging, heat, oxidative stress, and heavy metals. An in-depth understanding of how plants respond to abiotic stress from the molecular side view is an important criterion for its actual management. Abiotic stress tolerance is a complicated phenomenon that includes many interacting steps such as signal recognition and an array of subsequent responses in a signal transduction pathway. This multitude of reactions necessitates evidence at the omics level to comprehend it properly. Enormous advance has been made in the field of omics in different areas such as genomics, transcriptomics, proteomics, metabolomics, phenomics, and ionomics. These advanced approaches generate multifaceted data that can shed some light on what is going on inside the plant cells. For instance, functional genomics deals with the relation between the genome and the phenotype, this relation is highly affected by environmental abiotic stress conditions. For effective analysis of the huge amount of data generated from the omics approaches, advancement in bioinformatics and computational tools have been exploited. This review summarizes the advances in omics tools, both traditional and recent, comprising QTL mapping for abiotic stress tolerance, genome-wide association studies (GWAS) and genomic selection (GS) used to examine the mechanisms of abiotic stress tolerance in plants.

Keywords: abiotic stress, omics, genomics, transcriptomics, ionomics, QTL mapping

1. Introduction

Abiotic stress tolerance is a complex molecular regulatory system. It is considered a multigenic response that involves stress-responsive gene expression, signal transduction, and sensing. This complex array of interplaying reactions happening inside the plant cells can be understood with the help of different omics tools. Plants, being sessile organisms, are continuously exposed to unavoidable abiotic stress factors, which highly affect plants, reduce productivity, and result in significant losses in crop yields of more than 60% [1]. Examples of the most influencing abiotic stress conditions on plants include salinity, drought, submergence, oxidative stress, and temperature change (heat and extreme cold) as well as environmental pollutants such as heavy metals, herbicides, and pesticides.

 IntechOpen

The phases of alarm, resistance, and exhaustion might be seen as the three primary phases of plant stress events and responses [2]. Each phase may induce the expression of a different set of genes.

Before the omics era, efforts were concentrated on the study of individual genes. Nowadays, advancement in omics tools, including genomics, transcriptomics, proteomics, and metabolomics enabled researchers to find the molecular interaction and relationships with signal cascades which connect specific signals obtained with specific molecules.

Each omics strategy has a limit that affects the approach's specificity or sensitivity, but employing a multi-omics approach may help to get around this restriction by giving a full picture of the changes that happen in plants under abiotic stress conditions at the genomic, transcriptomic, proteomic, metabolomic, and phenomics side of view. In the following parts, we will be looking in some detail at each of these points of view.

Over the past 20 years, many plant genomes have been sequenced, thanks to advancements in genomics and next-generation sequencing. Modern genomics, including marker-assisted selection (MAS), genome-wide association studies (GWAS), genomic selection (GS), transgenic breeding, and gene editing, as well as transcriptomics, proteomics, and next-generation sequencing, offer a variety of novel, potent methods for the study of plant species. This review investigates the general effects of significant abiotic stresses on plants, as well as the adaptive mechanisms and the interplay of omics approaches including genomic, transcriptomic, and proteomic procedures employed by researchers to lessen these difficulties and address abiotic stresses. Understanding the response of a plant to an abiotic stress is the first step toward breeding stress-tolerant varieties. To provide a comprehensive image of the abiotic stress tolerance mechanism in plants, a detailed review of each omics technique and multi-omics integration has been covered in the following sections, thus helping in plant abiotic stress tolerance as well as breeding enhancement (**Figure 1**).

1.1 Genomics approach

Scanning the genomes of many plant species for abiotic stress tolerance, specifically the crop species, allows us to predict stress-related genes. There are numerous opportunities for functional investigations of abiotic stress sensitive genes and tolerance mechanisms, thanks to RNA-Seq, random and targeted mutagenesis, gene shifting, complementation, and synthetic promoter trapping techniques [3].

Recent advances in major genomic tools such as marker-assisted selection (MAS), single nucleotide polymorphisms (SNPs), next-generation sequencing (NGS), and genome-wide association studies (GWAS) have a great impact to study genetic diversity and emphasize it in case genetic diversity is low. The complicated biological mechanisms and pathways regulating abiotic stress tolerance in rapeseed have been extensively studied using omics, GE tools, and molecular breeding strategies. To reveal various molecular, physiologic, and metabolic pathways related to abiotic stress tolerance in rapeseed, multi-dimensional omics studies have gathered many datasets at the transcriptome, proteome, and metabolome level. This has opened new research avenues to understand complex traits such as abiotic stress tolerance [4].

Deoxyribonucleic acid (DNA) markers are applied to scan the genomes to screen desired loci and transgene; SNPs have been discovered in many plants using QTL mapping for the desired trait. Genes linked to various abiotic stresses were identified

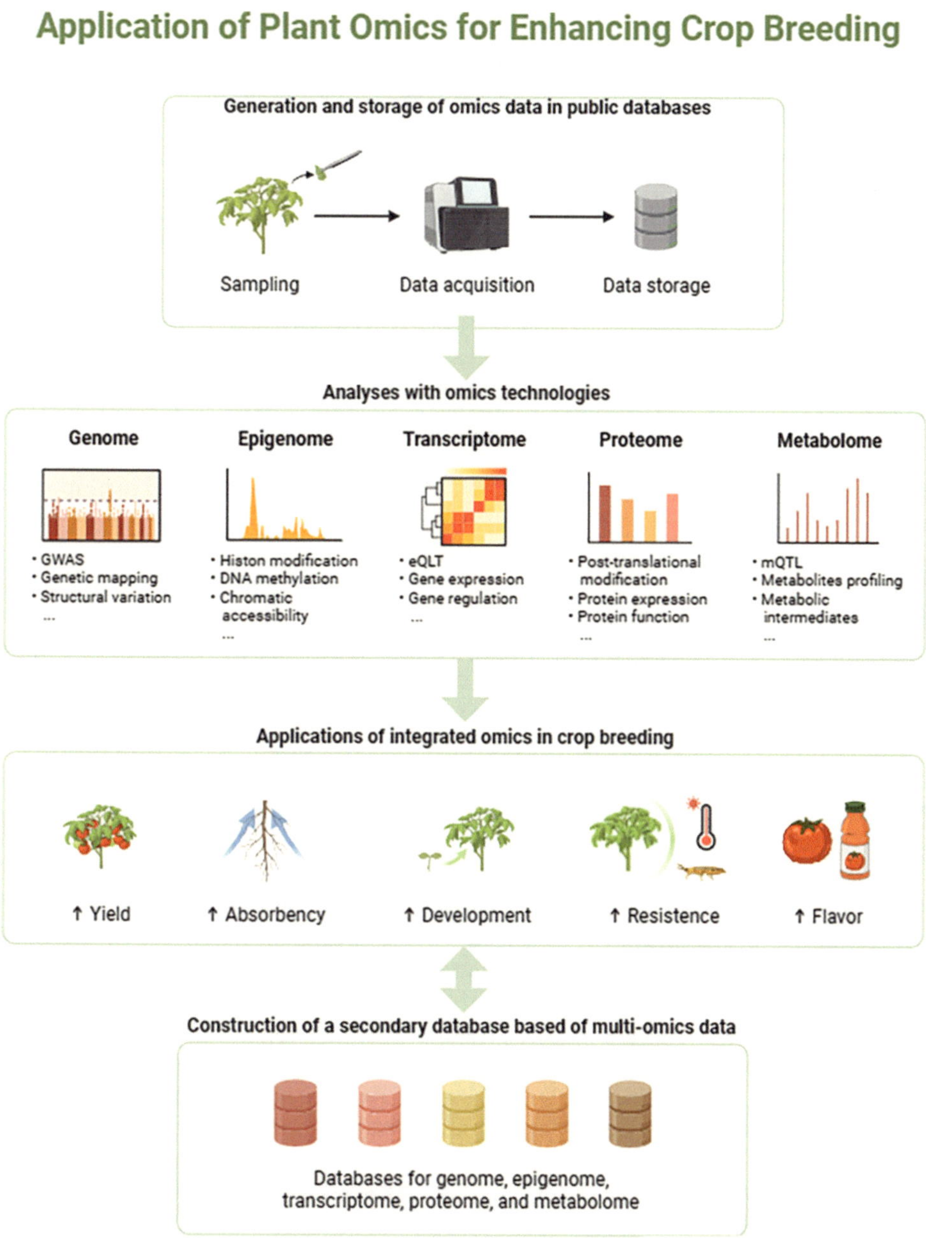

Figure 1.
Integrative multi-omics techniques to provide plants with abiotic stress tolerance. The diagram was produced using BioRender (https://biorender.com/).

in rapeseed breeding programs. Finding SNPs for various abiotic stress-related QTLs is made possible by high-throughput sequencing techniques and rapeseed EST construction [5].

Stress tolerance results from multiple complex responses and mechanisms. Assessment of these mechanisms and stress-related QTLs has assisted plant breeders to develop elite genotypes with stress tolerance. Accurate depictions of the genetic pathways in crop plants that regulate abiotic stress responses will be possible with the utilization of numerous "omics data". Combining this information with pan-genomics may improve the precision of genome-assisted breeding, boost crop breeding effectiveness, and open new opportunities for genome editing-based abiotic stress tolerance development [6].

1.2 Transcriptomic approach

Large-scale genomic sequence information has been revealed in many plants due to the advancements in transcriptome analysis methods. A comprehensive picture of the transcriptomes can be obtained by identifying partial or whole cDNA sequences. There are three primary databases that house the available ESTs.

These ESTs are organized by NCBI, TIGR, and Sputnik, which have fully characterized gene sequences. Microarrays and cDNA were frequently utilized to pinpoint gene expression profiles in abiotically stressed Arabidopsis, potato, rice, sorghum, maize, and wheat [7].

The introduction of genome sequencing technologies and the subsequent decline in their price have revolutionized the field of omics studies and hence the comprehension of plant metabolic pathway identification [8].

By using transcriptome investigations, it is possible to find molecular markers that are connected to widespread reactions in a variety of plant species; these molecular markers are utilized to comprehend the way of action of many agents such as biostimulants [9] and genes differentially expressed in each case. Alterations in gene expression have been studied as the intensity of RNA signals observed in plants in response to different treatments.

A comprehensive picture of the transcriptomes can be obtained by identifying partial or complete cDNA sequences. The available ESTs are arranged with fully characterized gene sequences in three major databases, namely NCBI, TIGR, and Sputnik. Researchers studying the gene expression profiles of stress tolerance-related traits in Arabidopsis and rice have greatly benefited from abiotic stress-related ESTs [7].

Transcriptome analysis is a dynamic method for assessing any increased gene expression over a given period [10]. Conventional techniques including cDNAs-AFLP, differential display-PCR (DD-PCR), and SSH were initially used to study transcriptome dynamics under drought conditions, but these techniques have poor resolution [11].

Digital gene expression (DGE) and microarray RNA expression profiles are also analyzed, thanks to advances in the relevant technology [12]. Microarray research demonstrated the differential expression of soybean and barley genes during the embryonic and reproductive stages of drought stress [13]. The Affymetrix gene chip array was used in a manner like this to identify the various gene expression levels in soybeans during dehydration stress [14].

1.3 Proteomic approach

Sequence, structural, functional, and expression proteomics are the four subcategories of proteomics. Proteomics is used to profile the total amount of expressed proteins in an organism [15]. Proteomics typically refers to two different sorts of studies: (1) the characterization of a proteome, which involves identifying all the proteins expressed in a particular cell, tissue, organ, organism, or population; and (2) other type of proteomics which is differential proteomics in which the proteome of a plant under control conditions is compared to the proteome of the same plant under study conditions, such as the exposure to a heavy metal or a water shortage, or in another example, the comparison of protein expression profiles between different varieties.

Structural proteomics involves determining the protein amino acid sequence using High-Performance Liquid Chromatography (HPLC). Using computational-based methods to deal with the huge number of natural protein structures, like modeling, experimental techniques like crystallization, nuclear magnetic resonance (NMR), X-ray diffraction of protein crystals, and electron microscopy, structural proteomics studies protein structure to better understand potential activities of proteins [16].

Large-scale proteomic research is being done on how plants react to salinity and drought stressors. To understand the stress-response mechanism in plants, proteomic techniques have been used at the level of the entire plant, the organ, and the subcellular level. Key proteomic research on plant species under salinity and drought stress has been previously summarized [7].

Proteomic investigations revealed that heat-shock proteins, nucleoside diphosphate kinase, RuBisCO, Cu-Zn superoxide dismutase (SOD), and 2-Cys-peroxiredoxin were shown to be substantially increased in sugar beetroot under drought stress. With the aid of 2D gel electrophoresis and image analysis tools, the changes in leaf proteins were investigated. Certain proteins displayed genotype-specific patterns of up- or downregulation. Liquid chromatography-tandem mass spectrometry (LC-MS/MS) was used to examine relevant protein spots. Rubisco and other additional proteins involved in signal transduction, oxidative stress, redox control, and chaperone functions were identified. Some of these proteins may offer a physiological benefit in arid conditions, making them desirable MAS targets [17]. A proteome investigation of maize under drought stress [18] revealed proteins involved in metabolism, photosynthesis, and stress responses. Branched-chain amino acid amino transferase 3 protein and zinc finger transcription factor oxidative stress 2 proteins played a significant role in drought stress responses in the plants that over-expressed ethylene response factor, according to a proteomic analysis of Arabidopsis under drought stress [19].

2. Metabolomic approach

Under abiotic stress, the plant metabolome needs to be reprogrammed to maintain its essential metabolic homeostasis. This takes place by producing protective metabolites that enhance stress tolerance. Flavonoids, terpenes, phenols, and alkaloids are typical stress-induced specialized metabolites that are produced in certain plant species, organs, tissues, and cells [20].

If we want to rationally adjust plant systems to maximize plant tolerance to various abiotic stresses, we must first understand how key metabolites are regulated. We provide a brief overview of the modern metabolomics approaches used to study plant metabolism in the sections that follow. We must first comprehend how important metabolites are regulated, if we hope to rationally modify plant systems to increase plant tolerance to various abiotic stimuli. Metabolite profiling might be targeted, aiming to identify specific responses or non-targeted, aiming to classify as many metabolites as possible.

Using a non-targeted metabolome profiling, it has been demonstrated that both primary and secondary metabolites play significant roles in how plants react to abiotic stress such as drought and salinity. Primary metabolites with significant roles in photosynthetic failure and osmotic readjustment have been identified by gas chromatography coupled to mass spectrometry, including sugars, amino acids, and Krebs cycle intermediates. While the secondary metabolites, such as coenzymes, regulatory compounds, and antioxidant scavengers, responded to other specific stress circumstances. In addition to establishing the phenotypic response of the plant and identifying stress-tolerant lines, studies of metabolites in response to abiotic stress are useful for elucidating the genetic and biochemical mechanisms underlying the stress situation [21]. There is a complicated crosstalk between different stresses, as plants are continuously exposed to more than one stress rather than to individual stresses, resulting in significant effects on key metabolic processes like those involving sugars, their phosphates, and sulfur-containing substances.

Atomic and molecular mass spectrometry (MS) and nuclear magnetic resonance (NMR) spectroscopy are the two analytical techniques most frequently utilized in metabolomics research [22]. The non-destructive and effective NMR-based metabolomic study can reveal the structural details of the metabolites [22].

Previous research has reviewed a list of significantly changed central metabolites under cold, drought, flooding, heat, and salinity stress from published studies [23]. Specific metabolites were found to be impacted by cold stressors, others by drought, by flooding, by heat, and by salinity, among the summarized metabolites. Some metabolites associated with stress displayed common responses to many abiotic stresses, while others only responded to few stresses.

Amino acids, the antioxidant phenolic compounds catechin and kaempferol, as well as the osmolytes raffinose and galactinol, exhibited increased abundance under drought stress, whereas metabolites involved in photorespiration, redox regulation, and carbon fixation showed decreased abundance under drought stress [24].

2.1 Ionomic approach

The term "ionome" refers to an organism's composition of minerals, nutrients, and trace elements, which serves as the inorganic part of cellular and organismal systems [25]. Ionomics, considered as the functional genomics of elements, is thus the quantitative and simultaneous measurement of the elemental composition of living organisms (metal, non-metal, and metalloid) and changes in this composition in response to physiological stimuli, developmental state, and genetic modifications [25]. Due to their numerous uses, minerals (both essential and non-essential) have recently been recognized as important players in the biology of plant stress. ICP-AES and ICP-MS are two examples of high-throughput elemental analysis technologies that can be used to analyze the compositions of metal, non-metal, and metalloid elements in living organisms [26].

High-throughput elemental analysis technologies must be used for ionomics and must be integrated with both genetic and bioinformatics techniques. Ionomics has the capacity to record data regarding the functional status of an organism under abiotic stress. Samples must be digested in trace metal grade acid to reveal their fundamental elements. Compared to proteins, transcripts, and metabolites, elements have the advantages that they are relatively cheap to measure and the whole class of components can be measured in a single run.

2.2 Phenomics approach

In the context of the organism, phenomics is defined as the process of characterizing phenotype using high-dimensional phenotype data [27].

However, the term "phenotype" refers to the entire phenotype, and plant phenotype can be affected by the underlying interactions between the genome, environment, and management [28, 29]. For this reason, the phenomenon of phenomics is also known as genotype-phenotype-environotype interactions [30].

With the use of phenomics pipelines and advancement of non-destructing sensing and imaging techniques, researchers may turn vast amounts of image and sensor data into knowledge, necessitating the use of novel data processing and modeling techniques. Together, these breakthroughs are accelerating the development of the next generation of crops that are more climate change-resilient and sustainable, and whose advantages are expected to extend beyond physiology to breeding and have a real-world influence on ongoing efforts to ensure global food security. Advancements in technologies have a great impact in the phenomics area. One such technology is the WinRoots technology, which allows for both the immediate capture of high-throughput and high-quality whole-plant phenotypic data as well as the high-throughput growth of crops under controlled soil stress conditions. It would facilitate precision crop breeding by integrating phenotypes, genotypes, and other omics data sources in a comprehensive manner to unravel the genetic regulatory networks and mechanisms underpinning traits related to soil stress [31].

3. QTL mapping for abiotic stress tolerance

Plant breeders are now better able to understand how plants tolerate abiotic stress, thanks to the quantitative trait locus (QTL) discovery and mapping procedure. Molecular markers and stable QTL are crucial for molecular breeding.

Recent research has emphasized methods and developments for abiotic stress tolerance breeding programs in cereals, the difficulties in introducing advantageous QTL using conventional breeding techniques like mutation breeding and marker-assisted selection (MAS), and the advancements made by new breeding techniques like genome-wide association studies (GWASs), the clustered regularly interspaced short palindromic repeat, and others [32].

Conventional breeding, often known as traditional breeding, is the act of developing new plant varieties using traditional techniques and biological procedures [33]. Pyramiding several types of resistance, molecular markers have been utilized since the 1990s to select superior hybrids, thus creating multi-line cultivars with enduring resilience to abiotic stressors using many resistance genes [34].

SSRs (Single sequence repeats) and SNPs (Single nucleotide polymorphisms) can detect single nucleotide changes at the whole genome level as well as DNA variations in populations that are closely linked to one another [35]. The most well-known polymerase chain reaction (PCR)-based markers are microsatellites or SSR markers. Because they are co-dominant, hypervariable, locus sensitive, and multi-allelic, these markers are frequently utilized among plant species for screening, characterizing, and assessing genetic diversity [36].

SNP detection and application have become comparatively simple with the development of NGS and high-throughput genotyping technology [37]. The identification of genomic regions targeted for breeding programs for cold tolerance and the huge number of markers to find QTL with economically significant traits have both been made possible using high-density SNP iSelect assays, using 2 probes per SNP to determine the relative intensity ratio of the 2 possible target alleles for any given locus (9 and 90 K) in wheat (*T. aestivum*). The linkage markers produced in this study should help with molecular-assisted breeding of wheat with cold tolerance [38].

4. Genome-wide association studies (GWAS)

Abiotic stress has an impact on agricultural production. Investigations are typically based on single genes/traits. However, plant stress is controlled not by one trait but rather by multiple traits. GWASs are an effective method for examining a variety of traits linked to a single or multiple stresses. New gene candidates or quantitative trait loci, responsible for abiotic stress, have been discovered through GWAS on a variety of plants and crops. Future crop breeders and designers will benefit greatly from the findings of these investigations to understand the genetics underlying complex phenotypic traits.

In this field, network maps have been developed to reveal these complex traits and how genes interact and coordinate. A network map developed in maize shows that duplicated genes may have functional divergence at various layers. The integrative network map provides a potential that can be used to predict gene function, analyze cross-regulating interactions, and identify the molecular processes underlying complex traits [39].

The development of suitable cultivars may be facilitated and accelerated by the discovery of molecular markers linked to important genes that regulate tolerance to many abiotic stress factors and thus improve the plant yield and quality. Studies of multigene-associated traits, particularly GWASs, make extensive use of improved statistical methods and increased computational efficiency. The findings of these studies on plants have given breeders and future breeding studies useful information on the interactions between genotype, phenotype, and environment [38].

Through GWAS SNPs, selected genotypes were genotyped using the Barley 9 K iSelect SNP Array and several likely candidates for genes that underlie cold acclimation stress and/or chlorophyll were found. The candidate genes for cold-related traits were found in nearby regions of significant SNPs, primarily on chromosomes 1H, 3H, and 6H. For example, abscisic acid (ABA) signaling, hydrolase activity, protein kinase, and transduction of environmental signal transduction at the post-translational modification levels are a few examples of candidate genes that participate in plant response to abiotic stress at the post-transcriptional modification level [40].

5. Genomic selection (GS)

Genomic selection (GS) is a relatively straightforward, reliable, and powerful approach that has been used to solve the problem of breeding complex polygenic traits such as abiotic stress. In GS, the breeding values of lines are predicted using their phenotypes and marker genotypes [41, 42]. GS is more efficient than Marker-assisted breeding (MAB), because it builds a prediction model using all available marker information simultaneously, preventing the effects of skewed markers. Another advantage is that the majority of the variation's small-effect QTL, including epistatic interaction effects, are captured by GS.

By using precise, high-throughput phenotyping and a wide base of selection, genomic selection can be used to select for genotypes that are tolerant to abiotic stress. The development of genomics, transcriptomics, and proteomics has led to the identification of numerous target genes that may reduce the impact of abiotic stress. Additionally, mutant loci that facilitate complex properties, such as yield, have been found, which are otherwise very challenging to understand.

Genomic selection can be helpful to precisely exercise genetic gain and selection needed for improvement of any plant trait [43] and hence has become a crucial tool that can use this data to model crop yield for quick and efficient selection under different stress conditions to meet the production challenges in a world, where the climate is continuously changing, thus causing many abiotic stresses affecting plant growth and yield. GS is currently utilized in the breeding of numerous crops reviewed in [44]. Genetic and environmental (GE) interactions play a major role in the accuracy of GS, but many studies have only attempted to estimate the main effects for each marker. In addition to studying G and E, these multi-environmental trials are crucial for plant breeding, because they will increase the number of breeding cycles per year.

Acknowledgements

This research has received no external funding. The authors would like to extend their gratitude to the reviewers and the editors.

Funding

The authors declare that there is no funding provided for this review.

Authors' contributions

A. Morsy conceived the paper and the gap in the literature that this research aims to address, N. El-Sherif (corresponding author) wrote the paper and revised the related literature related to the manuscript. Both authors revised the final manuscript.

Competing interests

The authors declare that there is no conflicting interest of any type that could influence this research.

Declarations

Ethics approval and consent to participate.

The manuscript has not included any data or material and ethics issues. The authors declare that they have no personal relationships that could have appeared to influence the work in this manuscript and that their participation is completely voluntary.

Availability of data and material

The authors ensure that their data is available all the time.

Consent for publication

The authors give their consent for publication of this review.

Author details

Amal Morsy and Nahla El-Sherif*
Faculty of Science, Botany Department, Ain Shams University, Cairo, Egypt

*Address all correspondence to: elsherif.nahla@sci.asu.edu.eg

References

[1] Singhal P, Jan AT, Azam M, Haq QMR. Plant abiotic stress: A prospective strategy of exploiting promoters as alternative to overcome the escalating burden. Frontiers in Life Science. 2016;**9**(1):1-12. DOI: 10.1080/21553769.2015.1077478

[2] Teklehaimanot Z. Review: Physiological plant ecology: Ecophysiology and stress physiology of functional groups, 4th ed. Forestry. 2004;**77**. DOI: 10.1093/forestry/77.4.365-a

[3] Ismail AM, Horie T. Genomics, physiology, and molecular breeding approaches for improving salt tolerance. Annual Review of Plant Biology. 2017;**68**. DOI: 10.1146/annurev-arplant-042916-040936

[4] Raza A, Razzaq A, Mehmood SS, Hussain MA, Wei S, He H, et al. Omics: The way forward to enhance abiotic stress tolerance in Brassica napus L. 2021;**12**(1):251-281

[5] Basunanda P, Radoev M, Ecke W, Friedt W, Becker HC, Snowdon RJ. Comparative mapping of quantitative trait loci involved in heterosis for seedling and yield traits in oilseed rape (*Brassica napus L.*). Theoretical and Applied Genetics. 2010;**120**(2):271-281. DOI: 10.1007/s00122-009-1133-z

[6] Derbyshire MC, Batley J, Edwards D. Use of multiple 'omics techniques to accelerate the breeding of abiotic stress tolerant crops. Current Plant Biology. 2022;**32**(44):100262

[7] Sayyar Khan M, Nawaz Khan M. Understanding plant responses to drought and salt stresses: Advances and challenges in "omics" approaches. In: Transgenic Crops - Emerging Trends and Future Perspectives. 2019

[8] Owen C, Patron NJ, Huang A, Osbourn A. Harnessing plant metabolic diversity. Current Opinion in Chemical Biology. 2017;**40**:24-30

[9] González-Morales S, Solís-Gaona S, Valdés-Caballero MV, Juárez-Maldonado A, Loredo-Treviño A, Benavides-Mendoza A. Transcriptomics of Biostimulation of Plants under Abiotic Stress. 2021. DOI: 10.3389/fgene.2021.583888

[10] El-Metwally S, Ouda OM, Helmy M. Next generation sequencing technologies and challenges in sequence assembly. In: Springer Briefs in Systems Biology. 2014

[11] Nataraja KN, Madhura Bhat G, Parvathi MS. OMICS: Modern tools for precise understanding of drought adaptation in plants. In: Plant OMICS and Crop Breeding. 2017

[12] Duque AS, De Almeida AM, Bernardes A, Marques J, Farinha AP, Santos D, et al. Abiotic stress responses in plants: Unraveling the complexity of genes and networks to survive. In: Abiotic Stress - Plant Responses and Applications in Agriculture. 2013. DOI: 10.5772/52779

[13] Le DT, Nishiyama R, Watanabe Y, Tanaka M, Seki M, Ham LH, et al. Differential gene expression in soybean leaf tissues at late developmental stages under drought stress revealed by genome-wide transcriptome analysis. PLoS One. 2012;7(11):e49522. DOI: 10.1371/journal.pone.0049522

[14] Khan F, Chai HH, Ajmera I, Hodgman C, Mayes S, Lu C. A transcriptomic comparison of two

bambara groundnut landraces under dehydration stress. Genes (Basel). 2017;**8**. DOI: 10.3390/genes8040121

[15] Mosa KA, Ismail A, Helmy M. Omics and system biology approaches in plant stress research. In: SpringerBriefs in Systems Biology. 2017. ISSN 2193-4746, ISSN 2193-4754 (electronic). ISBN 978-3-319-59377-7, ISBN 978-3-319-59379-1 (eBook). DOI: 10.1007/978-3-319-59379-1. Library of Congress Control Number: 2017943691

[16] Heal JW, Bartlett GJ, Wood CW, Thomson AR, Woolfson DN. Applying graph theory to protein structures: An atlas of coiled coils. Bioinformatics. 2018;**34**. DOI: 10.1093/bioinformatics/bty347

[17] Hajheidari M, Abdollahian-Noghabi M, Askari H, Heidari M, Sadeghian SY, Ober ES, et al. Proteome analysis of sugar beet leaves under drought stress. Proteomics. 2005;**5**(4):950-960

[18] Kim SG, Lee JS, Kim JT, Kwon YS, Bae DW, Bae HH, et al. Physiological and proteomic analysis of the response to drought stress in an inbred Korean maize line. Plant Omics. 2015;**8**(2):159-168

[19] Scarpeci TE, Frea VS, Zanor MI, Valle EM. Overexpression of AtERF019 delays plant growth and senescence, and improves drought tolerance in Arabidopsis. Journal of Experimental Botany. 2017;**68**(3):erw429. DOI: 10.1093/jxb/erw429

[20] Khare S, Singh NB, Singh A, Hussain I, Niharika K, Yadav V. Plant secondary metabolites synthesis and their regulations under biotic and abiotic constraints. Journal of Plant Biology. 2020;**63**(11)

[21] Hong J, Yang L, Zhang D, Shi J. Plant metabolomics: An indispensable system biology tool for plant science. International Journal of Molecular Sciences. 2016;**17**(6):767

[22] Dunn WB, Bailey NJC, Johnson HE. Measuring the Metabolome: Current Analytical Technologies. 2005

[23] Xu Y, Fu X. Reprogramming of Plant Central Metabolism in Response to Abiotic Stresses: A Metabolomics View. 2022

[24] Barchet GLH, Dauwe R, Guy RD, Schroeder WR, Soolanayakanahally RY, Campbell MM, et al. Investigating the drought-stress response of hybrid poplar genotypes by metabolite profiling. Tree Physiology. 2014;**34**(11). DOI: 10.1093/treephys/tpt080

[25] Salt DE, Baxter I, Lahner B. Ionomics and the study of the plant ionome. Annual Review of Plant Biology. 2008;**59**

[26] Shariatipour N, Shams Z, Heidari B. The role of ionomics as functional genomics of elements in plant abiotic stress tolerance. In: Omics Analysis of Plants under Abiotic Stress. Apple Academic Press (AAP); 2022. DOI: 10.1201/9781003282761-5

[27] Houle D, Govindaraju DR, Omholt S. Phenomics: The next challenge. Nature Reviews Genetics. 2010;**11**(12):855-866

[28] Gjuvsland AB, Vik JO, Beard DA, Hunter PJ, Omholt SW. Bridging the genotype-phenotype gap: What does it take? The Journal of Physiology. 2013;**591**(Pt 8)

[29] Großkinsky DK, Syaifullah SJ, Roitsch T. Integration of multi-omics techniques and physiological phenotyping within a holistic phenomics approach to study senescence in model and crop plants. Journal of Experimental Botany. 2018;**69**(4):825-844

[30] Song P, Wang J, Guo X, Yang W, Zhao C. High-throughput phenotyping: Breaking through the bottleneck in future crop breeding. The Crop Journal. 2021;**9**(S9)

[31] Zhang Y, Zhang W, Cao Q, Zheng X, Yang J, Xue T, et al. WinRoots: A high-throughput cultivation and phenotyping system for plant Phenomics studies under soil stress. Frontiers in Plant Science. 2022;**12**. DOI: 10.3389/fpls.2021.794020

[32] Raj SRG, Nadarajah K. QTL and candidate genes: Techniques and advancement in abiotic stress resistance breeding of major cereals. International Journal of Molecular Sciences. 2023;**24**(1):6

[33] Acquaah G. Conventional plant breeding principles and techniques. In: Advances in Plant Breeding Strategies: Breeding, Biotechnology and Molecular Tools. 2016. DOI: 10.1007/978-3-319-22521-0_5

[34] Kamboj D, Kumar S, Mishra CN, Srivastava P, Singh G, Singh GP. Marker assisted breeding in cereals: Progress made and challenges in India. Journal of Cereal Research. 2020;**12**. DOI: 10.25174/2582-2675/2020/104208

[35] da Silva Linge C, Antanaviciute L, Abdelghafar A, Arús P, Bassi D, Rossini L, et al. High-density multi-population consensus genetic linkage map for peach. PLoS One. 2018;**13**. DOI: 10.1371/journal.pone.0207724

[36] Jasim Aljumaili S, Rafii MY, Latif MA, Sakimin SZ, Arolu IW, Miah G. Genetic diversity of aromatic rice germplasm revealed by SSR markers. BioMed Research International. 2018;**2018**. DOI: 10.1155/2018/7658032

[37] Varshney RK, Kudapa H, Roorkiwal M, Thudi M, Pandey MK, Saxena RK, et al. Advances in genetics and molecular breeding of three legume crops of semi-arid tropics using next-generation sequencing and high-throughput genotyping technologies. Journal of Biosciences. 2012;**37**. DOI: 10.1007/s12038-012-9228-0

[38] Zhao Y, Li J, Zhao R, Xu K, Xiao Y, Zhang S, et al. Genome-wide association study reveals the genetic basis of cold tolerance in wheat. Molecular Breeding. 2020;**40**. DOI: 10.1007/s11032-020-01115-x

[39] Han L, Zhong W, Qian J, Jin M, Tian P, Zhu W, et al. A multi-omics integrative network map of maize. Nature Genetics. 2023;**55**. DOI: 10.1038/s41588-022-01262-1

[40] Elakhdar A, Slaski JJ, Kubo T, Hamwieh A, Ramirez GH, Beattie AD, et al. Genome-wide association analysis provides insights into the genetic basis of photosynthetic responses to low-temperature stress in spring barley. Frontiers in Plant Science. 2023;**14**

[41] Heffner EL, Sorrells ME, Jannink JL. Genomic selection for crop improvement. Crop Science. Jan 2009;**49**(1)

[42] Wang X, Xu Y, Hu Z, Xu C. Genomic selection methods for crop improvement: Current status and prospects. The Crop Journal. 2018;**6**(4)

[43] Penna S, Jain SM. Fruit crop improvement with genome editing, in vitro and transgenic approaches. Horticulturae. 2023;**9**(1):58

[44] Deshmukh R, Sonah H, Patil G, Chen W, Prince S, Mutava R. Integrating omic approaches for abiotic stress tolerance in soybean. Frontiers in Plant Science. 2014;**5**(244):244

Chapter 11

Grafting in Pepper to Overcome Drought, Salinity, and High Temperature

Yaiza G. Padilla, Ramón Gisbert-Mullor, Salvador López-Galarza and Ángeles Calatayud

Abstract

Since the twentieth century, pepper production and consumption have increased worldwide. However, pepper-harvested area decreases every year, which is partly associated with climate change effects such as extreme temperatures, salinity, and drought. These abiotic stresses affect pepper plants by limiting photosynthesis, growth, and development; increasing reactive oxygen species (ROS); and blocking metabolic processes, among others, leading to reduced production and fruit quality. Grafting rises as an effective technique to cultivate in unfavorable environmental conditions, because crops yields increase when tolerant rootstocks are employed due to vanished stress perception in the scion. Tolerant rootstocks favor water and nutrients uptake, photosynthesis maintenance, antioxidant system and hormonal signaling activation, and gene expression regulation, facilitated by the bidirectional signal transmission between rootstock and scion. This chapter summarizes the latest advances in pepper abiotic stress mitigation by grafting: how tolerance is achieved with the help of tolerant pepper rootstocks under heat, salt, and water stress.

Keywords: *Capsicum annuum*, grafting, abiotic stress, rootstock, thermal stress, salt stress, water stress

1. Introduction

The genetic pool of *Capsicum annuum* includes lots of artificial and natural hybrids, breeding lines, cultivars, and landraces [1], as well as the most relevant commercial pepper varieties, such as bell peppers. In line with Ref. [2], the harvested pepper area worldwide in 2021 was 2.055.310 ha, with 36286.643 tons of pepper production all around the world. For continents, Asia was the first producer (67.9%), followed by the Americas (11.5%) and Europe (10.8%), in 2021 [2]. In the last two decades, pepper production has grown, but farmland areas diminish annually. This scenario can be partly owing to climate change because it accelerates land degradation, caused by drought, desertification, extreme temperatures, among other socio-environmental constraints [3].

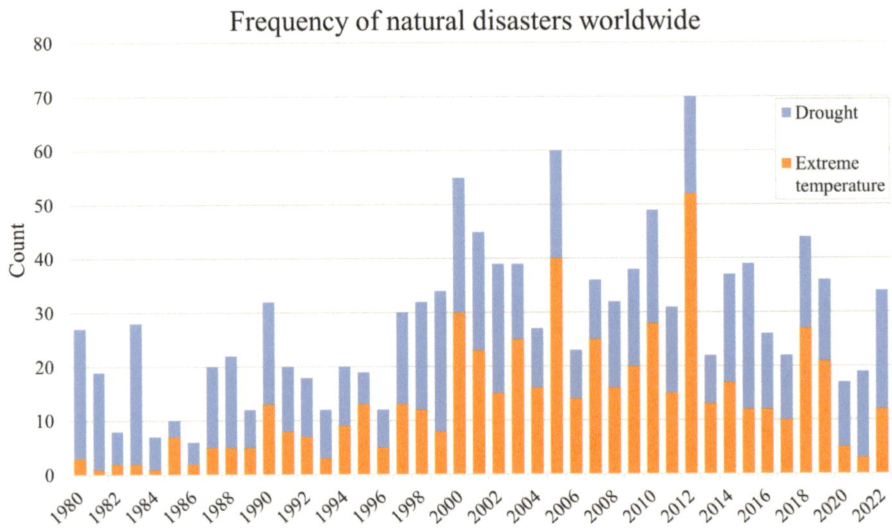

Figure 1.
Frequency of natural disasters in the world (1980–2022). International Monetary Fund indicators dataset: Climate-related disasters frequency (2022). Blue bars refer to drought events, and orange bars depict extreme temperature events.

Pepper is normally eaten fresh or processed as a spice or vegetable, and can also be used as cosmetic and medical applications from its extract, and also for ornamental purposes [4]. Global pepper success can be partly due to its nutritional composition, which comprises fiber, tocopherols like vitamin E, ascorbic acid (vitamin C), minerals, carotenoids, and flavonoids. Its vitamin C concentration is sufficiently high to be made available to consumers [5, 6].

One serious threat to the world's agriculture is abiotic stress, and it is the first source of crop productivity loss that, for major crops, can be as high as 50% [7, 8]. This stress includes extreme temperatures, salinity, and drought, among others, whose prevalence and severity rise with today's expected climate change conditions (**Figure 1**) [9]. Indeed, salinity and drought are expected to transform more than half the world's arable land into unproductive land by 2050 [8].

Certain strategies that aim to solve the problems caused by abiotic stress could include either conventional or molecular breeding. Nevertheless, tolerance to abiotic stress results from numerous quantitative characters controlled by several minor genes [8, 10]. Ameliorating abiotic stress by grafting on Solanaceae crops is an effective strategy, especially for heat stress, drought, and salinity conditions [11]. With horticultural crops, the basis of grafting for cultivation under abiotic stress conditions is rootstocks' ability to survive undesirable stressors, in other words, their resistance [12]. Tolerant pepper rootstocks could attenuate the impact of abiotic stresses. Hence, the mechanisms implicated in this tolerance response should be studied.

2. Salinity

Salinity stress markedly limits plant growth and productivity, and influences water quality, soil biodiversity, and soil erosion [13]. Of every salt type that brings about salinity stress, plants must cope with the most soluble and ubiquitous form:

sodium chloride (NaCl) [14]. All development stages from germination to reproduction are affected by salt stress because it causes wide-ranging symptoms: more succulent and thicker but smaller leaves that are abscised with necrosis on roots and shoots, and consequences for several physiological processes such as transpiration and photosynthesis. This stress interferes with the metabolism of phytohormones and other metabolic pathways, and can affect protein functions and disturb genes [15].

Plants generally respond to salinity in two phases: by one first quick response, which begins a few minutes after being exposed to salinity and lasts some days. It is known as the osmotic phase, when stomatal closure is promoted and young leaf growth is inhibited, followed by a delayed response, which can last for days, or even weeks. It is called the ionic phase, and it aims to stop metabolic processes, induce mature leaf senescence, and even cause cell death [14, 16].

The osmotic phase that is driven by salinity destroys water potential homeostasis due to the excessive concentration of ions in soil and/or water outside roots [17]. To successfully prevail osmotic stress, plants carry out osmotic adjustment to lower the osmotic potential in the cytosol. Next, they activate water transport systems and synthesize compatible solutes. All this involves less energy use for plant growth and for survival under stress conditions [18, 19]. Compatible solutes are also known as osmoprotectants or osmolytes, which consist of a substantial quantity of compounds, including sugar alcohols (myo-inositol), sugars (raffinose, trehalose), quaternary ammonium compounds (glycine betaine), and amino acids (proline) that operate to diminish oxidative damage from drought stress [20, 21].

The imbalance of ionic distribution and ionic homeostasis owing to not only the cellular toxicity of sodium (Na^+) and chloride (Cl^-) ions that accumulate but also an unbalanced K^+/Na^+ (potassium ion/sodium ion) ratio inside plants occurs in the ionic stress phase [15, 22, 23]. To reduce ionic stress, plants attempt to avoid high Cl^- and Na^+ concentrations in the cytoplasm *via* two distinct mechanisms: transferring salt from the cytoplasm to vacuoles (sequestration) and/or preventing salt uptake by roots and it being transported across plants (exclusion) [18].

As salt stress triggers secondary stresses, mainly oxidative stress, to restore the reactive oxygen species (ROS) balance, plants develop detoxification strategies [24] to promote the synthesis of antioxidants and antioxidant enzymes activities to scavenge ROS, regulate ROS-responsive genes expression, and balance ROS generation [19, 24]. To overcome salinity, a molecular response occurs that involves salt-responsive genes, which carry out lots of functions, such as molecular chaperones (HSPs), ion homeostasis or transport (*SOS*, *NHX1*, *H^+-ATPase*), and transcription factors (*DREB*s), among others [25]. This response is frequently mediated by Ca^{2+} accumulation sensing. This, in turn, begins complex signal transduction pathways and/or abscisic acid (ABA) because these genes can contain ABA-responsive elements [26].

In circumstances in which peppers face salt stress, the authors note less growth with salinity because of the toxic effect of ions and the limited capacity to make osmotic adjustments [23]. According to Suarez et al. [27], tolerant pepper cultivars limit the root-to-shoot Na^+ flux. This involves the lesser gene expression of ion transporters and lower Na^+ concentrations in leaves. These authors conclude that in pepper plants' response to salt stress, Na^+ exclusion is a crucial trait, with wide variability among genotypes related to salt tolerance [27]. The work of Penella et al. [28] establishes that one reliable trait for choosing pepper cultivars for their salt stress tolerance is maintaining the net photosynthetic rate.

In order to collect more in-depth knowledge about the tolerance mechanisms in pepper, the authors investigate the response of two pepper genotypes with different salt tolerances [29]. They find that the lower relative water content (RWC) in the tolerant genotype is lesser than in the sensitive genotype. The tolerant genotype maintains the chlorophyll concentration; the increase in lipid peroxidation is not so severe, and the increases in both superoxide dismutase activity and glutathione content are more marked than in the sensitive genotype [29]. The authors in Ref. [30] detect that tolerant pepper cultivars have less disturbed photosynthesis, a higher proline content, and a bigger leaf/and root biomass than the sensitive genotype under salinity conditions.

3. Heat stress

As **Figure 1** shows, extreme temperature events are becoming increasingly frequent due to climate change. Heat stress is complex stress of variable intensity and duration that is frequently deemed to be heat shock when temperatures are 10–15°C above the normal crop development range for sufficiently long periods to irreversibly cause plant damage [31]. Rises in temperature can be transitory (heat wave or heat shock) or constant (prolonged heat stress), and can result in detrimental effects for plant development, plant growth, and yield losses, and can even cause plants to die in drastic circumstances [32, 33]. Thus, heat stress can impact plants at physiological, biochemical, reproductive, and morpho-anatomical levels [34].

Plants exposed to high temperatures can suffer cellular damage, including interference with enzyme functions by modifying protein activity or alterations to lipid membrane integrity by changing fluidity [35]. The impact of high temperatures on plant development depends on the growth stage, that is, delayed seed germination, which can lead to completely inhibited germination, as determined by heat stress intensity and plant species [31]. In succeeding stages, chloroplasts become extremely vulnerable to rising temperatures, and damage to chloroplasts signifies CO_2 assimilation, electron transport between both photosystems, and disturbed chlorophyll biosynthesis, among other photosynthesis-related processes [35]. Heat stress interferes with other relevant processes such as water relations, respiration, and primary/secondary metabolisms, which trigger changes in protein regulation and gene expression toward plant tolerance [31, 35].

Plant strategies to successfully overcome heat stress include changing leaf orientation to prevent direct solar radiation; the promotion of cooling *via* transpiration; ROS scavenging by means of antioxidants synthesis; preserving plant membrane stability *via* alteration to its lipid composition; the activation of heat response-related factors and genes, including heat shock factors (HSFs) and heat shock proteins (HSPs); the generation of compatible osmolytes; and several signaling strategies such as mitogen-activated protein kinase (MAPK), chaperones, and calcium-dependent protein kinase (CDPK) cascades [31, 34, 36].

In HSP terms, the pepper heat stress response is investigated in Ref. [37]. The authors put forward the principal role in thermotolerance for the *Hsp70-1* gene. This gene is induced by putrescine, hydrogen peroxide (H_2O_2), and calcium ion (Ca^{2+}). This induction is more marked with heat stress. The authors of Guo et al. [38] confirm the implication of HSP70s in the pepper heat stress response *via* the regulation of stress-related genes expression, plus their connection with plant development and growth. The glutathione metabolism has been related to thermotolerance in pepper because the increase in the metabolites and genes involved on this pathway in the tolerant pepper genotype under heat stress is greater than that in the sensitive genotype [39].

The authors of Rajametov et al. [40] suggest that the photosynthetic rate is maintained *via* improved transpiration, along with a higher proline concentration, as an effective strategy for thermotolerant peppers in the seedling stage.

For the reproductive stage, some authors have indicated a relation between flower abscission with heat stress and ethylene production [41]. In other words, flower abscission depends on flowers' susceptibility to ethylene, which generates these stress conditions. Some studies note and confirm the effects of high temperatures on male pepper gametophytes, which are more sensitive than female gametophyte, and not only during their formation and development but also in the later germination and pollen tube growth stages [42, 43]. Fruit set at high temperature can be affected during fruit development. This leads to fruit weighing less and being smaller in size, and is closely related to pollination efficiency and subsequent ovule fertilization, which determine the number of seeds per fruit [44, 45]. In pepper fruit, some authors have run multi-omic analyses (proteomics, metabolomics, and transcriptomics) and noted how heat stress inhibits secondary metabolite synthesis, that is, ascorbic acid and capsaicin [46]. Three transcription factors (TFs) have been found to be involved in capsaicin synthesis regulation under heat stress conditions [46].

4. Drought

Water is an essential resource. It is threatened by an ever-growing population and urban areas emerging in arid areas under climate change conditions. In turn, the span, prevalence, and frequency of droughts all increase (**Figure 1**) [47]. FAO [47] indicates that agriculture is the most, and the first, sector to be affected by drought. Drought impacts on crops can result from many factors, including high light intensity, scarce rainfall, extreme (high and low) temperatures, or salinity [48].

The drought impact on plants has several adverse effects that influence all plant levels: biochemical, morphological, physiological, molecular, and cellular. Ultimately, this implies smaller crop yields [49]. Drought-related morphological alterations mostly include delayed growth and shorter plant height, which are consequences of slower cell expansion on roots, stems, and senescing leaves, whose number and size diminish [49, 50].

When the CO_2 assimilation rate lowers, it might be accompanied by a reduction in photosynthesis and transpiration [49–51]. This is when phytohormones play a role in the drought response, particularly ABA, whose synthesis is promoted under drought conditions in roots, and is transported to leaves and shoots to induce mainly stomatal closure and to avoid transpiration-related water loss [48].

Changes in water relations favor root growth over shoot growth due to rapid drops in the water potential (Ψ_w) in roots. This promotes water uptake, while the Ψ_w in leaves lowers at a slower rate [52]. In connection with rapid Ψ_w decline to allow water to enter, osmotic adjustment is normally made by accumulation and osmoprotectant synthesis, which make the concentration of the solutes inside cells rise [53].

Oxidative damage takes place by drought exposure and is generated by excess ROS. All this brings about harmful effects, such as DNA nicking, photosynthesis restriction, lipid peroxidation, and protein denaturation, among others [50]. Plants attempt to counteract oxidative damage by antioxidant systems being activated, which involve enzymatic activities such as ascorbate peroxidase (APX), monodehydroascorbate reductase (MDHAR), catalase (CAT), dehydroascorbate reductase (DHAR), superoxide dismutase (SOD), glutathione peroxidase (GPX), glutathione reductase (GR), glutathione S-transferase (GST), ferritin, thioredoxins, glutaredoxins, as well as antioxidant compounds (i.e., carotenoids, glutathione, ascorbic acid, α-tocopherol flavonoids, and phenolic acids) [54, 55].

ROS can act as secondary messengers during drought signaling with other secondary messengers like ABA and Ca^{2+} to regulate the expression of the drought-induced proteins and drought-responsive genes implicated on signaling pathways [56]. Certain TFs like *WRKY*, dehydration-responsive element binding (*DREB*), *NAC*, *MYB*, *HSF*s, and basic leucine zipper (*bZIP*) are key regulators of the downstream genes involved in the drought molecular response [57]. Some drought-responsive proteins are aquaporins, which are water channels that transport metal ions, small solutes and gases, dehydrins (promote osmolyte accumulation, water retention, ROS scavenging, and photosynthetic activity), and late embryogenesis abundant (LEA) proteins (protect plant membranes to avoid water loss, among other functions) [58–60].

In the study by Delfine et al. [61], drought effects are studied on pepper plants. Here the authors report photosynthetic limitations *via* lesser stomatal conductance (g_s) and lower leaf Ψ_w which lead to reductions in plant growth, plant productivity, and fruit quality in relation to well-irrigated control plants. In the study by Hu et al. [62], the authors offer similar results in two pepper cultivars under drought conditions: a lower photosynthetic rate, increases in ROS, and then antioxidant enzymes APX and SOD being activated.

The authors of Ref. [63] describe the behavior of two pepper cultivars with differing degrees of tolerance. They also note opposite responses to progressive drought stress. Cultivar Shanshu-2001 better tolerates drought and obtains higher growth and yield values. This is probably the result of a higher RWC in leaves, lower lipid peroxidation, and a prolonged rise in the proline concentration and electrolyte leakage than cultivar Nongchengjiao-2, which apparently shows a sensitivity response [63]. It is noteworthy that the more tolerant cultivar performs greater constitutive SOD, CAT, and peroxidase activities than the more sensitive cultivar, which may confer better drought resistance ability [63].

Some authors have investigated the response of two pepper genotypes (one tolerant, one sensitive) under water stress and report comparable results: The RWC and photosynthetic rate are higher and g_s is lower in the tolerant genotype [64]; moreover, there are changes in gene expression; in other words, aquaporin genes are downregulated in the sensitive genotype and upregulated in the tolerant one. Some works have studied the transcriptomic response of water-stressed pepper plants, and Park et al. [65] have identified seven rapidly induced cDNAs. One (*Ca-LEAL1*) encodes for an LEA-like protein and is induced by ABA application. The authors in Ref. [66] identify and describe a *DREB*-like gene, which is rapidly induced by water stress in pepper roots, and they put forward a role in transcription activation.

Drought affects pepper fruit in several ways: delayed ripening, less production, and modified volatile and bioactive compounds. As such alterations depend on the plant stage, these mechanisms are required to adapt to drought conditions [67].

5. Grafting technique to overcome abiotic stresses

Grafting as an agricultural technique involves joining two different plant parts: the scion, which provides the shoot, and the rootstock, which supplies the root system (**Figure 2**). The scion and the rootstock are connected *via* the union of their vascular systems to become a single plant after being joined [69].

Figure 2.
Grafted pepper seedlings [68].

For more than 2000 years, grafting has been an ancient technique applied to fruit trees. In the last century, this practice was extended to vegetable crops [70]. Grafting in vegetables began in Korea and Japan to restore crop loss, which was caused by soilborne pathogens and led to grave diseases, intensified by serial cropping [71]. In the 1950s, tunnels and greenhouses became popular, and most farmers performed protected cultivation to prolong crop seasons and left the rotating crop system behind. Therefore, grafting a variety onto tolerant rootstocks is required to control soilborne infections, among other diseases [70].

Grafting has many other uses than soil disease mitigation applications, of which cultivation performed under adverse environmental conditions (biotic-abiotic stresses) is highlighted, as are improved crop yields that result from applying vigorous rootstocks [71, 72]. In the present day, most research centers on environmental stress resistance and vegetable grafting physiology, as well as on two families, that is, Solanaceae and Cucurbitaceae, which are cultivated as grafted plants in North America, Europe, and Asia [73, 74].

Grafting amends the scion stress perception of an unfavorable environment, that is, a big robust root system from rootstocks, which involves a better water relations balance, better ability for water and nutrients uptake in relation to the latter, enhanced photosynthetic activity and antioxidant system, hormonal signaling network activation, and, finally, long-distance signal transmission by mRNAs, small RNAs, and proteins [75]. There is a general consensus that bidirectional genetic exchange between the rootstock and scion occurs in grafting plants by the grafting union [76]. Long-distance signals particularly appear to be a promising strategy for modifying physiological functions and regulating rootstock-scion relations [74, 77] to move toward higher production.

5.1 Grafting in pepper to counteract abiotic stress

Abiotic stress amelioration by grafting on Solanaceae crops has been proven to be an effective strategy, mainly under salinity, heat stress, and drought conditions. Regarding *salinity*, the authors in Ref. [78] reveal that grafted plants mitigate growth reduction owing to enhanced photosynthesis and antioxidant enzyme activities.

On peppers that face salt stress, some authors attribute grafted plant tolerance to rootstocks' ability to maintain ion homeostasis in the scion. This leads to less disrupted photosynthesis, lipid peroxidation, and nitrate reductase activity on leaves, as reflected by bigger marketable fruit yields [79]. In Ref. [80], the authors observe that the impact of salinity is milder on the biomass of the plants grafted onto the NIBER® rootstock (F1 hybrid), with 32–80% higher yields than the ungrafted pepper plants previously employed in experiments under salinity field conditions for 3 years. This lesser impact is associated with lower reductions in nitrate reductase, photosynthetic rates, and stomatal conductance compared to the ungrafted and self-grafted plants. These authors also point out a signaling role for H_2O_2, which activates antioxidant mechanisms, as reflected by increased proline and phenols contents [80].

Finally, some other authors reveal that the response of grafted pepper plants to salinity is rootstock-dependent. They indicate three differing strategies: (i) The Creonte rootstock (from Ruiter seeds) increases fruit yield and the reproductive/vegetative ratio is not disturbed thanks to better biomass distribution, and also to opposite cytokinins and ethylene precursor regulation; (ii) the Terrano rootstock (from Syngenta) stunts plant height by non-gibberellins accumulation and induces ABA to maintain the reproductive and vegetative biomass; and (iii) the Atlante rootstock (from Ramiro Arnedo) and its invigorating effect that result from improving the ABA/cytokinins balance [81].

The impact of *heat stress* on pepper grafted plants is investigated in Ref. [82]. These authors indicate that employing the Creonte rootstock (from Ruiter seeds) under heat stress allows higher fruit yields and lowers the percentage of sunscalded fruits, whereas the photosynthetic rate rises. In a controlled environment, other authors have studied the response of grafted pepper to high temperatures. They have verified that marketable pepper yields increase (> 50%) grafting onto a thermotolerant rootstock under greenhouse conditions *versus* the ungrafted variety [43, 83]. These higher yields depend on mechanisms in the vegetative (i.e., undisturbed chlorophyll and carotenoids, more ascorbic acid, and less electrolyte leakage) and reproductive (enhanced pollen germination and fruit set and improved anthers proline content) stages.

Heat stress also affects pepper roots growth, and a bigger number of roots appeared on the peppers grafted onto a hybrid rootstock bred toward abiotic stress tolerance [84]. The combination of the variety with the hybrid rootstock obtained better results at the two evaluated upper depths (0–15 cm and 15–30 cm) compared to the self-grafted and ungrafted variety under heat stress [84].

With *drought*, grafted pepper responses under water stress induced by PEG are studied in Ref. [85]. For all the grafted plants, the authors observe diminished photosynthetic activity and NO_3^- uptake and transport to leaves. The photosynthesis of the plants grafted onto tolerant pepper accessions is less disturbed. The authors propose a protective role of osmotic adjustment and proline accumulation to overcome water stress [85]. Some other authors report a protective role for the tolerant pepper rootstock tested under short-term water stress against oxidative stress for being capable of alleviating oxidative stress in the scion and might balance the C and N biomass distribution, which results in larger crop yields [86].

Oxidative stress mitigation by the antioxidant system and its relation with photosynthesis maintenance with water stress in pepper grafted plants is studied in Ref. [87]. The authors show that pepper plants grafted onto tolerant hybrid rootstock H92 under water stress are able to avoid oxidative damage and sustain the photosynthetic rate thanks to enhanced proline and ascorbic acid contents.

The work in Ref. [88] puts the NIBER® pepper tolerant rootstock under deficit irrigation conditions for consecutive years to the test. It reports the milder impact of water stress on the plants grafted onto the NIBER® rootstock in photosynthesis, fruit production, and plant biomass terms (marketable yield losses were up to 50% lower) *versus* the ungrafted variety. With other studies, the authors confirm their previous results obtained in the pepper plants grafted onto the NIBER® rootstock (50% higher marketable yield under water stress) and they note better root biomass distribution compared to the ungrafted variety [89].

Regarding roots implication in pepper water stress tolerance, in Ref. [90], the authors describe constitutive differences in the roots gene expression between pepper-sensitive accession A10 and pepper-tolerant rootstock NIBER®. These differences under the control conditions were associated with the detoxification system and ABA induction in the NIBER® rootstock. Under short-term water stress conditions in the NIBER® rootstock, TFs like *DREB*s and *MYC*, chaperones, and aquaporins were upregulated, while the concentration of auxins, abscisic acid, jasmonic acid, trehalose, raffinose, and spermidine increased [90]. The authors point out that these mechanisms, together with lower oxidized glutathione, indicate the ability to prevent oxidative damage and water stress tolerance in the very early phase of stress.

The signal transmission from roots to shoots, and *vice versa*, is key for water stress tolerance achievement [91]. The phytohormones network regulation of roots and shoots during the water stress response in pepper grafted plants is studied in Ref. [92]. The authors quantify a higher ABA concentration in the variety grafted onto NIBER® roots at 4 h after water stress and suggest the transport of this hormone from roots to shoots because the ABA concentration lowered in the roots and rose in the leaves of this plant combination after 48 h of water stress. Moreover, the authors link the ABA levels at 48 h with the higher WUE (water use efficiency) in the variety grafted onto NIBER® under water stress, which is favored by major stomata closure in relation to the self-grafted variety. Simultaneously, the concentrations of the ethylene precursor ACC, jasmonic acid, and salicylic acid increased in this plant combination under water stress, which the authors explain by their role in abiotic stress signaling and tolerance [92].

6. Conclusion

In summary, grafting is a proven effective strategy for mitigating abiotic stress effects on horticultural plants, specifically on pepper plants. In this chapter, some abiotic stress tolerance mechanisms of grafted pepper plants are presented, which lead to better production and fruit quality (**Figure 3**). Tolerant rootstocks are able to modify the stress perception of the grafted variety by means of different tolerance strategies to overcome high temperature, salinity, or drought. Given a trend toward sustainable approaches to fight climate change, the grafting technique has a lot to offer. Recent efforts have been made to study the molecular mechanisms involved in grafting-acquired tolerance, but a lot of work remains to be done.

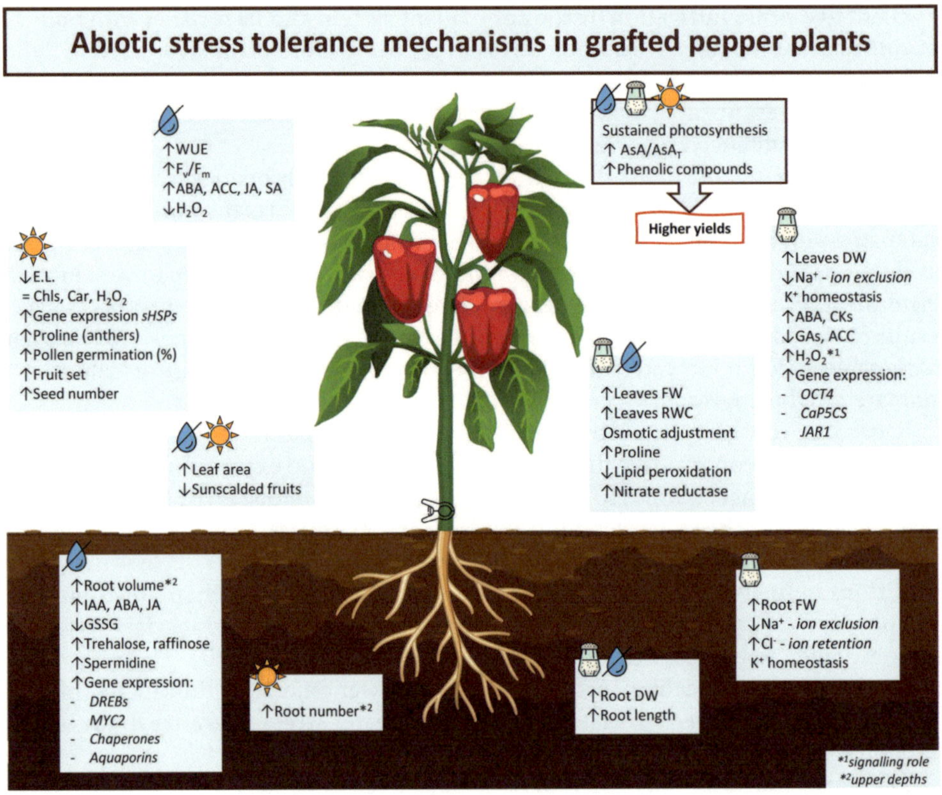

Figure 3.
Abiotic stress tolerance mechanisms in roots and leaves of grafted pepper plants. The drop icon depicts drought stress; the sun icon depicts high temperature stress, and the salt shaker icon refers to salt stress. The blue squares include the tolerance mechanisms found in one stress condition (one icon) or common to more than one stress condition individually (two or three icons). Arrows depict an increase (up) or decrease (down) under stress conditions in relation to control conditions, and the equal sign reflects no changes. This figure was designed using Freepik illustrations.

Acknowledgements

This work has been financed by Grant PID2020-118824RR-C21 funded by MCIN/AEI/10.13039/501100011033. Ramón Gisbert-Mullor is a beneficiary of a doctoral fellowship (FPU-MEFP (Spain)). Yaiza Gara Padilla is a beneficiary of grant PRE2018-086374 funded by MCIN/AEI/10.13039/501100011033 and, as appropriate, by "ESF Investing in your future." Parts of this chapter were previously published in the doctoral thesis by the same author: Yaiza G. Padilla. Study of the Physiological, Metabolomic and Transcriptional Changes Mediated by Rootstocks to Explain the Water Stress Tolerance of Grafted Pepper Plants. 2023. Universitat Politècnica de València. Available from: https://riunet.upv.es/handle/10251/199992.

Author details

Yaiza G. Padilla[1*], Ramón Gisbert-Mullor[2], Salvador López-Galarza[2] and Ángeles Calatayud[1]

1 Horticulture Department, Citriculture and Crop Production Center, Valencian Institute for Agricultural Research (IVIA), Moncada, Valencia, Spain

2 Crop Production Department, Polythecnic University of Valencia (UPV), Camí de Vera, Valencia, Spain

*Address all correspondence to: padilla_yai@gva.es

References

[1] Barchenger DW, Naresh P, Kumar S. Genetic resources of capsicum. In: The Capsicum Genome. Cham: Springer; 2019. pp. 9-23. DOI: 10.1007/978-3-319-97217-6_2

[2] FAO. FAOSTAT-crops and livestock products [Internet]. 2023. Available from: https://www.fao.org/faostat/en/#data/QCL/visualize [Accessed: 30 May 2023]

[3] Climate change – food and agriculture organization of the United Nations [Internet]. Available from: https://www.fao.org/climate-change/en/ [Accessed: 24 October 2022]

[4] Morris WL, Taylor MA. The Solanaceous vegetable crops: Potato, tomato, pepper, and eggplant. Encyclopedia of Applied Plant Sciences. 2017;**3**:55-58. DOI: 10.1016/B978-0-12-394807-6.00129-5

[5] Olatunji TL, Afolayan AJ. The suitability of chili pepper (*Capsicum annuum* L.) for alleviating human micronutrient dietary deficiencies: A review. Food Science & Nutrition. 2018;**6**(8):2239-2251. DOI: 10.1002/fsn3.790

[6] Martínez-Ispizua E, Martínez-Cuenca MR, Marsal JI, Díez MJ, Soler S, Valcárcel JV, et al. Bioactive compounds and antioxidant capacity of Valencian pepper landraces. Molecules. 2021;**26**(4):1031. DOI: 10.3390/MOLECULES26041031

[7] Sah SK, Reddy KR, Li J. Abscisic acid and abiotic stress tolerance in crop plants. Frontiers in Plant Science. 2016;**7**:571. DOI: 10.3389/fpls.2016.00571

[8] Wang W, Vinocur B, Altman A. Plant responses to drought, salinity and extreme temperatures: Towards genetic engineering for stress tolerance. Planta. 2003;**218**(1):1-14. DOI: 10.1007/s00425-003-1105-5

[9] IMF. Climate change indicators dashboard [Internet]. 2022. Available from: https://climatedata.imf.org/ [Accessed: 06 August 2023]

[10] Athar HR, Ashraf M. Strategies for crop improvement against salinity and drought stress: An overview. In: Salinity and Water Stress. Dordrecht: Springer; 2009. pp. 1-16. DOI: 10.1007/978-1-4020-9065-3_1

[11] Rouphael Y, Venema JH, Edelstein M, Savvas D, Colla G, Ntatsi G, et al. Grafting as a tool for tolerance of abiotic stress. In: Vegetable Grafting: Principles and Practices. UK: CABI; 2017. pp. 171-215. DOI: 10.1079/9781780648972.0171

[12] Rivero RM, Ruiz J, Romero L. Role of grafting in horticultural plants under stress conditions. Journal of Food, Agriculture and Environment. 2003;**1**(1):70-74. ISSN: 1459-0255

[13] FAO. Soil salinity – global soil partnership [Internet]. 2023. Available from: https://www.fao.org/global-soil-partnership/areas-of-work/soil-salinity/en/ [Accessed: 13 June 2023]

[14] Munns R, Tester M. Mechanisms of salinity tolerance. Annual Review of Plant Biology. 2008;**59**(1):651-681. DOI: 10.1146/annurev.arplant.59.032607.092911

[15] Al MM, Khan AL, Muneer S. Silicon in horticultural crops: Cross-talk, signaling, and tolerance mechanism under salinity stress.

Plants. 2020;**9**(4):460. DOI: 10.3390/PLANTS9040460

[16] Isayenkov SV, Maathuis FJM. Plant salinity stress: Many unanswered questions remain. Frontiers in Plant Science. 2019;**10**:435515. DOI: 10.3389/FPLS.2019.00080

[17] Munns R. Comparative physiology of salt and water stress. Plant, Cell & Environment. 2002;**25**(2):239-250. DOI: 10.1046/j.0016-8025.2001.00808.x

[18] Munns R. Plant adaptations to salt and water stress. In: Advances in Botanical Research. Oxford, UK: Academic Press; 2011. pp. 1-32. DOI: 10.1016/B978-0-12-387692-8.00001-1

[19] Zhao S, Zhang Q, Liu M, Zhou H, Ma C, Wang P. Regulation of plant responses to salt stress. International Journal of Molecular Sciences. 2021;**22**(9):4609. DOI: 10.3390/ijms22094609

[20] Ejaz S, Fahad S, Anjum MA, Nawaz A, Naz S, Hussain S, et al. Role of Osmolytes in the Mechanisms of Antioxidant Defense of Plants. Cham: Springer; 2020. pp. 95-117. DOI: 10.1007/978-3-030-38881-2_4

[21] Jogawat A. Osmolytes and their role in abiotic stress tolerance in plants. In: Molecular Plant Abiotic Stress. New Jersey, USA: Wiley; 2019. pp. 91-104. DOI: 10.1002/9781119463665.ch5

[22] Miura K, Tada Y. Regulation of water, salinity, and cold stress responses by salicylic acid. Frontiers in Plant Science. 2014;**5**:4. DOI: 10.3389/fpls.2014.00004

[23] Navarro JM, Garrido C, Martínez V, Carvajal M. Water relations and xylem transport of nutrients in pepper plants grown under two different salts stress regimes. Plant Growth Regulation. 2003;**41**(3):237-245. DOI: 10.1023/B:GROW.0000007515.72795.c5

[24] Yang Y, Guo Y. Unraveling salt stress signaling in plants. Journal of Integrative Plant Biology. 2018;**60**(9):796-804. DOI: 10.1111/jipb.12689

[25] Gupta B, Huang B. Mechanism of salinity tolerance in plants: Physiological, biochemical, and molecular characterization. International Journal of Genomics. 2014;**701596**:1-18. DOI: 10.1155/2014/701596

[26] Chaudhry UK, Gökçe ZNÖ, Gökçe AF. The influence of salinity stress on plants and their molecular mechanisms. In: International Electronic Conference on Plant Sciences (IECPS 2021); 1-15 December 2021; Online. Vol. 11, No. 1. Basel, Switzerland: MDPI; 2021. p. 31. DOI: 10.3390/IECPS2021-12017

[27] Suarez DL, Celis N, Ferreira JFS, Reynolds T, Sandhu D. Linking genetic determinants with salinity tolerance and ion relationships in eggplant, tomato and pepper. Scientific Reports. 2021;**11**(1):16298. DOI: 10.1038/s41598-021-95506-5

[28] Penella C, Nebauer SG, Lopéz-Galarza S, San Bautista A, Gorbe E, Calatayud A. Evaluation for salt stress tolerance of pepper genotypes to be used as rootstocks. Journal of Food, Agriculture and Environment. 2013;**11**(3-4):1101-1107. DOI: 10.17221/163/2013-hortsci

[29] Aktas H, Abak K, Eker S. Anti-oxidative responses of salt-tolerant and salt-sensitive pepper (*Capsicum annuum* L.) genotypes grown under salt stress. The Journal of Horticultural Science and Biotechnology. 2012;**87**(4):360-366. DOI: 10.1080/14620316.2012.11512877

[30] López-Serrano L, Calatayud Á, López-Galarza S, Serrano R, Bueso E. Uncovering salt tolerance mechanisms in pepper plants: A physiological and transcriptomic approach. BMC Plant Biology. 2021;**21**(1):169. DOI: 10.1186/s12870-021-02938-2

[31] Wahid A, Gelani S, Ashraf M, Foolad M. Heat tolerance in plants: An overview. Environmental and Experimental Botany. 2007;**61**(3):199-223. DOI: 10.1016/j.envexpbot.2007.05.011

[32] Kotak S, Larkindale J, Lee U, von Koskull-Döring P, Vierling E, Scharf KD. Complexity of the heat stress response in plants. Current Opinion in Plant Biology. 2007;**10**(3):310-316. DOI: 10.1016/j.pbi.2007.04.011

[33] Raza A, Charagh S, Abbas S, Hassan MU, Saeed F, Haider S, et al. Assessment of proline function in higher plants under extreme temperatures. Plant Biology. 2023;**25**(3):379-395. DOI: 10.1111/plb.13510

[34] Nadeem M, Li J, Wang M, Shah L, Lu S, Wang X, et al. Unraveling field crops sensitivity to heat stress: Mechanisms, approaches, and future prospects. Agronomy. 2018;**8**(7):128. DOI: 10.3390/agronomy8070128

[35] Hu S, Ding Y, Zhu C. Sensitivity and responses of chloroplasts to heat stress in plants. Frontiers in Plant Science. 2020;**11**:375. DOI: 10.3389/fpls.2020.00375

[36] Panikulangara TJ, Eggers-Schumacher G, Wunderlich M, Stransky H, Schöffl F, Galactinol synthase1. A novel heat shock factor target gene responsible for heat-induced synthesis of raffinose family oligosaccharides in *Arabidopsis*. Plant Physiology. 2004;**136**(2):3148-3158. DOI: 10.1104/pp.104.042606

[37] Guo M, Zhai YF, Lu JP, Chai L, Chai WG, Gong ZH, et al. Characterization of *CaHsp70-1*, a pepper heat-shock protein gene in response to heat stress and some regulation exogenous substances in *Capsicum annuum* L. International Journal of Molecular Sciences. 2014;**15**(11):19741-19759. DOI: 10.3390/ijms151119741

[38] Guo M, Liu JH, Ma X, Zhai YF, Gong ZH, Lu MH. Genome-wide analysis of the *Hsp70* family genes in pepper (*Capsicum annuum* L.) and functional identification of *CaHsp70-2* involvement in heat stress. Plant Science. 2016;**252**:246-256. DOI: 10.1016/j.plantsci.2016.07.001

[39] Wang J, Lv J, Liu Z, Liu Y, Song J, Ma Y, et al. Integration of Transcriptomics and metabolomics for pepper (*Capsicum annuum* L.) in response to heat stress. International Journal of Molecular Sciences. 2019;**20**(20):5042. DOI: 10.3390/ijms20205042

[40] Rajametov SN, Yang EY, Cho MC, Chae SY, Jeong HB, Chae WB. Heat-tolerant hot pepper exhibits constant photosynthesis via increased transpiration rate, high proline content and fast recovery in heat stress condition. Scientific Reports. 2021;**11**(1):14328. DOI: 10.1038/s41598-021-93697-5

[41] Aloni B, Karni L, Zaidman Z, Riov Y, Huberman M, Goren R. The susceptibility of pepper (*Capsicum annuum*) to heat induced flower abscission: Possible involvement of ethylene. Journal of Horticultural Science. 1994;**69**(5):923-928. DOI: 10.1080/14620316.1994.11516528

[42] Aloni B, Peet M, Pharr M, Karni L. The effect of high temperature and high atmospheric CO_2 on carbohydrate changes in bell pepper (*Capsicum annuum*) pollen in relation

to its germination. Physiologia Plantarum. 2001;**112**(4):505-512. DOI: 10.1034/j.1399-3054.2001.1120407.x

[43] Gisbert-Mullor R, Padilla YG, Calatayud Á, López-Galarza S. Rootstock-mediated physiological and fruit set responses in pepper under heat stress. Scientia Horticulturae. 2023;**309**:111699. DOI: 10.1016/j.scienta.2022.111699

[44] Marcelis LFM, Baan Hofman-Eijer LR. Effects of seed number on competition and dominance among fruits in *Capsicum annuum* L. Annals of Botany. 1997;**79**(6):687-693. DOI: 10.1006/ANBO.1997.0398

[45] Pagamas P, Nawata E. Sensitive stages of fruit and seed development of chili pepper (*Capsicum annuum* L. var. Shishito) exposed to high-temperature stress. Scientia Horticulturae. 2008;**117**(1):21-25. DOI: 10.1016/J.SCIENTA.2008.03.017

[46] Liu C, Luo S, Zhao Y, Miao Y, Wang Q, Ye L, et al. Multiomics analyses reveal high temperature-induced molecular regulation of ascorbic acid and capsaicin biosynthesis in pepper fruits. Environmental and Experimental Botany. 2022;**201**:104941. DOI: 10.1016/J.ENVEXPBOT.2022.104941

[47] FAO. Drought and agriculture. Food and agriculture organization [Internet] 2023. Available from: https://www.fao.org/land-water/water/drought/droughtandag/en/ [Accessed: 13 December 2021]

[48] Salehi-Lisar SY, Bakhshayeshan-Agdam H. Drought stress in plants: Causes, consequences, and tolerance. In: Drought Stress Tolerance in Plants. Vol. 1. Cham: Springer International Publishing; 2016. pp. 1-16. DOI: 10.1007/978-3-319-28899-4_1

[49] Anjum SA, Xie XY, Wang LC, Saleem MF, Man C, Lei W. Morphological, physiological and biochemical responses of plants to drought stress. African Journal of Agricultural Research. 2011;**6**(9):2026-2032. DOI: 10.5897/AJAR10.027

[50] Yang X, Lu M, Wang Y, Wang Y, Liu Z, Chen S. Response mechanism of plants to drought stress. Horticulturae. 2021;7(3):50. DOI: 10.3390/horticulturae7030050

[51] Flexas J, Bota J, Cifre J, Escalona J, Galmes J, Gulias J, et al. Understanding down-regulation of photosynthesis under water stress: Future prospects and searching for physiological tools for irrigation management. The Annals of Applied Biology. 2004;**144**(3):273-283. DOI: 10.1111/j.1744-7348.2004.tb00343.x

[52] Hsiao TC, Xu L. Sensitivity of growth of roots versus leaves to water stress: Biophysical analysis and relation to water transport. Journal of Experimental Botany. 2000;**51**(350):1595-1616. DOI: 10.1093/jexbot/51.350.1595

[53] Ozturk M, Turkyilmaz Unal B, García-Caparrós P, Khursheed A, Gul A, Hasanuzzaman M. Osmoregulation and its actions during the drought stress in plants. Physiologia Plantarum. 2021;**172**(2):1321-1335. DOI: 10.1111/ppl.13297

[54] Hasanuzzaman M, Bhuyan MHMB, Zulfiqar F, Raza A, Mohsin SM, Mahmud JA, et al. Reactive oxygen species and antioxidant defense in plants under abiotic stress: Revisiting the crucial role of a universal defense regulator. Antioxidants. 2020;**9**(8):681. DOI: 10.3390/antiox9080681

[55] Reddy AR, Chaitanya KV, Vivekanandan M. Drought-induced responses of photosynthesis and

antioxidant metabolism in higher plants. Journal of Plant Physiology. 2004;**161**(11):1189-1202. DOI: 10.1016/j.jplph.2004.01.013

[56] Kaur G, Asthir B. Molecular responses to drought stress in plants. Biologia Plantarum. 2017;**61**(2):201-209. DOI: 10.1007/s10535-016-0700-9

[57] Manna M, Thakur T, Chirom O, Mandlik R, Deshmukh R, Salvi P. Transcription factors as key molecular target to strengthen the drought stress tolerance in plants. Physiologia Plantarum. 2021;**172**(2):847-868. DOI: 10.1111/ppl.13268

[58] Afzal Z, Howton TC, Sun Y, Mukhtar MS. The roles of Aquaporins in plant stress responses. Journal of Developmental Biology. 2016;**4**(1):9. DOI: 10.3390/JDB4010009

[59] Aslam M, Maqbool MA, Cengiz R. Mechanisms of drought resistance. In: Drought Stress in Maize (*Zea mays* L.). Cham: Springer; 2015. pp. 19-36. DOI: 10.1007/978-3-319-25442-5_3

[60] Riyazuddin R, Nisha N, Singh K, Verma R, Gupta R. Involvement of dehydrin proteins in mitigating the negative effects of drought stress in plants. Plant Cell Reports. 2021;**41**(3):519-533. DOI: 10.1007/s00299-021-02720-6

[61] Delfine S, Tognetti R, Loreto F, Alvino A. Physiological and growth responses to water stress in field-grown bell pepper (*Capsicum annuum* L.). The Journal of Horticultural Science and Biotechnology. 2002;**77**(6):697-704. DOI: 10.1080/14620316.2002.11511559

[62] Hu WH, Xiao YA, Zeng JJ, Hu XH. Photosynthesis, respiration and antioxidant enzymes in pepper leaves under drought and heat stresses. Biologia Plantarum. 2010;**54**(4):761-765. DOI: 10.1007/s10535-010-0137-5

[63] Anjum SA, Farooq M, Xie X, yu, Liu X jian, Ijaz MF. Antioxidant defense system and proline accumulation enables hot pepper to perform better under drought. Scientia Horticulturae. 2012;**140**:66-73. DOI: 10.1016/j.scienta.2012.03.028

[64] Sahitya UL, Krishna MSR, Suneetha P. Integrated approaches to study the drought tolerance mechanism in hot pepper (*Capsicum annuum* L.). Physiology and Molecular Biology of Plants. 2019;**25**(3):637-647. DOI: 10.1007/s12298-019-00655-7

[65] Park JA, Cho SK, Kim JE, Chung HS, Hong JP, Hwang B, et al. Isolation of cDNAs differentially expressed in response to drought stress and characterization of the *Ca-LEAL1* gene encoding a new family of atypical LEA-like protein homologue in hot pepper (*Capsicum annuum* L. cv. Pukang). Plant Science. 2003;**165**(3):471-481. DOI: 10.1016/S0168-9452(03)00165-1

[66] Hong JP, Kim WT. Isolation and functional characterization of the *Ca-DREBLP1* gene encoding a dehydration-responsive element binding-factor-like protein 1 in hot pepper (*Capsicum annuum* L. cv. Pukang). Planta. 2005;**220**(6):875-888. DOI: 10.1007/s00425-004-1412-5

[67] Borràs D, Plazas M, Moglia A, Lanteri S. The influence of acute water stresses on the biochemical composition of bell pepper (*Capsicum annuum* L.) berries. Journal of the Science of Food and Agriculture. 2021;**101**(11):4724-4734. DOI: 10.1002/jsfa.11118

[68] López-Serrano L. Unravelling the Physiological and Genetic Adaptation of Grafted Pepper under Saline and Hydric

Stresses [thesis]. Universitat Politècnica de València; 2021. Available from: https://riunet.upv.es/handle/10251/162875

[69] Hu B, Beartrack M. Ferguson College of Agriculture. Introduction to Vegetable Grafting [Internet] 2021. Available from: https://extension.okstate.edu/fact-sheets/introduction-to-vegetable-grafting.html [Accessed: 19 June 2023]

[70] Bie Z, Nawaz MA, Huang Yuan HY, Lee JungMyung LJ, Colla G. Introduction to vegetable grafting. In: Colla G, Pérez-Alfocea F, Schwarz D, editors. Vegetable Grafting: Principles and Practices. UK: CABI; 2017. pp. 1-21. DOI: 10.1079/9781780648972.0001

[71] Lee JM, Kubota C, Tsao SJ, Bie Z, Echevarria PH, Morra L, et al. Current status of vegetable grafting: Diffusion, grafting techniques, automation. Scientia Horticulturae. 2010;**127**(2):93-105. DOI: 10.1016/j.scienta.2010.08.003

[72] Kyriacou MC, Rouphael Y, Colla G, Zrenner R, Schwarz D. Vegetable grafting: The implications of a growing agronomic imperative for vegetable fruit quality and nutritive value. Frontiers in Plant Science. 2017;**8**:741. DOI: 10.3389/fpls.2017.00741

[73] Gaion LA, Braz LT, Carvalho RF. Grafting in vegetable crops: A great technique for agriculture. International Journal of Vegetable Science. 2018;**24**(1):85-102. DOI: 10.1080/19315260.2017.1357062

[74] Goldschmidt EE. Plant grafting: New mechanisms, evolutionary implications. Frontiers in Plant Science. 2014;**5**:727. DOI: 10.3389/fpls.2014.00727

[75] Rouphael Y, Kyriacou MC, Colla G. Vegetable grafting: A toolbox for securing yield stability under multiple stress conditions. Frontiers in Plant Science. 2018;**8**:2255. DOI: 10.3389/fpls.2017.02255

[76] Tsaballa A, Xanthopoulou A, Madesis P, Tsaftaris A, Nianiou-Obeidat I. Vegetable grafting from a molecular point of view: The involvement of epigenetics in rootstock-Scion interactions. Frontiers in Plant Science. 2021;**11**:621999. DOI: 10.3389/FPLS.2020.621999

[77] Lu X, Liu W, Wang T, Zhang J, Li X, Zhang W. Systemic long-distance signaling and communication between rootstock and Scion in grafted vegetables. Front Plant Science. 2020;**11**. DOI: 10.3389/FPLS.2020.00460

[78] He Y, Zhu Z, Yang J, Ni X, Zhu B. Grafting increases the salt tolerance of tomato by improvement of photosynthesis and enhancement of antioxidant enzymes activity. Environmental and Experimental Botany. 2009;**66**(2):270-278. DOI: 10.1016/j.envexpbot.2009.02.007

[79] Penella C, Nebauer SG, Quiñones A, San Bautista A, López-Galarza S, Calatayud A. Some rootstocks improve pepper tolerance to mild salinity through ionic regulation. Plant Science. 2015;**230**:12-22. DOI: 10.1016/j.plantsci.2014.10.007

[80] López-Serrano L, Canet-Sanchis G, Selak GV, Penella C, San Bautista A, López-Galarza S, et al. Physiological characterization of a pepper hybrid rootstock designed to cope with salinity stress. Plant Physiology and Biochemistry. 2020;**148**:207-219. DOI: 10.1016/j.plaphy.2020.01.016

[81] Gálvez A, Albacete A, Martínez-Andújar C, Del Amor FM, López-Marín J. Contrasting rootstock-mediated growth and yield responses in salinized pepper plants (*Capsicum annuum* L.) are

associated with changes in the hormonal balance. International Journal of Molecular Sciences. 2021;**22**(7):3297. DOI: 10.3390/IJMS22073297

[82] López-Marín J, González A, Pérez-Alfocea F, Egea-Gilabert C, Fernández JA. Grafting is an efficient alternative to shading screens to alleviate thermal stress in greenhouse-grown sweet pepper. Scientia Horticulturae. 2013;**149**:39-46. DOI: 10.1016/j.scienta.2012.02.034

[83] Gisbert-Mullor R, Padilla YG, Martínez-Cuenca MR, López-Galarza S, Calatayud Á. Suitable rootstocks can alleviate the effects of heat stress on pepper plants. Scientia Horticulturae. 2021;**290**:110529. DOI: 10.1016/j.scienta.2021.110529

[84] Aidoo MK, Sherman T, Ephrath JE, Fait A, Rachmilevitch S, Lazarovitch N. Grafting as a method to increase the tolerance response of bell pepper to extreme temperatures. Vadose Zone Journal. 2017;**17**(1):170006. DOI: 10.2136/vzj2017.01.0006

[85] Penella C, Nebauer SG, Bautista AS, López-Galarza S, Calatayud Á. Rootstock alleviates PEG-induced water stress in grafted pepper seedlings: Physiological responses. Journal of Plant Physiology. 2014;**171**(10):842-851. DOI: 10.1016/j.jplph.2014.01.013

[86] López-Serrano L, Canet-Sanchis G, Vuletin Selak G, Penella C, San Bautista A, López-Galarza S, et al. Pepper rootstock and Scion physiological responses under drought stress. Frontiers in Plant Science. 2019;**10**:38. DOI: 10.3389/fpls.2019.00038

[87] Padilla YG, Gisbert-Mullor R, López-Serrano L, López-Galarza S, Calatayud Á. Grafting enhances pepper water stress tolerance by improving photosynthesis and antioxidant defense systems. Antioxidants. 2021;**10**(4):576. DOI: 10.3390/ANTIOX10040576

[88] Gisbert-Mullor R, Pascual-Seva N, Martínez-Gimeno MA, López-Serrano L, Badal Marín E, Pérez-Pérez JG, et al. Grafting onto an appropriate rootstock reduces the impact on yield and quality of controlled deficit irrigated pepper crops. Agronomy. 2020;**10**(10):1529. DOI: 10.3390/agronomy10101529

[89] Gisbert-Mullor R, Martín-García R, Bažon Zidarić I, Pascual-Seva N, Pascual B, Padilla YG, et al. A water stress–tolerant pepper rootstock improves the behavior of pepper plants under deficit irrigation through root biomass distribution and physiological adaptation. Horticulturae. 2023;**9**(3):362. DOI: 10.3390/horticulturae9030362

[90] Padilla YG, Gisbert-Mullor R, Bueso E, Zhang L, Forment J, Lucini L, et al. New insights into short-term water stress tolerance through transcriptomic and metabolomic analyses on pepper roots. Plant Science. 2023;**333**:111731. DOI: 10.1016/J.PLANTSCI.2023.111731

[91] Albacete AA, Martínez-Andújar C, Pérez-Alfocea F. Hormonal and metabolic regulation of source–sink relations under salinity and drought: From plant survival to crop yield stability. Biotechnology Advances. 2014;**32**(1):12-30. DOI: 10.1016/j.biotechadv.2013.10.005

[92] Padilla YG, Gisbert-Mullor R, López-Galarza S, Albacete A, Martínez-Melgarejo PA, Calatayud Á. Short-term water stress responses of grafted pepper plants are associated with changes in the hormonal balance. Frontiers in Plant Science. 2023;**14**:1412. DOI: 10.3389/FPLS.2023.1170021

Chapter 12

Potato Genomics, Transcriptomics, and miRNomics under Abiotic Stressors

Beyazıt Abdurrahman Şanlı, Zahide Neslihan Öztürk and Orkun Gencer

Abstract

Potato (*Solanum tuberosum* L.) is one of the essential non-cereal crops with noticeably greater production and consumption rates across the world. Because of the diverse range of utilization of nutritious tubers, potatoes can be used as an alternative food to address regional, national, and global food security issues compromised by global climate change. Since potato yield and quality are dramatically affected by abiotic stress conditions, the development of potato plants showing greater performance after being exposed to individual or combined stress treatments has become intriguing for the past decade. In this book chapter, recent studies and/or approaches associated with potato genomics, transcriptomics, and identification of miRNAs are summarized to discuss the response of potatoes to abiotic stress factors in different regulatory levels. Such a summary should encompass the importance and identification of factors for the development of potato plants under current and potential exacerbating effects caused by abiotic stress.

Keywords: potato, genomics, transcriptomics, miRNomics, abiotic stress

1. Introduction

Global climate change stimulates significant threats to sustainable agricultural production by altering climate patterns, causing large-scale detrimental impacts around the globe such as excessive heat, prolonged drought, severe hurricanes, and melting glaciers [1]. As a result, the frequency of unfavorable environmental conditions seriously undermining plant growth, development, and productivity increases each and every year [2]. Unfortunately, detrimental consequences caused by abiotic stresses on plant productivity and survival are likely to intensify as greenhouse gas emissions continue over a long period of time worldwide [1, 3]. Scientists have anticipated that ambient CO_2 concentrations will double by the end of the twenty-first century to 700 ppm, accompanied by a surge in global air temperatures of 0.3–4.8°C [1]. Hence, the expected rise in air temperature may exacerbate the evapotranspiration effect, resulting in water scarcity in the soil and agricultural drought [4, 5]. Additionally, the adverse influences of climate change on soil content are expected to boost soil salinity because of increasing sea levels and saltwater intrusion, particularly along coastal regions [6, 7]. While the estimated surge in atmospheric CO_2 levels is suggested to improve productivity in some crops including wheat, maize, rice, and soybean [8], yield losses resulting from high temperatures and water scarcity may outweigh the positive effect of the CO_2 increase [9]. A multitude of factors have also intensified the burden on plants including the growing increase rate in the world's population, the reduced arable area caused by soil degradation, urbanization, and the increased need for animal feeding which entails the development of tolerant plants adapted to projected scenarios in the future and improved yields of food crops to fulfill individual, regional, national, and global food security requirements [10, 11].

Cultivated in more than a hundred countries and consumed by over a billion people worldwide in 2021, potato (*Solanum tuberosum*) is one of the major non-cereal staples with the highest harvested area (**Figure 1**) and production quantity in China (**Figure 2**) [12]. Although China, India, and Ukraine have a significant role in potato production, Kuwait, New Zealand, and the United States of America rank in the top three places based on yield in the world (**Figure 3**).

Along with a shallow root system, potato is known as a cool season crop that grows well in temperate locations under mild temperatures. Due to the high nutritional content of potato tubers including carbohydrates, vitamins, proteins, antioxidants, and minerals, attempts toward increasing potato crop productivity may help to fulfill the nutritional needs of the growing worldwide population under the adverse effects of climate change [13]. Physiological processes during potato tuber initiation and maturation that substantially shapes the quality and productivity of potato include carbohydrate synthesis via the photosynthesis process in the leaves, the transport of sucrose to the tubers, and the conversion of sucrose to starch in the tubers [14]. Depending upon the cultivar, duration, type, and mode of stress (either as individual and/or combined), the potato has been considerably influenced at morphological, physiological, and molecular levels as well as post-harvest storage.

Abiotic stresses, often known as drought, heat, salinity, heavy metals, and nutrient deficiency, that interfere with metabolic processes in potatoes are most likely to damage tuber productivity and quality by impairing during plant growth, development, and tuberization stages [15, 16]. Potato yield across the world is anticipated

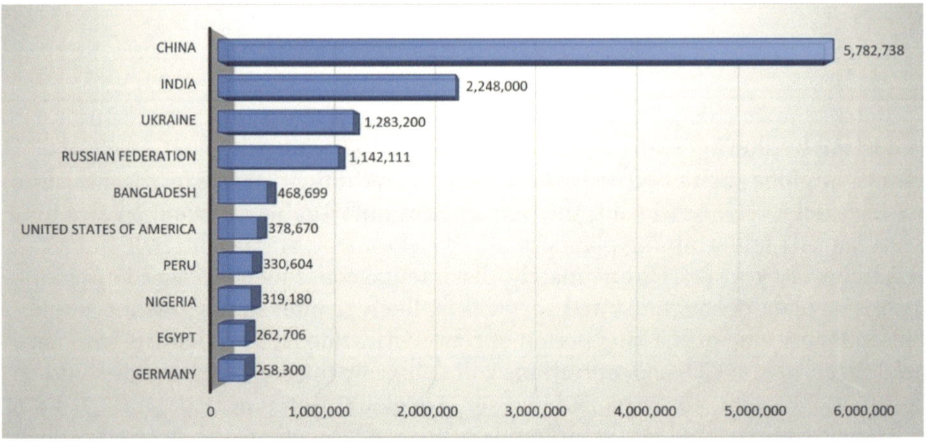

Figure 1.
Harvested area of potato around the world (Ha) [12]. Figure constructed by Orkun GENCER.

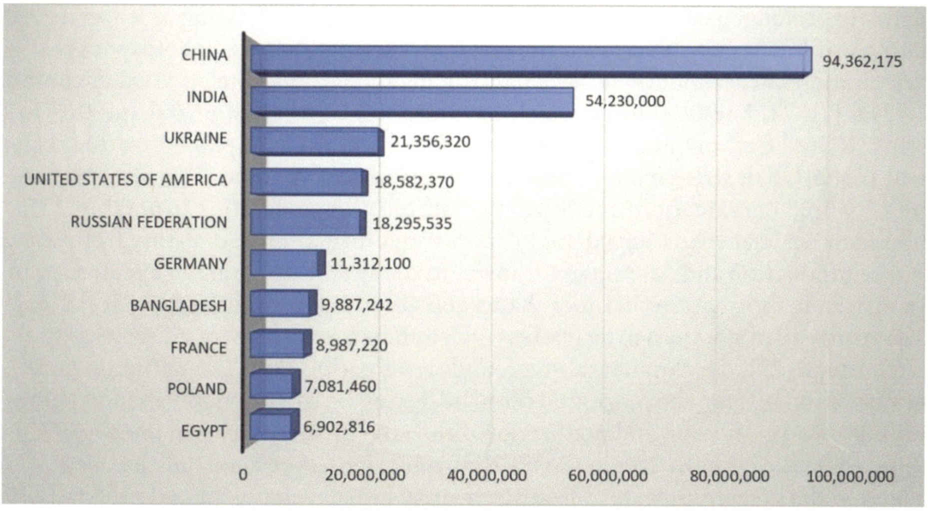

Figure 2.
Production quantity of potato around the world (Tonnes) [12]. Figure constructed by Orkun GENCER.

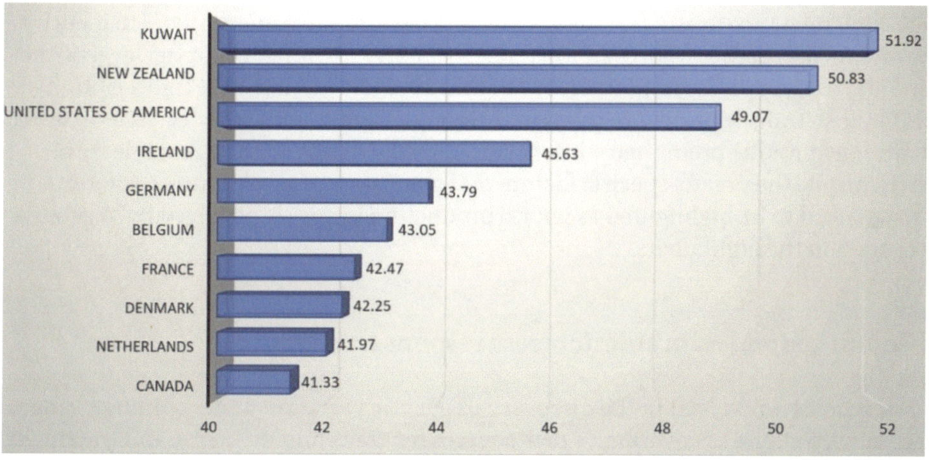

Figure 3.
Yield of potato around the world (Tonnes/Ha) [12]. Figure constructed by Orkun GENCER.

to fall markedly by 2055 as a consequence of the greenhouse effect and frequent occurrence of drought stress [17]. In addition to those, another study has predicted that global potato production would fall by 18–32% between 2040 and 2069 resulting from abiotic stresses linked to global climate change [18]. Although the severity of yield loss resulting from abiotic stress is dependent on several factors including the type of genotype, stress type/period, and developmental stage of the potato [19], the earlier exposure to drought stress, for example, indicates the more detrimental impacts during tuber formation, bulking and maturation due to a lower rate of carbon assimilation and then reduced partitioning of photosynthesis products to tubers [20]. Similar to drought stress, the severity of salinity influences on potato plants differs depending upon cultivar and salt concentration. The negative impacts of salt stress are analogous to those of drought based on morphological and physiological aspects including prolonged emergence of seedlings, dehydration/yellowing of leaves, accelerating leaf senescence, decreased tuber size/dry weight, reduced photosynthesis/transpiration rate, reduction in stomatal conductance, and less relative water content in leaves [21, 22] because both cause water stress, which forces stomatal aperture to shrink, K^+ and Ca^{2+} mineral deficiency in soil and inhibition of plant growth/development [23, 24]. Salt stress has also been shown to increase Na^+ and Cl^- concentration, intensify lipid peroxidation, and lower the rate of callus generation in potatoes [25, 26]. Likewise, heat stress considerably limits tuber formation and quality by limiting sucrose production and subsequent transfer to the stolon, resulting in a reduction in dry matter, malformations in tuber shape, and abnormal secondary growth [15, 16]. The sensitivity of potatoes to heat stress is also shown to be heavily influenced by cultivar type [27], developmental stage, and stress period [28]. The optimum temperature demands for above-ground biomass during potato development and below-ground tuber production and maturation differ; the former thrives in the 20–25°C range, while the latter thrives in the 15–20°C range [29]. A recent study has also suggested the greater impacts of heat stress on potato development and productivity when applied earlier during tuber formation as well as inhibition of tuber production temperatures over 25°C using six potato cultivars [30].

Considering the various negative direct or indirect impacts on potato quality and yield as aforementioned above, it is therefore highly essential to obtain potato lines with high tolerance to abiotic stress. Thus, the first step to achieve this is to identify the factors that confer improved tolerance to abiotic stress factors at the level of the genome, transcriptome, and microRNAs that are involved in either transcriptional and/or post-transcriptional regulation. Although a great number of studies on other plants may provide preliminary information for the discovery of stress tolerance mechanisms, the potato-specific factors that possess crucial roles under abiotic stresses need to be highlighted as potato production is mostly achieved by vegetative propagation through tubers.

2. Potato genomics on abiotic stress response/tolerance

Genomics is involved in the discovery of genetic variation underpinning changes in phenotypes, the uncovering of new sources for traits and diversity, and the elucidation of molecular pathways regulated by stress conditions. Over time, new breeding targets have been specified due to advances in genomic technologies including third-generation sequencing to obtain reference genomes with a greater quality, investigation of non-coding parts of the genome to illuminate the function and utilization of pangenome for the analysis of the structural variation at species-wide level [31]. Besides, some other approaches have also contributed to genomic studies such as machine learning, short breeding, and high-throughput phenotyping.

Globally, the most extensively grown potato varieties are *Solanum tuberosum* subsp. tuberosum, but there are also other potato species with different ploidy levels, including diploids, triploids, pentaploids, and hexaploids [32, 33]. Along with polyploid species, tetraploid cultivars are self-compatible, whereas diploids are often self-incompatible [34]. Due to complicated genetic inheritance and highly heterozygous characteristics of the *S. tuberosum* subsp. Tuberosum, diploids developed from tetraploid genotypes by anther/pollen culture or generated by interspecific hybridization with specific genotypes are widely utilized as parents for mapping analyzes [35]. Yet, the self-incompatibility nature of diploid potatoes hinders the production of pure lines, necessitating the use of F1-hybrid populations to employ genome mapping in potatoes as opposed to typical homozygous mapping approaches [36].

The Potato Genome Sequencing Consortium (PGSC), composed of 26 international institutes from 14 countries, successfully sequenced the doubled monoploid (DM) potato genome of *S. tuberosum* group Phureja, with the identification of 39,031 protein-coding genes [34]. Subsequently, another study refined the DM potato assembly more precisely by arranging scaffolds and pseudomolecules in the genome [37]. Lately, a chromosome-scale long-read reference assembly was created in potatoes [38]. More than 100 different potato species have now been sequenced or re-sequenced, primarily with the help of Illumina platforms including the *wild S. commersonii*, tuber-bearing Solanum species, *Solanum chacoense* "M6", a somatic hybrid produced from *Solanum pinnatisectum* as well as cultivated potato taxa using Illumina and long-read PacBio platforms [39–43]. The identification of novel genes, markers, and haplotypes has been stimulated by the ongoing improvement of sequencing technologies and bioinformatic tools to develop our understanding of potato biology [44].

The genotyping-by-sequencing (GBS) method has recently become one of the most popular approaches as it attempts to illuminate genome-wide association studies and genetic diversity in a variety of crops, including potatoes [45, 46]. GBS has been successfully used in potatoes for the construction of physical mapping related to tuber yield and starch content/yield [47], to identify and characterize single nucleotide polymorphisms (SNPs) particularly linked to simple traits [48], genetic variation as well as population structure studies [49]. On the other hand, an SNP array-based high-throughput genotyping system has been developed and utilized in a wide variety of genomic studies in potatoes including the determination of novel genes and SNPs against late blight resistance (*Phytophthora infestans*) [50], identification/characterization of the genetic diversity and the construction of population structure for Andigenum germplasm [51], determination of haplotype-specific SNPs against wart resistance [52], and deciphering QTLs against common scab resistance in Canadian potato germplasm [53]. Moreover, QTL mapping has been effectively applied to identify genomic areas under abiotic stress, including drought [54], heat [55], and salinity [56]. One study has explored 28 QTLs specifically related to drought treatment, 17 QTLs under recovery treatment, and two QTLs in control conditions using a diploid potato population [57]. A few years later, it was suggested a total of 45 QTLs were linked to morphological, physiological, and agronomical traits upon exposure to terminal drought and well-watered application, of which 26 presumably attributed to drought tolerance [58]. Similarly, the identification of QTLs related to heat stress has revealed that allelic variation in the promoter region of heat shock cognate 70 (HSC70) corresponds with high-temperature tolerance by shielding photooxidative damage against PSII across a diverse variety of wild potato relatives [59].

Along with genomic QTLs, numerous additional classes of QTLs, including expression QTL (eQTL), proteomic QTL (pQTL), phenomic QTL (phQTL), and Meta-analysis QTL (MQTL) are also considered important tools in potato breeding. An eQTL attempts to uncover genetic variations that influence the expression of a particular gene(s) [60], whereas a pQTL correlates protein content and amount with genetic variability [61]. Besides, phQTL are chromosomal areas that include loci conferring to genetic variability in phenotypes [62]. Meta-analysis of QTLs (MQTL) involves QTL data from a wide range of studies, in various locations, years and genetic backgrounds, providing identification of stable and reliable QTLs [62]. Recent modern NGS-driven QTL mapping strategies namely, QTL-seq, individual population resequencing, and MutMap, which uses bi-parental mapping and mutant populations, could be utilized for effective detection of major QTLs controlling agronomic traits in potatoes. Consequently, with the progressive advances in the QTL mapping approach and sequencing technology such as high-throughput genotyping (HTG) and high-throughput digital phenotyping (HTP) as well as the development of specific software such as GWASpoly designed for tetraploid potato [63], more precise and accurate marker-trait association studies for traits including biotic/abiotic stress, quality, and productivity pave the way for the speeding up potato breeding cycles from over 10 years to around 4 years [64].

3. Recent transcriptomic studies of potato under abiotic stress

Transcriptomics is the investigation of the transcriptome, the overall set of RNA transcripts created by the genome, under certain conditions or in a single cell via high-throughput technologies such as RNA sequencing (RNA-seq) and microarray [65]. Transcriptome analysis facilitates the determination of genes that express differentially in particular cell types or reactions to different treatments. While genomics uncovers the entire set of genetic information such as genome structure and mapping, transcriptomics reveals all RNA-related information including their expression levels, roles, locations, and transports/degradations to understand the relationship between a given phenotype and transcriptome.

A great number of attempts at transcriptomic analysis have been undertaken in potatoes upon abiotic stress treatments including heat [66, 67], drought [68, 69], heat combined with drought [70], salt [71], a combination of drought and salt [72], and nitrogen deficiency [73, 74]. Another attempt has constructed cDNA libraries from normal potatoes grown at 20°C day/18°C night and those exposed to 3 days of heat treatment at 35°C day/28°C night [66]. A total of 1420 differentially expressed genes (DEGs) were discovered, of which 771 genes were considerably upregulated, whereas 649 were substantially downregulated. The expression profiles of 12 selected genes also coincided with the sequencing results. As shown by gene ontology analysis, these DEGs were grouped into 49 distinct Gene Ontology (GO) categories, revealing the functional variety of the high-temperature stress-responsive genes. Although StHsp26-CP and StHsp70 levels were significantly increased upon 3 days of heat application, the majority of potato heat transcription factors (StHsfs) and heat shock proteins (StHsps) were not effectively expressed according to the expression analyzes of DEGs. In a comprehensive study examining the potato transcriptome and metabolome, researchers have revealed the upregulation of 448 genes and downregulation of 918 genes when the heat-treated group was compared with the control plants [67]. After the application of short stress (6 h), a total of 160 genes have been upregulated and 538 genes have been downregulated, while 130 genes have been upregulated and 94 genes have been downregulated in response to extended heat stress (3 days). Differentially expressed genes have been closely associated with photosynthesis, protein denaturation, RNA processing, and cell wall damage under heat stress. Moreover, differentially expressed compounds related to amino acid production and accumulation of secondary metabolites including L-proline/tyrosine production and synthesis of flavonol-derivative molecules have largely been upregulated in response to heat stress, indicating that these compounds may have a remarkable role in improved heat tolerance. Meanwhile, it has been suggested to reduce some secondary metabolites such as stevioside under short heat stress and suppress amino acid production such as histidine as well as jasmonic acid (JA) hormone synthesis under prolonged heat stress. A study that utilized EMS-mutagenized potato plants showed an increased tolerance to drought treatment than the wild-type (WT) Desiree plants [68]. Moreover, alterations in transcripts induced by drought treatment in drought-tolerant (DR) and WT plants were analyzed using the de novo assembly approach. Total RNA was extracted from one-week-old WT and DR plants after being exposed to polyethylene glycol-8000 at different time intervals including 0, 6, 12, 24, and 48 h for RNA sequencing, respectively. Sixty one thousand one hundred transcripts and 5118 DEGs showing up or downregulation were discovered following the pairwise analyzes of WT and DR plants. Under the drought condition, transcriptome results have suggested a total of 909, 977, 1181, 1225, and 826 DEGs between WT and DR plants at 0, 6, 12, 24, and 48 h. As a result of KEGG enrichment analysis, photosynthetic-antenna protein and endoplasmic reticulum protein processing have been suggested to have major functions in the drought tolerance mechanism of the DR plant. With drought-tolerant potato landrace, scientists have examined the transcriptome response to drought, rehydration, and re-dehydration stress under consecutive treatments of dehydration, rehydration, and re-dehydration [69]. While, drought-responsive genes have been primarily associated with photosynthesis, regulation of lipid and sugar metabolism, signal transduction, wax formation, rigidity of cell wall, and osmoregulation. A total of 178 transcription factors were differentially expressed in drought-tolerant landrace JSY potato following drought stress, representing 4.7% of the drought-response genes (3764) and mainly related to 11 gene families. The majority of them were upregulated apart from the bHLH family, which was downregulated under drought stress. It has also been shown that the recovery of potato by regulating the gene expressions are closely associated with tuberin production, flavonoid, lipid, and sugar metabolism, detoxification mechanism as well as the reverse expression of the majority of drought-responsive genes. Moreover, upregulation and downregulation of transcriptional factor families, such as increased expression of the majority of ethylene responsive factor (ERF) transcription factor families, have been found more upon re-drought as compared to initial drought, indicating the previous drought treatment might induce a favorable response to the upcoming drought by drought hardening effect. In addition to drought-responsive genes regulating cell structure and components, drought hardening triggers the intensification of gene expressions that play important roles in photosynthesis, sugar metabolism, proteases and protease inhibitors, flavonoid production, transporters, and signal transduction. Instead of the individual stress treatment, researchers have tested the effects of drought, heat, and simultaneous drought and heat in a growing chamber using five different potato lines [70]. Interestingly, drought and combined stress have triggered an increase in chlorophyll content as well as typical responses including reduced leaf size and plant height, disruption in the membrane structure and decreasing relative water content (RWC) under abiotic stress conditions. They have also suggested that the concurrent treatment of drought and heat stress most likely had a higher impact on the physiological traits, causing wilting earlier and severe effects on growth rate as compared to a single treatment. A study has examined salt stress-responsive genes in Longshu No.5, a salt-tolerant tetraploid potato genotype [71]. A total of 5508 DEGs have been found in response to salt stress. As a result of GO and KEGG analysis, DEGs were found to be involved in the groups of nucleic acid binding, transporter function, ion movement, binding, kinase function, and oxidative phosphorylation. The highly differential expression of genes that function in the ion transport signaling, in particular, implies that modulation of this signaling cascade is crucial upon salt treatment in potatoes. In another recent study, researchers have predicted the effects of drought, salinity, and simultaneous drought and salinity on agricultural traits using maize and potato using a remote sensing approach in a wide range of geographical regions in the Netherlands [72]. They have discovered that both stressors (individual and co-occurring) had a considerable impact on the physiological traits including leaf area index, chlorophyll content, fraction of vegetation cover, relative water content, and amount of photosynthetically active radiation absorbed by plants in both groups, with drought being more pronounced effect than salt. Although, it has been revealed aggravating effects on the traits under combined stress conditions, the degree of impact level was substantially dependent on the growth and development season.

The number of studies is quite limited related to the simultaneous treatment of heat and salt in potatoes, however, its effects at morphological, physiological, and molecular levels have been analyzed to understand cross-talking mechanisms in tomato [75]. It has been found that the combination of stresses has affected profoundly in shoots and roots as well as dry weight compared to individual stress. Similarly, tomato plants exposed to salt (100 mM NaCl) or combined stress (42°C; 4 h/day) have shown a significant rise in Na^+ levels in both roots and shoots, of which under combined stress (21 days of co-exposure) being more impacted than exposed to salinity stress alone. Moreover, inhibition of nutrient absorption proteins caused by heat stress has stimulated decreased uptake of Ca^{2+} and K^+, imbalancing salt overly sensitive (SOS) pathway accompanied by the increased risk of toxicity which ultimately leads to considerable damage to plant development. Proline has been suggested to have important roles under a combination of heat and salt stress due to osmoprotectant and ROS scavenging function as well as energy source molecule under nutrient deficiency conditions, producing 30 ATP equivalents per oxidation of proline. Concerning the overall findings, it has been shown that activation of many alternative oxidases (AOX) enzymes and increased amount of antioxidant metabolites have overcome oxidative damage and build-up of reactive oxygen species upon stress treatments.

4. MiRNomics in potato

MiRNAs are small, single-stranded, and non-coding RNAs that modulate gene expression at either the transcriptional or post-transcriptional stages in plants by perfectly binding to the messenger RNA (mRNA) molecule of the target gene (**Figure 4**).

In general, miRNAs have a length of 20–24 nucleotides and are classified as species-specific or conserved between species in plants. They have been involved in a wide range of processes such as the flowering period, growth development, hormone regulation, signaling pathways, homeostasis maintenance, and in response to biotic/abiotic stresses in plants [76]. As a result of the emerging studies, it has been suggested that miRNAs show different expression profiles depending on the type and duration of stress and the type of specific tissue in plant organisms [77–79]. Therefore, it is thought that miRNA-based molecular editing studies can be utilized in enhanced tolerance to abiotic stress in plants as well as an increase in yield and quality of agricultural products (**Figure 5**).

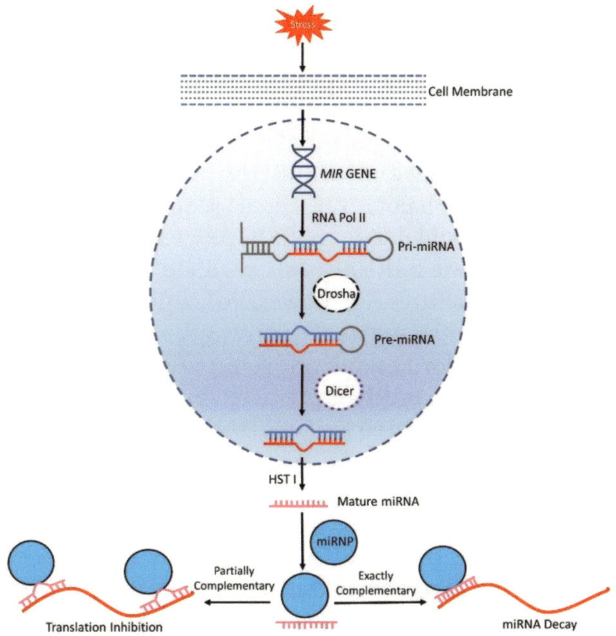

Figure 4.
General miRNA biogenesis pathway in plants. Figure constructed by Orkun GENCER.

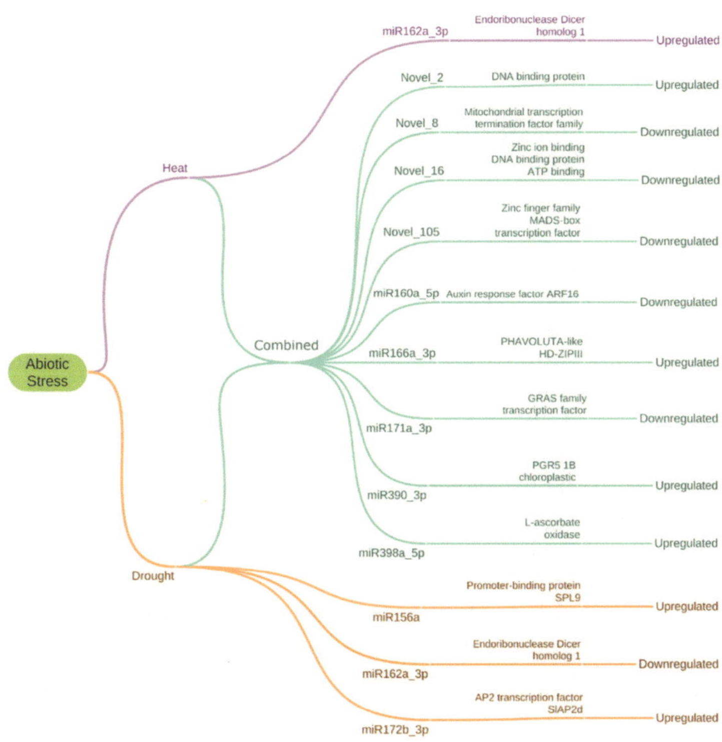

Figure 5.
Abiotic stress-responsive miRNAs in potato. Figure constructed by Orkun GENCER.

4.1 Identification of conserved and novel miRNAs in potato

Attempts to identify miRNAs and their target genes associated with potato growth/development and in response to abiotic stress, have begun in the last decade. As such, scientists have identified 48 potential miRNAs in *Solanum tuberosum* using an in silico approach by comparing known miRNAs from other organisms such as *Arabidopsis thaliana* and rice with potato expressed sequence tag (EST), genome survey sequence (GSS), and nucleoid databases [80]. As a result of bioinformatic analysis, they have reported that these miRNAs could regulate 186 potential target genes involved in above-ground structures e.g., flower, leaf, and stem development, root development, signal transduction, metabolic pathways, and stress response. RT-PCR analyzes have been conducted for 12 randomly selected out of 48 candidate miRNAs to validate the presence of potato miRNAs predicted by the bioinformatics approach [80]. Altogether, it has been discovered that while certain miRNAs have been expressed in all plant tissues, others have displayed temporal or spatial expression during development stages. Similarly, another in silico approach was used to predict miRNAs in potatoes where 71 putative miRNAs from 48 families have been identified [81]. It is noteworthy that 65 of the 71 miRNAs were predicted for the first time. Only 7 of the 71 potential miRNAs have been chosen for verification, and their expression levels in multiple potato tissues have been investigated using real-time PCR. The findings reveal that all seven miRNAs have been effectively amplified and their expression patterns have varied among tissues [81]. It has been postulated the highly variable expression level of each miRNA in various tissues might be linked to miRNA roles in the modulation of the vegetative development in potatoes. Another study has identified a total of 202 candidate miRNAs corresponding to 78 families using a newly modified comparative genome comparison approach in potatoes [82]. It has been discovered that 54 of the 78 families were novel in potatoes as well as 24 miRNA families reported in earlier studies. This group has analyzed the expressions of 12 miRNAs in young leaf, immature flower, and mature flower tissues. As a result, they have observed that a vast majority of the miRNAs have been expressed in all three tissues and one has no expression in young leaf tissue. Ref. [82] have estimated that the identified miRNAs target a total of 1094 genes, encoding transcription factors, regulating stress response, signal transduction, and other metabolic stages. According to GO analysis, these targets participate in 545 biological processes, 28 of which are associated with potato defensive systems, and biological metabolisms including carbon, starch, lipid, and primary/lateral root formation. Pathway enrichment study using KEGG indicated that the identified miRNAs are involved in 98 metabolic pathways including sucrose, fatty acid, amino acid metabolism, and carbon fixation as well as plant hormone synthesis. In another study to predict miRNAs and target genes in various agricultural plants belonging to the Solanaceae family, 22 miRNAs and 221 target genes were found by analyzing ESTs in potatoes [83]. Researchers continued to have identified 259 miRNAs belonging to 159 miRNA families in potatoes using next-generation and potato genome sequencing [84, 85]. Researchers have reported that only 28 of these miRNA families have been classified as conserved miRNA families in other plants, while the others have been suggested to be potato-specific confirmed by RNA gel blot hybridization. It has been revealed that the putative target genes are transcription factors and genes involved in defense mechanisms, signal transduction, ion regulation, flowering, and tuber production. In another study, 89 conserved miRNAs from 33 families involving 147 potato-specific and 112 potato-specific candidate miRNAs were discovered using samples from three distinct tissues and four different phases of tuber formation [84]. The results obtained suggested that the majority of conserved miRNAs target various important transcription factor families including SBP, GRAS, YA, NAC, and ARF highlighting the importance of conserved miRNAs in critical biological processes. According to the GO analyzes, the great majority of those have been implicated in oxidation-reduction activities, metabolism-related processes, defensive responses, and transportation as well as signal transduction, implying that the majority of potential targets of potato-specific miRNA have not overlapped with conserved miRNA targets. Furthermore, the expression analyzes by qRT-PCR results have indicated that certain miRNAs have been expressed based on tissue type, while the expression level of others has differed markedly depending upon the developmental stage of the potato, such as tuberization. In another study, a scientist identified 120 new miRNAs belonging to 110 families with a comparative genomics approach [86]. It has been shown that the expression of only 10 randomly selected miRNAs by quantitative RT-PCR. According to the findings, 433 putative genes targeted by 120 miRNAs are involved in signal transduction, modulation of growth/development, transcription factors, structural functions, and biotic stress response. Recently, a study group has identified the differentially expressed miRNAs and corresponding target genes that function in the build-up of anthocyanin during purple potato development by small RNA and degradome sequencing. A total of 275 differentially expressed miRNAs have been revealed from sRNA libraries. It has been found through interaction analysis that 37 miRNAs and 23 target genes were involved in anthocyanin-responsive miRNA-mRNA modules. Various miRNAs control the major enzymes involved in anthocyanin production by regulating structural and regulatory genes in purple potatoes. In addition, it has been proposed that identified miRNAs are crucial in the coloring and accumulation of anthocyanins in potatoes [87].

4.2 Identification of conserved and novel miRNAs under abiotic stress

miRNA studies in potato in response to abiotic stress conditions has a quite recent history compared to agriculturally essential crops including wheat [88], rice [89], and maize [90]. Lately, a growing number of studies have identified known and novel miRNAs under abiotic stresses to elucidate their functions in a multitude of metabolic pathways, which ultimately confer the development of potato plants with enhanced tolerance to individual or combined stress conditions. The majority of stress treatments on potatoes are related to the identification of drought-responsive miRNAs, while the studies on the identification of heat, cold, or salt-responsive miRNAs are quite limited.

4.2.1 Drought-responsive miRNAs in potato

Under drought stress, a study has identified three drought-responsive miRNAs in potatoes including stu-miR396, stu-miR156a, and stu-miR157a by comparative deep-sequencing, and sequence alignment [91]. They have found that the expression of each miRNA has differentially changed when exposed to 15% polyethylene glycol (PEG-6000) or an air-drying method over time. It has been suggested that stu-miR156a and stu-miR157a have 68 and 71 potential mRNA targets which have significant functions in the developmental process or cell differentiation, respectively. The same group has discovered drought-responsive miRNAs in potatoes, including stu-miR171a, stu-miR171b, and stu-miR171c that target GRAS family transcription factors, as well as their expression patterns and target mRNAs under drought stress.

It has also been suggested that drought-induced miRNAs in potatoes also contain families of miR159, miR164, miR166, miR390, miR395, miR397, miR398, miR408, and miR482 [92]. In another study, researchers have shown altered expression of the miRNAs stu-miR172c, stu-miR172d, and stu-miR172e, and their targets by adding 15% polyethylene glycol (PEG 6000) or air-drying approach [93]. Contrary to stu-miR172e which shows a lack of transcription factor target, 8 of 13 identified targets for stu-miRNA172c and 6 of 11 putative stu-miRNA172d targets belong to transcription factor gene families that have vital functions in the developmental pattern or cell differentiation. A total of 11 miRNAs grouped into six different miRNA families have been predicted in the modulation of the gen expressions responsible for the build-up in proline amount in potatoes under drought stress [94]. Ten of 11 anticipated miRNAs have effectively been identified using qRT-PCR, encompassing nine downregulated miRNAs and one upregulated miRNA. It was discovered that miR172, miR396a, miR396c, and miR4233 may control the pyrroline-5-carboxylate synthase (P5CS) gene expression, whereas miR2673 and miR6461 may modulate the pyrroline-5-carboxylate reductase (P5CR) and proline dehydrogenase (PRODH) gene expressions, as suggested by expression and functional analyzes. Ref. [95] have specifically discovered three potato miR159s including stu-miR159a, stu-miR159b, and stu-miR159c, and corresponding target genes including StGAMyb-like1, StGAMyb-like2.1, and StGAMyb-like2.2. As a result of the expression analysis, the reduced expression of stu-miR159s and increased expression of GAMyb-like transcription factors in potatoes suggest that the potato adaptation might be improved by overexpression of GAMyb-like transcription factor family genes induced by stu-miR159s under drought stress treatment. Another study to investigate potato miRNAs implicated in the regulation of drought tolerance discovered 458 known and 674 new miRNAs in control samples, and 471 known and 566 new miRNAs in drought-treated samples [85]. Drought treatment has been performed using 15% PEG treatment once a day for 20 days. Conserved miRNAs whose expression varies more than twice under drought stress were selected, and a decrease in the expression of 100 miRNAs and an increase in the expression of 99 miRNAs have been explored in this study. As a consequence of PEG treatment, it has been found that downregulation of 100 and upregulation of 99 miRNAs, whereas upregulation of 119 novel miRNAs and downregulation of 151 novel miRNAs have been observed based on differential expression analyzes. Four miRNAs including miR811, miR814, miR835, miR4398, and their targets MYB transcription factor, hydroxyproline-rich glycoprotein, aquaporin, and WRKY transcription factor have been suggested to be involved in the modulation of drought-related genes based on expression analysis, target prediction, and annotation.

4.2.2 Potato cold responsive miRNAs

Scientists, for the first time, have identified genome-wide level miRNAs and corresponding targets in potatoes exposed to cold stress [96]. In this regard, small RNA and degradome libraries have been created by potato tubers held at 4 and 20°C for 23 days. As a result of deep sequencing and whole-genome sequence analysis, 53 known and 60 novel miRNAs have been explored. Degradome analyzes have revealed 70 miRNA targets, demonstrating that miRNAs have been involved in the modulation of gene expression in post-harvest tubers. The targets have notable functions in a multitude of biological processes such as regulation of carbohydrate metabolism, stimulus-response, transcriptional control, and signal transduction. Subsequent expression profiling has displayed unique expression patterns of 11 miRNAs/miRNA*s and 34 targets for two particular potato genotypes to cold storage treatment regarding the decrease in sugar deposition, indicating a potential role for miRNAs in cold-induced sweetening. Recently, another study group has reported the profiling of miRNA and target mRNAs in potato plants, Favorita var., using high-throughput sequencing under low temperature stress (0°C 12 h) by creating two RNA libraries and six mRNA libraries [97]. There were 294 known and 211 novel miRNAs identified, where 24 miRNAs showed higher expression in the low temperature-treated group than controls, and 80 miRNAs displayed lower expression in treated plants compared to control plants. Furthermore, it has been suggested that 3298 DEGs with 1629 upregulated and 1669 downregulated expression levels function in response to cold and drought response, and regulation of secondary metabolism. It has been also speculated that contrasting patterns of expression of two miRNAs have contributed to improved cold tolerance in potatoes by modulating peroxisome metabolism. Similar to this study, another study has investigated the impacts of low temperatures on miRNA profiling in potatoes exposed to various low temperatures, with freezing (−2°C), chilling (0°C), and critical growth temperature (2°C) [98]. Three hundred seven known miRNAs from 73 sRNA families and 211 novel miRNAs have been characterized in the study. The majority of miRNAs function under low temperature, and drought as well as disease stress where certain conserved miRNAs, such as stu-miR530, stu-miR156d, and stu-miR167b, have been discovered in response to low temperature in potatoes for the first time. It has been shown that 442 miRNA target genes showed differential expression most of which function in the carotenoid synthesis, regulation of circadian rhythm, and signal transduction prompting that numerous low temperature responsive pathways are possible; however, abscisic acid along with gibberellin has been suggested to have a coordinating role in a multitude of metabolism pathways under low temperature stress.

4.2.3 Potato heat-responsive miRNAs

Although heat stress severely disturbs potato yield and quality as compared to drought or cold stress, attempts to reveal heat stress-responsive miRNAs in potato is still meager. Over the past 15 years, one study has revealed 202 putative potato miRNAs belonging to 78 families by improved comparative genome method as well as 1094 target genes that are involved in the modulation of signal transduction, and a wide range of metabolic activities such as carbohydrate and lipid metabolisms, and stress response [82]. Of these families, miR414, miR1132, miR1435, miR1530, and miR1533 have been suggested in response to heat stress in potatoes by targeting genes encoding heat-shock proteins that function in the chloroplast or nucleus. A recent conducted study has analyzed the various roles of unknown function (DUF221) proteins in potatoes [99]. It has also been shown that DUF22 genes are associated with miRNA families upon heat treatment in potatoes. Among the explored 10 StDDP (*Solanum tuberosum* DUF domain containing) genes, StDDP4 has been targeted by 10 members of the stu-miR395. Three members of miR319-3p, stu-miR319a-3p, and stu-miR319b have targeted StDDP1 and StDDP5 genes, respectively. StDDP7 has been targeted by stu-miR172a-5 while StDDP2 and StDDP1 have been targeted by stu-miR172b-5p and stu-miR172d-3p, respectively. Moreover, more than one miRNA has been predicted to target the StDDP5, StDDP4, StDDP2, and StDDP1 genes along with StDDP6 targeted by stu-miR8033-3p. Additionally, it has been shown that high-temperature stress elevated the expression of StDDP2, StDDP5, and StDDP7 in potatoes yet the expressions of corresponding miRNAs have not been validated in this study.

4.2.4 Salt/alkaline responsive miRNAs in potato

After the identification of miR319 upregulation on the improved performance under salinity stress [100, 101], researchers have shown that transgenic potato plants constructed by overexpression of miR319 have displayed enhanced water conservation and cell membrane stability as compared with control plants under salt stress [102]. It has also been observed that accumulation of Na + in transgenic plants has been less than as compared to control plants upon salt treatment. Another similar study has used parallel screening of both noncoding and coding RNA transcriptomes to identify and characterize salt stress-responsive miRNA and corresponding target genes in the Andigena potato cultivar, Sullu. An inverse association was revealed between salinity stress-regulated miRNAs including miR166 targeting HD-ZIP-Phabulosa/Phavulota network and miR159 targeting Myb101 genes [103]. They have also postulated that the miR159-Myb101 pair could be critical for managing the stress-induced premature switch to the reproductive stage as well as regulating vegetative development whereas, miR166-HD-ZIP-Phabulosa/Phavulota network might be closely related to growth control under salt stress through modulating vegetative dormancy via interacting with defense-related pathways. Recently, a study group has analyzed alkali stress-responsive miRNAs and their targets in potatoes which differ from salt stress [104]. The findings of the miRNA sequencing study revealed 168 differentially expressed miRNAs upon alkali stress treatment, of which 21 known miRNAs and 110 novel miRNAs were constructed by 12 small RNA libraries. Besides, the mRNA sequencing data revealed 5731 differentially expressed mRNAs using the same number of mRNA libraries. MiRNA-mRNA integrated analysis yielded 33 miRNA-target gene pairings made up of 20 differentially expressed miRNAs and 33 differentially expressed mRNAs. It has been suggested that some mechanisms function under alkali stress in potatoes including the phenylpropanoid biosynthesis pathway, regulation of starch/sucrose pathway, and hormone-mediated signal transduction. Additionally, the negative impacts of miR4243-x and novel-m064-5p on shikimate O-hydroxycinnamoyltransferase (HCT) and sucrose-phosphate synthase (SPS) genes have been linked to the potato response under alkaline conditions.

4.2.5 Potato combined abiotic stress-responsive miRNAs

A recent study to identify miRNAs in contrasting potato cultivars under a simultaneous treatment of drought and heat stresses has shown 447 conserved miRNAs and 315 novel miRNAs from 10 libraries using similarity and structural analysis [79]. It has been indicated that miRNA expression profile varies based on the stress treatment alone or combined, and the type of cultivar, tolerant or sensitive (tolerant-Unica/sensitive-Russet Burbank cultivar). For example, a total of 156 miRNAs with differential expression in sensitive cultivars have been observed, 22 of which were unique to heat and drought combination stress. Besides, 111 miRNAs were differentially expressed in tolerant cultivars under drought stress, with 16 miRNAs being drought-specific based on multiple comparison analyzes. They have also analyzed eight miRNAs that function in either specific or common modulation under drought combined with heat stress. Six miRNAs were suggested downregulated in both potato cultivars under simultaneous heat and drought treatment and individual heat stress, whereas two were found to be upregulated in this study. Upon drought stress, however, four miRNAs were downregulated, whereas one was upregulated in both cultivars. These miRNAs have depicted in the modulation of metabolic processes including a total of 200 genes regulating ribonucleotide, purine ribonucleotide, adenyl nucleotide, carbohydrate derivative binding, purine nucleotide, 236 genes for nitrogen compound metabolism, more than 135 genes for ATP binding in sensitive cultivar under drought stress. Moreover, 213 genes have been closely linked to small molecules and anion binding as well as nucleotide phosphate binding in sensitive cultivars exposed to heat treatment. Upon heat and drought treatment, 200 genes have been found participating in the cyclic compound formation and the heterocycle metabolic process in the tolerant cultivar.

Following the identification/characterization of miRNAs upon abiotic stress conditions, the functions of those miRNAs were analyzed through overexpressing in potato plants when exposed to individual or combined stress treatment. For instance, a study has explored the upregulation of miR160a-5p and downregulation of auxin response factor (ARF16) under a combination of heat and drought stress in potato plants (Unica-tolerant and Russet Burbank-sensitive cultivar) using overexpression approach [105]. They have also indicated that transgenics have greater morphological (leaf number/size and stem size), physiological (gaseous exchange traits, chlorophyll content, and canopy temperature), and biological traits (less accumulation of H_2O_2) than control plants. In addition to these findings, researchers have proposed improved the physio-biochemical performances of transgenic potato plants obtained by over-expressing Novel_105 miRNA targeting E3 ubiquitin-protein ligase gene (XBAT35), which presumably results in the inhibition of cell death to some extent under single or combined stress condition [106]. Similarly, the role of miR172b-3p has been analyzed using two contrasting potato cultivars under drought combined with heat stress [107]. It has been reported that overexpression of miR172b-3p in transgenic lines inhibited ERTF RAP2-7-like expression, resulting in increased carbon fixation performance under a combination of drought and heat treatment.

4.3 Other studies to identify miRNAs in potato

Other than known stress conditions, there are also some studies for the identification of potato miRNAs to specific environmental or growth conditions. For example, a study has identified miRNAs in plants that are sensitive to a biocompatible stress resistance inducer which is a potential approach focused on generating induced resistance by using ecologically friendly chemicals [108]. As a result of next-generation sequencing (NGS) analysis on potato leaves exposed to potassium phosphite chemical, it has been discovered 25 differentially expressed miRNAs, 14 of which have already marked in miRBase and 11 of which have been mapped to the potato genome as possible novel miRNAs. Based on miRNA target prediction analysis, it has been implied that those genes are involved in disease resistance, transcription factors, and in response to oxidative stress. In addition to these studies, researchers have characterized miRNAs in potato-grown aeroponic environments under the application of high and low nitrogen regimes [109]. Along with 119 conserved miRNAs from 41 miRNA families, a total of 1002 potential new miRNAs have been identified in potatoes. It has been observed that 52 conserved and 404 putative novel miRNAs, have been differently expressed in roots exposed to low levels of nitrogen application, while 54 conserved miRNAs and 628 putative novel miRNAs have been differentially expressed in shoots. GO study revealed the vast majority of the 34,135 putative targets were linked to biological activities, followed by molecular functions and cellular components. The roles of miR397 and miR398 in potatoes under low nitrogen stress have been displayed followed by target validations.

5. Conclusion

Improved stress tolerance in plants against dynamic environmental conditions is a major focus in agriculture due to the adverse effects of climate change on plant growth, development, and yield. Because potatoes are a susceptible plant to various abiotic stress conditions, the application of one or more stress treatments results in negative impacts on potato morphology and physiology during developmental stages as well as yield and tuber quality during/after harvesting.

Classical breeding methods have been used for many years in potato breeding. Moreover, genomics, transcriptomics, and miRNomics have significantly contributed to the efficiency of potato breeding efforts. Utilization of genomic studies includes the use of both cultivated and wild potato genotypes as a reservoir of desirable quality traits/allelic variety, as well as the ability to modify whole chromosomal sets, providing sexual hybridization a potent technique for producing novel and desirable genotypes. Therefore, several parameters that impede potato genetic advancement such as tetrasomic inheritance, a high incidence of heterozygosity, and incompatibility in potatoes might be partially overcome in potato breeding studies. Over time, genomics provides all genetic information in potatoes, such as genome structure and mapping while transcriptomics reveals all RNA-related information including stress-responsive genes depending upon expression levels/functions/locations in order to comprehend the link between a particular phenotype and transcriptome. Moreover, miRNA-based editing can be used as an alternative approach to advance potato tolerance under such individual or combined abiotic stresses since miRNAs have not only vital but also diversified functions in a multitude of metabolic processes and under assorted abiotic stress factors such as drought, cold, heat, salinity, and mineral deficiency. Because of the unique roles of miRNAs resulting from spatiotemporal ability and type of stress treatment, identification/characterization of miRNAs under individual and combined abiotic stress might be the first step in understanding the function of miRNAs in a wide range of metabolic pathways in potatoes to confer improved tolerance against the ever-mounting effects of abiotic stress in the future. These advancements in potato genomic, transcriptomic and miRNomics are highly crucial to keep up with the high demand for potatoes across the world due to the ever-growing human population and the reduction of potato production areas.

Funding

This study received no external funding.

Conflict of interest

The authors declare no conflict of interest.

Author details

Beyazıt Abdurrahman Şanlı*, Zahide Neslihan Öztürk and Orkun Gencer
Department of Agricultural Genetic Engineering, Ayhan Şahenk Faculty of Agricultural Sciences and Technologies, Niğde Ömer Halisdemir University, Niğde, Turkey

*Address all correspondence to: beyazitsanlitblt@gmail.com

References

[1] Pachauri R, Meyer L, editors. Climate Change 2014: Synthesis Report. Contribution of Working Groups I, II and III to the Fifth Assessment Report of the Intergovernmental Panel on Climate Change. Geneva, Switzerland: The Intergovernmental Panel on Climate Change (IPCC); 2014. p. 151

[2] Lesk C, Rowhani P, Ramankutty N. Influence of extreme weather disasters on global crop production. Nature. 2016;**529**(7584):84-87. DOI: 10.1038/nature16467

[3] DeLucia EH, Nabity PD, Zavala JA, Berenbaum MR. Climate change: Resetting plant-insect interactions. Plant Physiology. 2012;**160**(4):1677-1685. DOI: 10.1104/pp.112.204750

[4] Hatfield JL, Boote KJ, Kimball BA, Ziska LH, Izaurralde RC, Ort D, et al. Climate impacts on agriculture: Implications for crop production. Agronomy Journal. 2011;**103**(2):351-370. DOI: 10.2134/agronj2010.0303

[5] Vandegeer R, Miller RE, Bain M, Gleadow RM, Cavagnaro TR. Drought adversely affects tuber development and nutritional quality of the staple crop cassava (*Manihot esculenta* Crantz). Functional Plant Biology. 2013;**40**(2):195-200. DOI: 10.1071/FP12179

[6] Rahmstorf S. A semi-empirical approach to projecting future sea-level rise. Science. 2007;**315**(5810):368-370. DOI: 10.1126/science.1135456

[7] Dasgupta S, Laplante B, Meisner C, Wheeler D, Yan J. The impact of sea level rise on developing countries: A comparative analysis. Climatic Change. 2009;**93**(3-4):379-388. DOI: 10.1007/s10584-008-9499-5

[8] Deryng D, Elliott J, Folberth C, Müller C, Pugh TAM, Boote KJ, et al. Regional disparities in the beneficial effects of rising CO_2 concentrations on crop water productivity. Nature Climate Change. 2016;**6**(8):786-790. DOI: 10.1038/nclimate2995

[9] Lobell DB, Gourdji SM. The influence of climate change on global crop productivity. Plant Physiology. 2012;**160**(4):1686-1697. DOI: 10.1104/pp.112.208298

[10] Murchie EH, Pinto M, Horton P. Agriculture and the new challenges for photosynthesis research. New Phytologist. 2009;**181**(3):532-552. DOI: 10.1111/j.1469-8137.2008.02705.x

[11] Zhu X-G, Long SP, Ort DR. Improving photosynthetic efficiency for greater yield. Annual Review of Plant Biology. 2010;**61**:235-261. DOI: 10.1146/annurev-arplant-042809-112206

[12] FAO. Faostat: FAO Statistical Databases [Internet]. Rome, Italy: Food & Agriculture Organization of the United Nations (FAO); 2021. Available from: http://www.fao.org/faostat/en/\#data/QC

[13] Birch PRJ, Bryan G, Fenton B, Gilroy EM, Hein I, Jones JT, et al. Crops that feed the world 8: Potato: Are the trends of increased global production sustainable? Food Security. 2012;**4**:477-508. DOI: 10.1007/s12571-012-0220-1

[14] Dahal K, Li X-Q, Tai H, Creelman A, Bizimungu B. Improving potato stress tolerance and tuber yield under a climate

change scenario-A current overview. Frontiers in Plant Science. 2019;**10**:563. DOI: 10.3389/fpls.2019.00563

[15] Tuteja N, Gill SS, Tiburcio AF, Tuteja R, editors. Improving Crop Resistance to Abiotic Stress. Vol. 1. Weinheim, Germany: John Wiley & Sons; 2012

[16] Wang-Pruski G, Schofield A. Potato: Improving crop productivity and abiotic stress tolerance. In: Improving Crop Resistance to Abiotic Stress. Weinheim, Germany: John Wiley & Sons; 2012. pp. 1121-1153

[17] Holden NM, Brereton AJ, Fealy R, Sweeney J. Possible change in Irish climate and its impact on barley and potato yields. Agricultural and Forest Meteorology. 2003;**116**(3-4):181-196. DOI: 10.1016/S0168-1923(03)00002-9

[18] Hijmans RJ. The effect of climate change on global potato production. American Journal of Potato Research. 2003;**80**:271-279. DOI: 10.1007/BF02855363

[19] Evers D, Lefèvre I, Legay S, Lamoureux D, Hausman J-F, Rosales ROG, et al. Identification of drought-responsive compounds in potato through a combined transcriptomic and targeted metabolite approach. Journal of Experimental Botany. 2010;**61**(9):2327-2343. DOI: 10.1093/jxb/erq060

[20] Obidiegwu JE, Bryan GJ, Jones HG, Prashar A. Coping with drought: Stress and adaptive responses in potato and perspectives for improvement. Frontiers in Plant Science. 2015;**6**:542. DOI: 10.3389/fpls.2015.00542

[21] Akhtar SS, Andersen MN, Liu F. Biochar mitigates salinity stress in potato. Journal of Agronomy and Crop Science. 2015;**201**(5):368-378. DOI: 10.1111/jac.12132

[22] Jaarsma R, de Vries RSM, de Boer AH. Effect of salt stress on growth, Na^+ accumulation and proline metabolism in potato (*Solanum tuberosum*) cultivars. PLoS One. 2013;**8**(3):e60183. DOI: 10.1371/journal.pone.0060183

[23] Ma Y, Dias MC, Freitas H. Drought and salinity stress responses and microbe-induced tolerance in plants. Frontiers in Plant Science. 2020;**11**:591911. DOI: 10.3389/fpls.2020.591911

[24] Aghaei K, Ehsanpour AA, Komatsu S. Potato responds to salt stress by increased activity of antioxidant enzymes. Journal of Integrative Plant Biology. 2009;**51**(12):1095-1103. DOI: 10.1111/j.1744-7909.2009.00886.x

[25] Queirós F, Fidalgo F, Santos I, Salema R. In vitro selection of salt tolerant cell lines in Solanum tuberosum L. Biologia Plantarum. 2007;**51**:728-734. DOI: 10.1007/s10535-007-0149-y

[26] Queirós F, Rodrigues JA, Almeida JM, Almeida DPF, Fidalgo F. Differential responses of the antioxidant defence system and ultrastructure in a salt-adapted potato cell line. Physiology and Biochemistry. 2011;**49**(12):1410-1419. DOI: 10.1016/j.plaphy.2011.09.020

[27] Tang R, Niu S, Zhang G, Chen G, Haroon M, Yang Q, et al. Physiological and growth responses of potato cultivars to heat stress. Botany. 2018;**96**(12):897-912. DOI: 10.1139/cjb-2018-0125

[28] Ahn Y-J, Claussen K, Lynn Zimmerman J. Genotypic differences in the heat-shock response and thermotolerance in four potato cultivars. Plant Science. 2004;**166**(4):901-911. DOI: 10.1016/j.plantsci.2003.11.027

[29] Van Dam J, Kooman PL, Struik PC. Effects of temperature and photoperiod

on early growth and final number of tubers in potato (*Solanum tuberosum* L.). Potato Research. 1996;**39**:51-62. DOI: 10.1007/BF02358206

[30] Rykaczewska K. The effect of high temperature occurring in subsequent stages of plant development on potato yield and tuber physiological defects. American Journal of Potato Research. 2015;**92**:339-349. DOI: 10.1007/s12230-015-9436-x

[31] Pourkheirandish M, Golicz AA, Bhalla PL, Singh MB. Global role of crop genomics in the face of climate change. Frontiers in Plant Science. 2020;**11**:922. DOI: 10.3389/fpls.2020.00922

[32] Watanabe K. Potato genetics, genomics, and applications. Breed Science. 2015;**65**(1):53-68. DOI: 10.1270/jsbbs.65.53

[33] Diambra LA. Genome sequence and analysis of the tuber crop potato. Nature. 2011;**475**:189-195. DOI: 10.1038/nature10158

[34] Aksoy E, Demirel U, Öztürk ZN, Çalişkan S, Çalişkan ME. Recent advances in potato genomics, transcriptomics, and transgenics under drought and heat stresses: A review. Turkish Journal of Botany. 2015;**39**(6):920-940. DOI: 10.3906/bot-1506-25

[35] Jansky SH, Charkowski AO, Douches DS, Gusmini G, Richael C, Bethke PC, et al. Reinventing potato as a diploid inbred line-based crop. Crop Science. 2016;**56**(4):1412-1422. DOI: 10.2135/cropsci2015.12.0740

[36] Simko I, Jansky S, Stephenson S, Spooner D. Genetics of resistance to pests and disease. In: Vreugdenhil D, Bradshaw J, Gebhardt C, Govers F, Mackerron DKL, Taylor MA, et al., editors. Potato Biology and Biotechnology. Amsterdam, Netherlands: Elsevier; 2007. pp. 117-155. DOI: 10.1016/b978-044451018-1/50049-x

[37] Sharma SK, Bolser D, de Boer J, Sønderkær M, Amoros W, Carboni MF, et al. Construction of reference chromosome-scale pseudomolecules for potato: Integrating the potato genome with genetic and physical maps. G3: Genes, Genomes, Genetics. 2013;**3**(11):2031-2047. DOI: 10.1534/g3.113.007153

[38] Pham GM, Hamilton JP, Wood JC, Burke JT, Zhao H, Vaillancourt B, et al. Construction of a chromosome-scale long-read reference genome assembly for potato. GigaScience. 2020;**9**(9):1-11. DOI: 10.1093/gigascience/giaa100

[39] Aversano R, Contaldi F, Ercolano MR, Grosso V, Iorizzo M, Tatino F, et al. The *Solanum commersonii* genome sequence provides insights into adaptation to stress conditions and genome evolution of wild potato relatives. The Plant Cell. 2015;**27**(4):954-968. DOI: 10.1105/tpc.114.135954

[40] Hardigan MA, Laimbeer FPE, Newton L, Crisovan E, Hamilton JP, Vaillancourt B, et al. Genome diversity of tuber-bearing Solanum uncovers complex evolutionary history and targets of domestication in the cultivated potato. Proceedings of the National Academy of Sciences. 2017;**114**(46):E9999-E10008. DOI: 10.1073/pnas.1714380114

[41] Leisner CP, Hamilton JP, Crisovan E, Manrique-Carpintero NC, Marand AP, Newton L, et al. Genome sequence of M6, a diploid inbred clone of the high-glycoalkaloid-producing tuber-bearing potato species Solanum chacoense, reveals residual heterozygosity. The Plant Journal. 2018;**94**(3):562-570. DOI: 10.1111/tpj.13857

[42] Tiwari JK, Rawat S, Luthra SK, Zinta R, Sahu S, Varshney S, et al. Genome sequence analysis provides insights on genomic variation and late blight resistance genes in potato somatic hybrid (parents and progeny). Molecular Biology Reports. 2021;**48**:623-635. DOI: 10.1007/s11033-020-06106-x

[43] Kyriakidou M, Achakkagari SR, Gálvez López JH, Zhu X, Tang CY, Tai HH, et al. Structural genome analysis in cultivated potato taxa. Theoretical and Applied Genetics. 2020;**133**:951-966. DOI: 10.1007/s00122-019-03519-6

[44] Zhou Q, Tang D, Huang W, Yang Z, Zhang Y, Hamilton JP, et al. Haplotype-resolved genome analyses of a heterozygous diploid potato. Nature Genetics. 2020;**52**(10):1018-1023. DOI: 10.1038/s41588-020-0699-x

[45] Uitdewilligen JGAML, Wolters A-MA, D'hoop BB, TJA B, RGF V, van Eck HJ. A next-generation sequencing method for genotyping-by-sequencing of highly heterozygous autotetraploid potato. PLoS One. 2013;**8**(5):e62355. DOI: 10.1371/journal.pone.0062355

[46] Bastien M, Boudhrioua C, Fortin G, Belzile F. Exploring the potential and limitations of genotyping-by-sequencing for SNP discovery and genotyping in tetraploid potato. Genome. 2018;**61**(6):449-456. DOI: 10.1139/gen-2017-0236

[47] Schönhals EM, Ding J, Ritter E, Paulo MJ, Cara N, Tacke E, et al. Physical mapping of QTL for tuber yield, starch content and starch yield in tetraploid potato (*Solanum tuberosum* L.) by means of genome wide genotyping by sequencing and the 8.3 K SolCAP SNP array. BMC Genomics. 2017;**18**(1):642. DOI: 10.1186/s12864-017-3979-9

[48] Caruana BM, Pembleton LW, Constable F, Rodoni B, Slater AT, Cogan NOI. Validation of genotyping by sequencing using transcriptomics for diversity and application of genomic selection in tetraploid potato. Frontiers in Plant Science. 2019;**10**:670. DOI: 10.3389/fpls.2019.00670

[49] Pandey J, Scheuring DC, Koym JW, Coombs J, Novy RG, Thompson AL, et al. Genetic diversity and population structure of advanced clones selected over forty years by a potato breeding program in the USA. Scientific Reports. 2021;**11**(1):8344. DOI: 10.1038/s41598-021-87284-x

[50] Mosquera T, Alvarez MF, Jiménez-Gómez JM, Muktar MS, Paulo MJ, Steinemann S, et al. Targeted and untargeted approaches unravel novel candidate genes and diagnostic SNPs for quantitative resistance of the potato (*Solanum tuberosum* L.) to Phytophthora infestans causing the late blight disease. PLoS One. 2016;**11**(6):e0156254. DOI: 10.1371/journal.pone.0156254

[51] Berdugo-Cely J, Valbuena RI, Sánchez-Betancourt E, Barrero LS, Yockteng R. Genetic diversity and association mapping in the Colombian central collection of *Solanum tuberosum* L. Andigenum group using SNPs markers. PLoS One. 2017;**12**(3):e0173039. DOI: 10.1371/journal.pone.0173039

[52] Prodhomme C, Vos PG, Paulo MJ, Tammes JE, Visser RGF, Vossen JH, et al. Distribution of P1(D1) wart disease resistance in potato germplasm and GWAS identification of haplotype-specific SNP markers. Theoretical and Applied Genetics. 2020;**133**:1859-1871. DOI: 10.1007/s00122-020-03559-3

[53] Yuan J, Bizimungu B, De Koeyer D, Rosyara U, Wen Z, Lagüe M. Genome-wide association study of resistance to potato common scab. Potato Research. 2020;**63**:253-266. DOI: 10.1007/s11540-019-09437-w

[54] Schumacher C, Thümecke S, Schilling F, Köhl K, Kopka J, Sprenger H, et al. Genome-wide approach to identify quantitative trait loci for drought tolerance in tetraploid potato (*Solanum tuberosum* L.). International Journal of Molecular Sciences. 2021;**22**(11):1-29. DOI: 10.3390/ijms22116123

[55] Trapero-Mozos A, Morris WL, Ducreux LJM, McLean K, Stephens J, Torrance L, et al. Engineering heat tolerance in potato by temperature-dependent expression of a specific allele of heat-shock COGNATE 70. Plant Biotechnology Journal. 2018;**16**(1):197-207. DOI: 10.1111/pbi.12760

[56] Chourasia KN, Lal MK, Tiwari RK, Dev D, Kardile HB, Patil VU, et al. Salinity stress in potato: Understanding physiological, biochemical and molecular responses. Life. 2021;**11**(6):1-24. DOI: 10.3390/life11060545

[57] Anithakumari AM, Nataraja KN, Visser RGF, van der Linden CG. Genetic dissection of drought tolerance and recovery potential by quantitative trait locus mapping of a diploid potato population. Molecular Breeding. 2012;**30**:1413-1429. DOI: 10.1007/s11032-012-9728-5

[58] Khan MA, Saravia D, Munive S, Lozano F, Farfan E, Eyzaguirre R, et al. Multiple QTLs linked to agro-morphological and physiological traits related to drought tolerance in potato. Plant Molecular Biology Reporter. 2015;**33**:1286-1298. DOI: 10.1007/s11105-014-0824-z

[59] Campbell R, Ducreux L, Cowan G, Young V, Chinoko G, Chitedze G, et al. Allelic variants of a potato heat shock COGNATE 70 gene confer improved tuber yield under a wide range of environmental conditions. Food and Energy Security. 2022;**12**(1):e377. DOI: 10.1002/fes3.377

[60] Nica AC, Dermitzakis ET. Expression quantitative trait loci: Present and future. Philosophical Transactions of the Royal Society B: Biological Sciences. 2013;**368**(1620):20120362. DOI: 10.1098/rstb.2012.0362

[61] Rodziewicz P, Chmielewska K, Sawikowska A, Marczak Ł, Łuczak M, Bednarek P, et al. Identification of drought responsive proteins and related proteomic QTLs in barley. Journal of Experimental Botany. 2019;**70**(10):2823-2837. DOI: 10.1093/jxb/erz075

[62] Mondal R, Kumar A, Gnanesh BN. Crop germplasm: Current challenges, physiological-molecular perspective, and advance strategies towards development of climate-resilient crops. Heliyon. 2023;**9**:e12973. DOI: 10.1016/j.heliyon.2023.e12973

[63] Rosyara UR, De Jong WS, Douches DS, Endelman JB. Software for genome-wide association studies in autopolyploids and its application to potato. Plant Genome. 2016;**9**(2):1-10. DOI: 10.3835/plantgenome2015.08.0073

[64] Slater AT, Wilson GM, Cogan NOI, Forster JW, Hayes BJ. Improving the analysis of low heritability complex traits for enhanced genetic gain in potato. Theoretical and Applied Genetics. 2014;**127**:809-820. DOI: 10.1007/s00122-013-2258-7

[65] Tyagi P, Singh D, Mathur S, Singh A, Ranjan R. Upcoming progress of transcriptomics studies on plants: An overview. Frontiers in Plant Science. 2022;**13**:1030890. DOI: 10.3389/fpls.2022.1030890

[66] Tang R, Gupta SK, Niu S, Li X-Q, Yang Q, Chen G, et al. Transcriptome analysis of heat stress response genes in potato leaves. Molecular Biology Reports.

2020;**47**:4311-4321. DOI: 10.1007/s11033-020-05485-5

[67] Liu B, Kong L, Zhang Y, Liao Y. Gene and metabolite integration analysis through transcriptome and metabolome brings new insight into heat stress tolerance in potato (*Solanum tuberosum* L.). Plants. 2021;**10**(1):103. DOI: 10.3390/plants10010103

[68] Moon K-B, Ahn D-J, Park J-S, Jung WY, Cho HS, Kim H-R, et al. Transcriptome profiling and characterization of drought-tolerant potato plant (*Solanum tuberosum* L.). Molecules and Cells. 2018;**41**(11):979-992. DOI: 10.14348/molcells.2018.0312

[69] Chen Y, Li C, Yi J, Yang Y, Lei C, Gong M. Transcriptome response to drought, rehydration and re-dehydration in potato. International Journal of Molecular Sciences. 2019;**21**(1):159. DOI: 10.3390/ijms21010159

[70] Handayani T, Watanabe K. The combination of drought and heat stress has a greater effect on potato plants than single stresses. Plant, Soil and Environment. 2020;**66**(4):175-182. DOI: 10.17221/126/2020-PSE

[71] Li Q, Qin Y, Hu X, Li G, Ding H, Xiong X, et al. Transcriptome analysis uncovers the gene expression profile of salt-stressed potato (*Solanum tuberosum* L.). Scientific Reports. 2020;**10**(1):5411. DOI: 10.1038/s41598-020-62057-0

[72] Wen W, Timmermans J, Chen Q, van Bodegom PM. Monitoring the combined effects of drought and salinity stress on crops using remote sensing in the Netherlands. Hydrology and Earth System Sciences. 2022;**26**(17):4537-4552. DOI: 10.5194/hess-26-4537-2022

[73] Tiwari JK, Buckseth T, Singh RK, Kumar M, Kant S. Prospects of improving nitrogen use efficiency in potato: Lessons from transgenics to genome editing strategies in plants. Frontiers in Plant Science. 2020;**11**:597481. DOI: 10.3389/fpls.2020.597481

[74] Tiwari JK, Buckseth T, Zinta R, Saraswati A, Singh RK, Rawat S, et al. Transcriptome analysis of potato shoots, roots and stolons under nitrogen stress. Scientific Reports. 2020;**10**(1):1152. DOI: 10.1038/s41598-020-58167-4

[75] Sousa B, Rodrigues F, Soares C, Martins M, Azenha M, Lino-Neto T, et al. Impact of combined heat and salt stresses on tomato plants—insights into nutrient uptake and redox homeostasis. Antioxidants. 2022;**11**(3):478. DOI: 10.3390/antiox11030478

[76] Mangrauthia SK, Agarwal S, Sailaja B, Madhav MS, Voleti SR. MicroRNAs and their role in salt stress response in plants. In: Salt Stress in Plants: Signalling, Omics and Adaptations. New York: Springer; 2013. pp. 15-46. DOI: 10.1007/978-1-4614-6108-1_2

[77] Song X, Li Y, Cao X, Qi Y. MicroRNAs and their regulatory roles in plant–environment interactions. Annual Review of Plant Biology. 2019;**70**:489-525. DOI: 10.1146/annurev-arplant-050718-100334

[78] Wani SH, Kumar V, Khare T, Tripathi P, Shah T, Ramakrishna C, et al. miRNA applications for engineering abiotic stress tolerance in plants. Biologia. 2020;**75**:1063-1081. DOI: 10.2478/s11756-019-00397-7

[79] Öztürk Gökçe ZN, Aksoy E, Bakhsh A, Demirel U, Çalışkan S, Çalışkan ME. Combined drought and heat stresses trigger different sets of miRNAs in contrasting potato cultivars. Functional &

Integrative Genomics. 2021;**21**(3-4):489-502. DOI: 10.1007/s10142-021-00793-w

[80] Zhang W, Luo Y, Gong X, Zeng W, Li S. Computational identification of 48 potato microRNAs and their targets. Computational Biology and Chemistry. 2009;**33**(1):84-93. DOI: 10.1016/j.compbiolchem.2008.07.006

[81] Yang W, Liu X, Zhang J, Feng J, Li C, Chen J. Prediction and validation of conservative microRNAs of *Solanum tuberosum* L. Molecular Biology Reports. 2010;**37**:3081-3087. DOI: 10.1007/s11033-009-9881-z

[82] Xie F, Frazier TP, Zhang B. Identification, characterization and expression analysis of MicroRNAs and their targets in the potato (*Solanum tuberosum*). Gene. 2011;**473**(1):8-22. DOI: 10.1016/j.gene.2010.09.007

[83] Kim HJ, Baek KH, Lee BW, Choi D, Hur CG. In silico identification and characterization of microRNAs and their putative target genes in Solanaceae plants. Genome. 2011;**54**(2):91-98. DOI: 10.1139/G10-104

[84] Lakhotia N, Joshi G, Bhardwaj AR, Katiyar-Agarwal S, Agarwal M, Jagannath A, et al. Identification and characterization of miRNAome in root, stem, leaf and tuber developmental stages of potato (*Solanum tuberosum* L.) by high-throughput sequencing. BMC Plant Biology. 2014;**14**(1):6. DOI: 10.1186/1471-2229-14-6

[85] Zhang N, Yang J, Wang Z, Wen Y, Wang J, He W, et al. Identification of novel and conserved microRNAs related to drought stress in potato by deep sequencing. PLoS One. 2014;**9**(4):e95489. DOI: 10.1371/journal.pone.0095489

[86] Din M, Barozai MYK, Baloch IA. Identification and functional analysis of new conserved microRNAs and their targets in potato (*Solanum tuberosum* L.). Turkish Journal of Botany. 2014;**38**(6):1199-1213. DOI: 10.3906/bot-1405-105

[87] Wu X, Ma Y, Wu J, Wang P, Zhang Z, Xie R, et al. Identification of microRNAs and their target genes related to the accumulation of anthocyanin in purple potato tubers (*Solanum tuberosum*). Plant Direct. 2022;**6**(7):e418. DOI: 10.1002/pld3.418

[88] Akdogan G, Tufekci ED, Uranbey S, Unver T. miRNA-based drought regulation in wheat. Functional & Integrative Genomics. 2016;**16**:221-233. DOI: 10.1007/s10142-015-0452-1

[89] Mittal D, Sharma N, Sharma V, Sopory SK, Sanan-Mishra N. Role of microRNAs in rice plant under salt stress. Annals of Applied Biology. 2016;**168**(1):2-18. DOI: 10.1111/aab.12241

[90] Pei L, Jin Z, Li K, Yin H, Wang J, Yang A. Identification and comparative analysis of low phosphate tolerance-associated microRNAs in two maize genotypes. Physiology and Biochemistry. 2013;**70**:221-234. DOI: 10.1016/j.plaphy.2013.05.043

[91] Hwang EW, Shin SJ, Kwon HB. Identification of MicroRNAs and their putative targets that respond to drought stress in *Solanum tuberosum*. Journal of the Korean Society for Applied Biological Chemistry. 2011;**54**:317-324. DOI: 10.3839/jksabc.2011.051

[92] Hwang EW, Shin SJ, Yu BK, Byun MO, Kwon HB. miR171 family members are involved in drought response in *Solanum tuberosum*. Journal of Plant Biology. 2011;**54**:43-48. DOI: 10.1007/s12374-010-9141-8

[93] Hwang EW, Shin SJ, Park SC, Jeong MJ, Kwon HB. Identification of miR172 family members and

their putative targets responding to drought stress in *Solanum tuberosum*. Genes & Genomics. 2011;**33**:105-110. DOI: 10.1007/s13258-010-0135-1

[94] Yang J, Zhang N, Ma C, Qu Y, Si H, Wang D. Prediction and verification of microRNAs related to proline accumulation under drought stress in potato. Computational Biology and Chemistry. 2013;**46**:48-54. DOI: 10.1016/j.compbiolchem.2013.04.006

[95] Yang LI, Wang C, Wang L, Xu C, Chen K. An efficient multiplex PCR assay for early detection of *Agrobacterium tumefaciens* in transgenic plant materials. Turkish Journal of Agriculture and Forestry. 2013;**37**(2):157-162. DOI: 10.3906/tar-1009-1265

[96] Ou Y, Liu X, Xie C, Zhang H, Lin Y, Li M, et al. Genome-wide identification of micrornas and their targets in cold-stored potato tubers by deep sequencing and degradome analysis. Plant Molecular Biology Reporter. 2015;**33**:584-597. DOI: 10.1007/s11105-014-0771-8

[97] Liao H, Wang Q, Zhang N, Fu Y, Wu G, Ren X, et al. High-throughput MicroRNA and mRNA sequencing reveals that MicroRNAs may be involved in peroxidase-mediated cold tolerance in potato. Plant Molecular Biology Reporter. 2021;**39**:577-594. DOI: 10.1007/s11105-020-01272-5

[98] Yan C, Zhang N, Wang Q, Fu Y, Wang F, Su Y, et al. The effect of low temperature stress on the leaves and MicroRNA expression of potato seedlings. Frontiers in Ecology and Evolution. 2021;**9**:727081. DOI: 10.3389/fevo.2021.727081

[99] Zaynab M, Peng J, Sharif Y, Albaqami M, Al-Yahyai R, Fatima M, et al. Genome-wide identification and expression profiling of DUF221 gene family provides new insights into abiotic stress responses in potato. Frontiers in Plant Science. 2021;**12**:3154. DOI: 10.3389/fpls.2021.804600

[100] Sunkar R, Zhu JK. Novel and stress-regulated microRNAs and other small RNAs from Arabidopsis. Plant Cell. 2004;**16**(8):2001-2019. DOI: 10.1105/tpc.104.022830

[101] Liu X, Yang L, Zhou X, Zhou M, Lu Y, Ma L, et al. Transgenic wheat expressing *Thinopyrum intermedium* MYB transcription factor TiMYB2R-1 shows enhanced resistance to the take-all disease. Journal of Experimental Botany. 2013;**64**(8):2243-2253. DOI: 10.1093/jxb/ert084

[102] Pieczynski M, Marczewski W, Hennig J, Dolata J, Bielewicz D, Piontek P, et al. Down-regulation of CBP80 gene expression as a strategy to engineer a drought-tolerant potato. Plant Biotechnology Journal. 2013;**11**(4):459-469. DOI: 10.1111/pbi.12032

[103] Kitazumi A, Kawahara Y, Onda TS, De Koeyer D, de los Reyes BG. Implications of miR166 and miR159 induction to the basal response mechanisms of an andigena potato (*Solanum tuberosum* subsp. andigena) to salinity stress, predicted from network models in Arabidopsis. Genome. 2015;**58**(1):13-24. DOI: 10.1139/gen-2015-0011

[104] Kang Y, Yang X, Liu Y, Shi M, Zhang W, Fan Y, et al. Integration of mRNA and miRNA analysis reveals the molecular mechanism of potato (*Solanum tuberosum* L.) response to alkali stress. International Journal of Biological Macromolecules. 2021;**182**:938-949. DOI: 10.1016/j.ijbiomac.2021.04.094

[105] Şanlı BA, Öztürk Gökçe ZN. Investigating effect of miR160 through overexpression in potato cultivars

under single or combination of heat and drought stresses. Plant Biotechnology Report. 2021;**15**(3):335-348. DOI: 10.1007/s11816-021-00677-2

[106] Yalçin M, Gökçe ZNÖ. Investigation of the effects of overexpression of Novel_105 miRNA in contrasting potato cultivars during separate and combined drought and heat stresses. Turkish Journal of Botany. 2021;**45**(5):397-411. DOI: 10.3906/bot-2103-39

[107] Asim A, Gökçe ZNÖ, Bakhsh A, Çayli İT, Aksoy E, Çalişkan S, et al. Individual and combined effect of drought and heat stresses in contrasting potato cultivars overexpressing miR172b-3p. Turkish Journal of Agriculture and Forestry. 2021;**45**(5):651-668. DOI: 10.3906/tar-2103-60

[108] Rey-Burusco MF, Daleo GR, Feldman ML. Identification of potassium phosphite responsive miRNAs and their targets in potato. PLoS One. 2019;**14**(9):e0222346. DOI: 10.1371/journal.pone.0222346

[109] Tiwari JK, Buckseth T, Zinta R, Saraswati A, Singh RK, Rawat S, et al. Genome-wide identification and characterization of microRNAs by small RNA sequencing for low nitrogen stress in potato. PLoS One. 2020;**15**(5):e0233076. DOI: 10.1371/journal.pone.0233076

Chapter 13

The Knowledge Landscape of Adenylyl Cyclases in Model Plant, *Arabidopsis thaliana*

Patience Chatukuta, Angela Sibanda-Makuvise, Tshegofatso Dikobe, Enetia Bobo, Katlego Sehlabane, Grace Mabadahanye, Neo Mametja, Mutsa Takundwa, David Kawadza and Oziniel Ruzvidzo

Abstract

One strategy for improving responses and adaptation systems of plants to stress is to target molecules involved in signaling and transduction of the stimuli effected by stresses. One such molecule is adenylyl cyclase (AC) – an enzyme that catalyzes the conversion of adenosine 5′-triphosphate (ATP) to the second messenger, 3′,5′-cyclic adenosine monophosphate (cAMP). cAMP, in turn, transduces signals in response to the various biotic and abiotic stress factors. Surprisingly, as far as five decades ago, attempts to isolate ACs and/or detect cAMP from the research model plant, *Arabidopsis thaliana*, were inconclusive or a matter of serious debates due to the absence of appropriate techniques or advanced technologies. This chapter, therefore, herein takes the reader on a journey from the 1970s to the present day, unraveling the challenges encountered, developments made, and successes realized in efforts and attempts to identify and characterize ACs in *A. thaliana*. The chapter covers from the early age of unsuccessful attempts to the more recent and successful advanced technologies such as the motif search approach, omics analysis and homologous cloning. Perspectives on the direction that future knowledge-building around this important group of plant proteins are also shared.

Keywords: *Arabidopsis thaliana*, adenylyl cyclase, plant signaling, cAMP, motif search

1. Introduction

In this chapter, we present part of our work that involves the identification and characterization of a special group of plant enzymes termed adenylyl cyclases (ACs) in the model plant *Arabidopsis thaliana*. These enzymes essentially catalyze the conversion of adenosine 5′-triphosphate (ATP) to the second messenger, 3′,5′-cyclic adenosine monophosphate (cAMP) [1–3]. The two molecules (ACs and cAMP) play crucial roles in many diverse and downstream biological processes of all living organisms, ranging from prokaryotes (e.g., *Escherichia coli*) to complex multicellular eukaryotes (e.g., *Homo sapiens*). In plants, ACs and cAMP play crucial roles linked to highly complex systems that assist in the maintenance and progression of cellular homeostasis. This is quite important considering the sessile nature of plants in the current rapidly fluctuating environments associated with a wide range of biotic and abiotic stress factors. Ideally, these two signaling molecules mediate several metabolic and physiological processes in plants including control of the cell cycle [4], activation of the phenylalanine ammonia-lyase (PAL) enzyme [5] and regulation of the phenylpropanoid pathway [6]. In addition, the molecules are also involved in stomatal movements [7], cation transport via voltage-independent channels (VICs) [8], pollen tube growth [9–12] as well as response to stressful conditions via the cyclic nucleotide-gated channels (CNGCs) [7, 13–16].

The chapter begins by introducing *A. thaliana* as a model organism for plant research followed by a narration of the journey that was followed from the 1970s to the present day, on the details of the challenges encountered, developments made, and successes realized with regard to the identification/discovering and characterization of ACs in this plant. The chapter also goes on to present the impact of such an adventure by highlighting related achievements made in other plants. Finally, the chapter closes by looking at future prospects by proposing possible engagements that could be pursued in this kind of an intriguing area of study in Plant Sciences.

2. *A. thaliana* as a model organism for plant research

A. thaliana is a weed-like plant that belongs to the Cruciferae family, commonly known as the mustard family. It is a dicot. The plant has become an essential model for plant scientists since the inception of the mid-1980s due to the availability of its whole genome sequence, molecular genetic markers and a large collection of its sequence-indexed DNA-insertion mutants [17]. Its different ecotypes have been collected from natural populations and are available for experimental analysis. The Columbia (Col-0) and Landsberg (Ler) ecotypes are the scientifically accepted standards for genetic and molecular studies [18]. The plant has a relatively small genome, which has led to a large base of mutants, genetic mapping and gene sequencing [19]. The plant is small in size, it self-pollinates and has a fast-growing capacity with a generation time of only 6 weeks (from germination to the first seeds under continuous light). A single plant can produce thousands of seeds within 2 months that can easily be screened in a single standard petri dish. Large numbers of the plants can also be grown in a very small space (as much as 100 plants in a pot or a small chamber) and such plants also being able to be easily or conveniently mutagenized by chemical mutagens or radiation [20]. As a result of all these attributes that are positive and amenable, *A. thaliana* has become an ideal model organism for plant research, particularly dicots, in laboratories.

3. The long journey to discover ACs and cAMP signaling in plants

3.1 Doubts and controversy on the possible existence of ACs or cAMP signaling in plants

Since the identification of cAMP in 1957 by Earl Wilbur Sutherland and TW Rall [21], wherein they studied the effect of adrenaline on the decomposition of glycogen into glucose in hepatocytes, a number of studies on ACs in animals and microorganisms have been performed [22–24]. Such studies included classification, sequencing, crystal structure analysis and functional characterization, ultimately leading to an unmasking of the catalytic mechanisms of this special group of enzymes [22–24]. However, despite all this impressive progress being made in other living organisms, the existence and/or functioning of ACs or cAMP in plants, still remained a matter of strong debate and serious controversy [25]. Firstly, because higher plants differ from other eukaryotic organisms in that they lack neurotransmitters and possess cell walls, therefore, the presence of plant cell walls would seem to compromise the believed and accepted mammalian AC/cAMP model [26]. Secondly, early studies of higher plant ACs were hampered by the fact that the proteins under study were mostly membrane-bound rather than soluble ones and therefore, their recovery and detection were somewhat very sneaky and difficult [27]. Thirdly, the attendant presence in crude biological extracts of active cyclic nucleotide phosphodiesterases, which hydrolyze cAMP also was a huge drawback [27]. Lastly, ACs represented a relatively very low proportion of plant cellular proteins for firm conclusions to be reached [27].

3.2 Hopes and continued debates

By the mid-1970s, both ACs and cAMP had been firmly established as important signaling molecules in animals and lower eukaryotes [1, 2, 28, 29]. Given this realization, it is not surprising that plant scientists continued to believe that this very same eukaryotic AC/cAMP signaling system was universal and therefore, operating in plants too [30]. Therefore, more work on plant AC/cAMP signaling continued. Nonetheless, further controversy and debates still ensued on these efforts, which perhaps can be best summarized by a concluding remark in one of the reviews of 1977, which stated as follows: "Our present knowledge, or rather ignorance, of cAMP in higher plants does not permit us to indulge in speculation on its function and thus to increase the disparity between available facts and conclusions, which are based solely on the conviction that plants, with respect to cAMP, should behave like animals or certain bacteria" [31]. As a result of this, plant scientists were therefore, strongly cautioned not to jump into conclusions even though they strongly believed that an AC/cAMP system existed in plants. The main reasons for this restraint were mostly technical. For instance, one criticism was mainly based on the fact that the reported results were merely either presumptive deductions from observed physiological effects of endogenously supplied cAMP or cAMP analogs, or that the conclusions were based solely onto insufficiently rigorous chromatographic identifications [31].

For instance, the demonstration of higher plant AC/cAMP activities had typically and traditionally been conducted through the use of either histochemical or biochemical procedures, whereby the histochemical procedures were predominantly based on the standard Wachstein-Meisel lead phosphate precipitation technique, which by that time, gave early indications of AC activity in *Zea mays*, *Pisum sativum*, *Alnus glutinosa*, *Vicia faba*, *Phaseolus vulgaris* and *Populus* species; while on the other hand, the biochemical approach proposed the presence of AC activity in *Medicago sativa*, *Spinacea oleracea*, *Ricinis communis*, *P. sativum*, *Verticillium albo-atrum* and *P. vulgaris* [26]. Moreover, some mass spectrometry-based analytical techniques of that period also indicated the presence of an AC/cAMP system in several plant species but still, the actual solid functional evidence was lacking [32].

Even though not very conclusive, the procedures discussed above principally revealed that cAMP levels in higher plants were relatively very low compared to other organisms, e.g., <20 pmol/g fresh weight compared to >250 pmol/g wet weight in animals [30, 33]. Hence, because of the low AC activity and barely detectable amounts of cAMP as well as the questionable experimental procedures used to identify cAMP and/or AC activity, the significance of cAMP in higher plants remained strongly doubted and/or even overlooked [34]. For this reason, it has also been so difficult to discern function(s) of ACs or cAMP in plants [35]. Consequently, in 1995, Assmann even went on as far as to declare that no plant AC gene had ever been cloned and, moreover, in 2006, Linder also asserted that no AC molecule had ever been conclusively identified in higher plants [30, 36]. In general, it has therefore, been commonly argued that until a plant AC is either cloned and/or its protein sufficiently purified to allow for microsequencing and a complete enzymological characterization, the relevance and validity of the AC activity and/or cAMP signaling reported in plants would always be open to serious or rather very heated debates [30].

3.3 Leads and indications

Despite the uncertainty and controversy of the mid-1970s, surrounding the AC/cAMP signaling system in plants, the notion of its existence still continued to stay alive. Firstly, because of the fact that both cell-permeant 8-BrcAMP and stimulation of *albeit* unknown ACs with forskolin could elicit concentration- and time-dependent biological responses such as increases in Ca^{2+} influx across the plasma membrane [37, 38]. Secondly, biochemical evidence, which included the finding that crude alfalfa (*M. sativa* L.) root extracts could show a calmodulin-dependent AC activity, was reported [39]. Thirdly, data from the whole-cell patch-clamp recordings of *V. faba* mesophyll protoplasts, convincingly revealed that outward K^+ current increased in a dose-dependent fashion following the intracellular application of cAMP and not AMP, cGMP or GMP [40]. Lastly, cAMP-dependent up-regulations of a calcium-permeable conductance, activated by hyperpolarization, were reported in guard and mesophyll cells of *A. thaliana* and *V. faba* [41].

3.4 Setting up the platform

The advent and era of high-throughput technologies (twenty-first century) eventually made it possible to assign functions to genes based on their homology, using the Basic Local Alignment Search Tool (BLAST). Apparently and according to Wong and Gehring in 2013 [42], higher plant ACs could not be identified using BLAST homology searches based on annotated ACs from prokaryotes, lower eukaryotes or animals because plant ACs are often part of complex multifunctional proteins with different domains and functions that are not conserved. Thus, BLAST searches with the known and experimentally confirmed ACs from other organisms could not return any plausible candidate(s). In addition, it was also noted that the pro-site signatures for class I and II ACs ((EYFG[SA]X(2)LWXLYK) and (YRNXW[NS]E[LIVM]RTLHFXG) respectively) are not present in the Arabidopsis proteome even if 2 mismatches were allowed [33]. Fortunately, between 2003 and 2011, six functional guanylyl cyclases (GCs) (structural analogs of ACs that convert guanosine 5′-triphosphate (GTP) to another second messenger, 3′,5′-cyclic guanosine monophosphate (cGMP)) from Arabidopsis were positively identified [43–47], using a 14 amino acid long search motif deduced from an alignment of conserved and functionally assigned amino acids

```
A:  --[RKS][YFW][CTGH][VIL][FV]X[DNA]X[VIL]X(4)[KR]X{1,3}[DE]--
      1          3                               14

B:  --[RK][YFW][DE][VIL][FV]X(8)[KR]X{1,3}[DE]--
      1         3                 14
```

Figure 1.
Catalytic motifs of nucleotide cyclases (NCs). (A) Core motif of experimentally tested GCs in plants, wherein the red residue at position 1 does hydrogen bonding with the guanine base, the blue residue at position 3 confers substrate specificity, and the red residue at position 14 stabilizes the transition GTP/cGMP. The Mg^{2+}/Mn^{2+}-binding site is 1–3 residues downstream of position 14 (green). (B) The derived motif specific for ACs, where at position 3, the residue [CTGH] has been substituted with [DE] to allow for ATP binding [33].

Gene ID	Protein code	Name of protein
At1g73980	AtTTM 1	Triphosphate metalloenzyme 1 protein
At1g26190	AtTTM 2	Triphosphate metalloenzyme 2 protein
At2g11890	AtTTM 3	Triphosphate metalloenzyme 3 protein
At3g14460	AtLRR	Leucine-rich repeat protein
At1g25240	AtENTH	Epsin N-terminal homology protein
At1g62590	AtPPR	Pentatricopeptide repeat protein
At1g68110	AtClAP	Clathrin assembly protein
At2g34780	AtMEE	Maternal effect embryo arrest protein
At3g02930	AtMTA	Microtubule assembly protein
At3g21465	AtAC	Adenylyl cyclase protein
At3g04220	AtTIR-NBS-LRR	Toll interleukin-like receptor-nucleotide-binding site-leucine-rich repeat protein
At3g18035	AtLHL	Linker histone-like protein
At3g28223	AtFb	F-box protein
At4g39756	AtKRFb	Kelch repeat-containing F-box protein

Table 1.
Arabidopsis proteins harboring the AC search motif: [RK][YFW][DE][VIL][FV]X(8)[KR]X(1,3)[DE] [33].

in the catalytic center of annotated GCs from other organisms (prokaryotes, lower eukaryotes or animals) (**Figure 1A**) [48]. Subsequent to the success of this approach, another related 14 amino acid-long search motif was then designed to target ACs (**Figure 1B**) [33]. Based on this new motif, a BLAST search of the Arabidopsis genome then managed to retrieve a total of 14 hits (**Table 1**) as potential AC candidates [33]. This thus convincingly provided a platform for the plausible identification or discovery of ACs in *A. thaliana* and ultimately the most targeted and sought AC/cAMP-dependent signaling system in plants.

3.5 Breakthrough and achievements

Using a wide array of web-based approaches and molecular techniques, we studied all the 14 candidate proteins in **Table 1** to determine if they were ACs. The methods used are detailed below.

3.5.1 Confirmation and validation of the presence of the AC motif in targeted protein candidates

Before any step was taken, it was necessary and important to first check and confirm presence of the AC motif in each of the targeted AC protein candidates. To do this, complete copy DNA (cDNA) and amino acid sequences of the targeted candidate were retrieved from The Arabidopsis Information Resource (TAIR) (https://www.arabidopsis.org/) followed by analysis of the amino acid sequence for presence of the motif [33], using the PROSITE database located within the Expert Protein Analysis System (ExPASy) proteomics server (https://www.expasy.org/). In addition, both the presence and location of the motif in each targeted candidate were also confirmed and validated by ACPred, available at http://gcpred.com/acpred/ [49]. (See example in **Figure A1**, Appendix A).

3.5.2 Computational analysis of the targeted protein candidates

Next, computational analysis of each of the targeted AC protein candidates was also undertaken to assess and determine the ability of its AC center to bind ATP and catalyze its subsequent conversion into cAMP. To do that, a 3-dimensional (3D) model of the targeted protein candidate was constructed by artificial intelligence using its AlphaFOLD beta version with low predicted error and very high confidence (pLDDT >90) [54]. This software uses a neural network-based model of artificial intelligence to predict protein structures from their amino acid sequences at an atomic level of accuracy. It first aligns the amino acid sequence input with sequences of known structures for pair-wise representation. The representation is then used to produce atomic coordinates for each residue, thus predicting the necessary rotation and then assembling a structured chain of amino acid residues. Its developers freely provide the source code for access to trained modelers and a script for predicting structures of novel input sequences [54]. In our case, the full-length amino acid sequence of each of the targeted protein candidates was submitted to the AlphaFOLD database followed by downloading of the model with the highest quality (based on C-scores). The downloaded model was then visualized and analyzed using UCSF ChimeraX next-generation molecular visualization program (v.1.10.1.) [55]. SeeSAR 3D (v.12.0.1) desktop modeling platform was then next used to perform docking of ATP (PubChem ID: 5957) to the AC center of the selected model via FlexX docking functionality [56, 57]. A structural alignment was then conducted by fragment assembly simulations based on iterative templates using the iterative threading assembly refinement (I-TASSER) server to match the selected model to an experimentally confirmed structure in the PDB library [56]. The model with the highest C-score was then analyzed using PyMOL (v.1.7.4.) (Schrödinger LLC, New York, USA) and ultimately adopted in the study. (See example in **Figure A1**, Appendix A).

3.5.3 Cloning and expression of the recombinant AC proteins

After determining that the AC center in the structural model of each of the targeted AC protein candidates was solvent-exposed, thus allowing for unimpeded substrate interactions and ultimately catalysis [57], we then went on to generate the respective protein candidate as a recombinant product so as to test it biochemically. To clone a fragment of each of the targeted AC protein candidates harboring the AC motif, total RNA was extracted from six-week-old *A. thaliana* ecotype Col-0 seedlings using the RNeasy plant mini kit, in combination with DNase 1 treatment, as instructed by the manufacturer (Qiagen, Crawley, UK). cDNA synthesis from total RNA and subsequent amplification of the targeted gene fragment from the cDNA were simultaneously performed in the presence of two sequence-specific primers for each protein candidate (**Table A1**, Appendix B), using a Verso 1-Step RT-PCR kit, in accordance with the manufacturer's instructions (Thermo Scientific, Rockford, USA). The resultant PCR product was then cloned into a pTrcHis2-TOPO expression vector via the TA cloning system (Invitrogen Corp., Carlsbad, USA) to make a pTrcHis2-TOPO:AC fusion expression construct with a C-terminus His purification tag.

To express the recombinant AC protein, competent *E. coli* BL21 Star pLysS cells (Invitrogen, Carlsbad, USA) were transformed (through heat shock at 42°C for 2 min) with the pCRT7/NT-TOPO:AC fusion construct and grown in double strength yeast-tryptone (2YT) media (16 g/L tryptone, 10 g/L yeast extract, 5 g/L NaCl and 4 g/L glucose; pH 7.0) containing 100 μg/ml ampicillin and 34 μg/ml chloramphenicol, on an orbital shaker (250 rpm) at 37°C. Protein expression was induced through the addition of isopropyl-β-D-thiogalactopyranoside (IPTG, Sigma-Aldrich Corp., Missouri, USA) at a final concentration of 1 mM and when the optical density (OD_{600}) of the cell culture had reached 0.5 (approximately 3 h). The culture was then left to grow for a further 3 h at 37°C. (See example in **Figure A1**, Appendix A).

3.5.4 Purification of the recombinant AC proteins

The resultant expressed recombinant AC protein (for each targeted AC candidate molecule) was purified by preparing a cleared cell lysate of the induced *E. coli* cells under non-native denaturing conditions (since most of those protein recombinants were primarily expressed in form of inclusion bodies (IBS)), whereby the harvested cells were resuspended in lysis buffer (8 M urea, 100 mM NaH_2PO_4, 10 mM Tris-Cl; pH 8.0, 500 mM NaCl, 20 mM β-mercaptoethanol, 7.5% (v/v) glycerol) at a quantitative ratio of 1 g pellet weight to every 10 ml buffer volume, mixed thoroughly using a mechanical stirrer at 24°C for 1 h and then centrifuged at 2500×*g* for 15 min. Supernatant was collected as the cleared lysate and transferred to 2 ml of 50% (w/v) nickel-nitriloacetic acid (Ni-NTA) slurry (Sigma-Aldrich Corp., Missouri, USA) that had been pre-equilibrated with 10 ml of lysis buffer and the lysate/slurry mixture then gently swirled on a rotary mixer for 1 h at 24°C. This step allowed for binding of the AC recombinant protein onto the Ni-NTA resin. The lysate-resin mixture was loaded into an empty XK16 column (Bio-Rad Laboratories Inc., California, USA) and allowed to settle and flow through discarded. The protein-bound resin was then washed three times with 30 ml of wash buffer (8 M urea, 100 mM NaH_2PO_4, 10 mM Tris-Cl; pH 8.0, 500 mM NaCl, 20 mM β-mercaptoethanol, 7.5% (v/v) glycerol, and 40 mM imidazole) to remove unbound proteins.

To refold the purified denatured recombinant AC protein, the washed protein-bound resin was equilibrated with 2 ml of gradient buffer (8 M urea, 200 mM NaCl, 50 mM Tris-Cl; pH 8.0, and 20 mM β-mercaptoethanol) before the column was connected to a Bio-Logic F40 Duo-Flow chromatography system (Bio-Rad Laboratories Inc., California, USA), programmed to run a linear refolding gradient. The refolding gradient was performed by linearly diluting the 8 M gradient buffer to 0 M urea concentration with a refolding buffer (200 mM NaCl, 50 mM Tris-Cl; pH 8.0, 500 mM glucose, 0.05% (w/v) polyethylene glycol, 4 mM reduced glutathione, 0.04 mM oxidized glutathione, 100 mM non-detergent sulfobetaine, and 0.5 mM phenylmethanesulfonylfluoride (PMSF)) over 10 h at a flow rate of 0.5 ml/min.

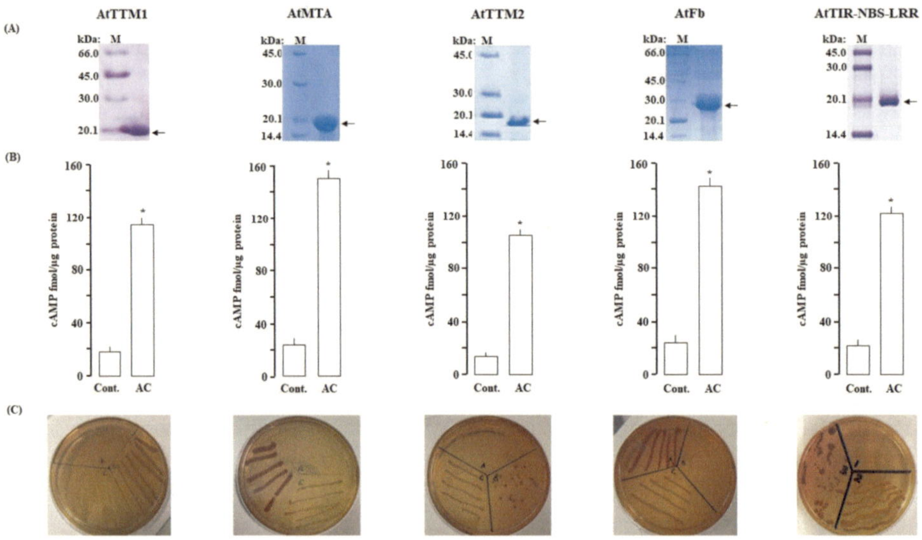

Figure 2.
Determination of the AC activity of A. thaliana *proteins. (A) Sodium dodecyl sulphate–polyacrylamide gel electrophoresis (SDS-PAGE) profiles of the Ni-NTA purified versions of AtTTM1, AtMTA, AtTTM2, AtFb and AtTIR-NBS-LRR recombinants, where (M) is the molecular weight marker and the arrow marking the expressed proteins. (B) cAMP levels generated by 5 µg recombinant in the presence of 1 mM ATP and 5 mM Mn^{2+}, and as determined by ELISA. Control reactions contained all other components except proteins. Data are mean values (n = 3) and error bars show the standard error (SE) of the mean. Asterisks denote values significantly different from those of control ($p < 0.05$) determined by the analysis of variance (ANOVA) and* post hoc *Student–Newman–Keuls (SNK) multiple range tests. (C) Ability of the recombinant AtTTM1, AtMTA, AtTTM2, AtFb and AtTIR-NBS-LRR proteins to complement* cyaA *mutation in* E. coli *SP850 to ferment lactose. All protein-expressing SP850* E. coli *cells showed a strong reddish color as if they were wild-type cells [59, 60] while* cyaA *mutant cells yielded yellowish colonies.*

After refolding, the resultant renatured recombinant AC protein was eluted in 2 ml of elution buffer (200 mM NaCl, 50 mM Tris-Cl; pH 8.0, 250 mM imidazole, 20% (v/v) glycerol, and 0.5 mM PMSF). The eluted native protein recombinant was then desalted and concentrated using a Spin-XUF filtration/concentration device with a molecular weight cut-off (MWCO) point of 3000 Da, in accordance with the manufacturer's instructions (Corning Corp., New York, USA). Protein concentration was then determined by both the Bradford method [58] and an ND2000 nanodrop spectrophotometer (Thermo Scientific Inc., Massachusetts, USA) before the recombinant protein was stored at −20°C for further downstream uses. (Refer to **Figure 2** below).

3.5.5 In vitro *assaying of the recombinant AC proteins*

The probable *in vitro* AC activity of the purified recombinant version of each of the targeted AC candidate molecules was tested by incubating 5 µg of the protein in 200 µl of 50 mM Tris-Cl (pH 8.0), containing 2 mM isobutylmethylxanthine (IBMX, Sigma-Aldrich Corp., Missouri, USA) (to inhibit phosphodiesterases), 5 mM Mn^{2+} (as co-factor) and 1 mM ATP (as substrate), followed by measurement of the generated cAMP by immunoassay. Background cAMP levels in control reactions were measured in tubes containing all the other components but no protein. All incubations were performed at room temperature (24°C) for 20 min and terminated by the addition of 10 mM EDTA followed by boiling for 3 min and cooling on ice for 2 min before centrifugation at 2500×*g* for 3 min. The resulting supernatants were in essence assayed for cAMP content using the cAMP-linked enzyme-linked immunosorbent assay (ELISA) kit, following its acetylation protocol and as described by the supplier's manual (Sigma-Aldrich Corp., Missouri, USA; code: CA201). The anti-cAMP antibody in this assaying system is highly specific for cAMP with an approximately 10^6 times lower affinity for cGMP. In all cases, each experiment was performed in triplicate ($n = 3$) using three different protein extracts that had been independently prepared. Data generated were then subjected to one-way analysis of variance (ANOVA) (Super-Anova, Statsgraphics Version 7, 1993; Statsgraphics Corp., The Plains, VI, USA) and wherever ANOVA revealed significant differences between treatments, means were separated by *post hoc* Student–Newman–Keuls (SNK) multiple range test ($p < 0.05$). (Refer to **Figure 2**).

3.5.6 Mass spectrometric analysis of the AC activities

To validate findings obtained by enzyme immunoassay, another additional method, tandem liquid chromatography-mass spectrometry (LC–MS/MS), was used. This method is capable of specifically and sensitively detecting cAMP levels at femtomolar concentrations. To do this, the acetylated cAMP samples from the immunoassay step were introduced into a Waters API Q-TOF Ultima mass spectrometer (Waters Microsep, Johannesburg, RSA) with a Waters Acquity UPLC at a flow rate of 180 ml/min. Separation was achieved in a Phenomenex Synergi (Torrance, CA) 4 μm Fusion-RP (250 × 2.0 mm) column when a gradient of solvent "A" (0.1% formic acid) and solvent "B" (100% acetonitrile) was applied over 18 min. During the first 7 min, the solvent composition was kept at 100% "A" followed by a linear gradient of up to 80% "B" for 3 min, and then a re-equilibration to the initial conditions. An electrospray ionization in the negative (W-) mode was used at a cone voltage of 35 V, to detect molecules and generate chromatograms. (See example in **Figure A1**, Appendix A).

3.5.7 Complementation testing of the recombinant AC proteins

To further confirm and validate the AC activities of the targeted and tested AC protein candidates, complementation testing was used. In this process, the *E. coli* mutant strain, SP850, (lam-, el4-, relA1, spoT1, *cya*A1400 (:kan), thi-1) [59, 60], deficient in the AC gene (*cya*A), was obtained from the *E. coli* Genetic Stock Centre (Yale University, New Haven, USA; accession No. 7200). The strain was prepared to be chemically competent followed by its transformation with the pTrcHis2-TOPO:AC fusion construct (through heat shock at 42°C for 2 min). The transformed bacteria together with the non-transformed cells were grown at 37°C in Luria-Bertani (LB) media containing kanamycin (15 μg/ml) up until their cell culture had reached an optical density (OD_{600}) of 0.5. Both groups of cells were streaked on MacConkey agar supplemented with 15 μg/ml kanamycin and 0.5 mM IPTG (Sigma-Aldrich Corp., Missouri, USA) (for transgene induction) before the streaked plates were incubated for 40 h at 37°C, for visual evaluation of the cultured cells on the media. After incubation, an ability of the induced transformed mutant cells to now ferment lactose would then be considered as an indication of the expressed recombinant protein's ability to generate cAMP from ATP, as a functional AC. As a result, the induced transformed cells would turn deep red or purple (just like wild-type cells) while the mutant control cells would remain yellowish or colorless [59, 60]. (Refer to **Figure 2**).

3.5.8 Phylogenetic analysis of the targeted protein candidates

Since classification of plant ACs has not yet been systematically undertaken and also considering that the relationship of plant ACs with the other currently existing six classes of ACs in animals and microorganisms is still very unclear, we assessed to determine if these plant proteins have any probable definitive groupings or clusters, particularly those ones harboring the AC search motif. In this regard therefore, full-length amino acid sequences of all plant proteins known to harbor the AC search motif were retrieved from the TAIR (https://www.arabidopsis.org/), NCBI (https://www.ncbi.nlm.nih.gov/) and Uniprot (https://www.uniprot.org/) websites, followed by construction of a phylogenetic tree through multiple sequence alignment and tree

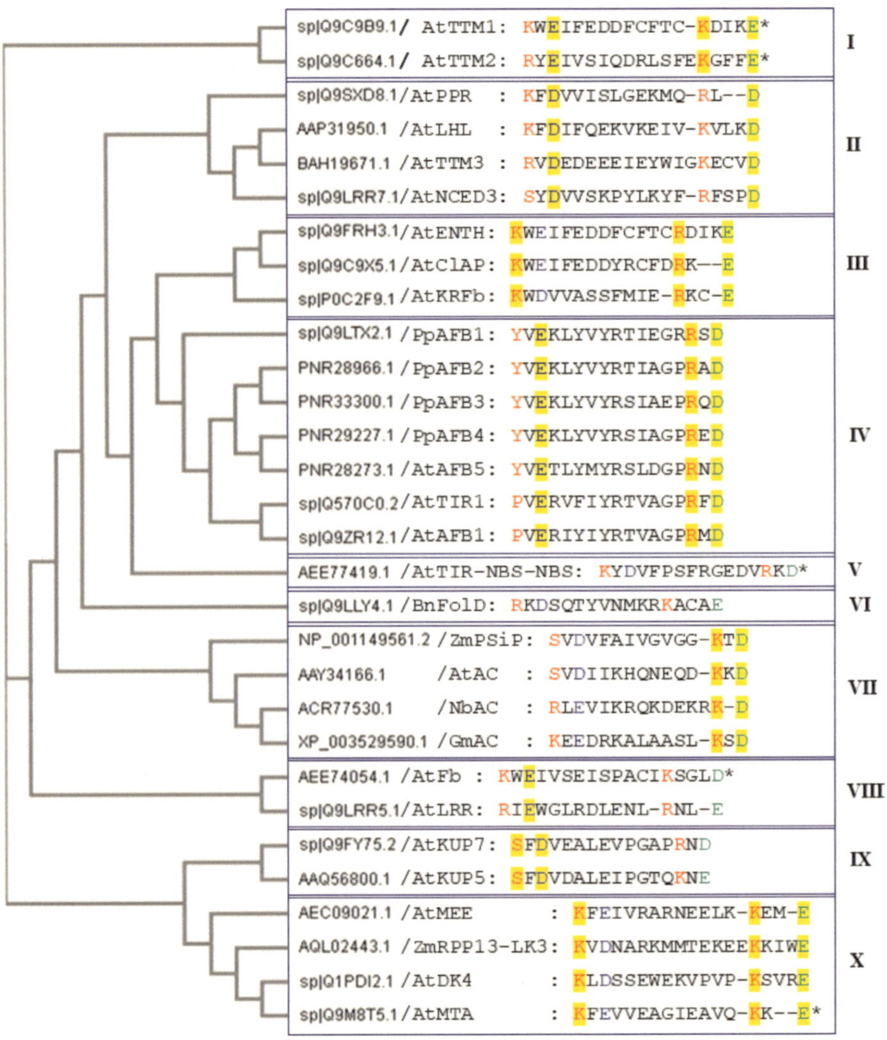

Figure 3.
Phylogenetic tree and core motif analysis of all plant proteins currently known to harbor the AC search motif. Specific key amino acid residues in the core AC motif appear to be conserved in a certain pattern per group (yellow highlight), something that possibly could be exploited and used to predict and/or assign groups for any newly identified AC candidate harboring the same AC core motif. Asterisks indicate AC candidates presented herein in this chapter for publication consideration.

generation using Clustal Omega (ClustalO) (https://www.ebi.ac.uk/Tools/msa/clustalo/). (Refer to **Figure 3**).

3.5.9 Bioinformatic analysis of the targeted protein candidates

Since in eukaryotes, it is widely accepted that proteins that are co-expressed often have related functions and linked coordinated regulatory systems [61–64], we then sought to explore and gain insights into the probable biological functions of each of our targeted protein candidates. This was achieved by subjecting each protein candidate to correlation expression analysis and stimulus-specific expression analysis so as to obtain its expression partners and then infer function(s). For correlation expression analysis, The Arabidopsis co-expression tool (ACT) (http://www.arabidopsis.leeds.ac.uk) [65] was used across all available microarray experiments, using the targeted protein candidate as the driver molecule and leaving the list limit of all other molecules blank to obtain a full correlational list. In this search, the top 50 co-expressed proteins (CEG50) were mostly considered, based on the Pearson correlation coefficient as a measure of similarity between them.

For stimulus-specific expression analysis, the expression profiles of the CEG50 together with the targeted protein candidate (AC-CEG50 were initially screened over all available ATH1:22 K arrays, Affymetrix public microarray data in the Genevestigator V3 (https://www.genevestigator.com) using the stimulus and mutation tools [66]. To obtain a greater resolution of the protein expression profiles, the normalized microarray data were subsequently downloaded and analyzed for experiments that were found to induce differential expression of the proteins. The data were downloaded from the following repository sites; GEO (NCBI) (http://www.ncbi.nlm.nih.gov/geo/) [67], the NASCArrays (http://affymetrix.arabidopsis.info/narrays/experimentbrowse.pl) [68], and the TAIR-ATGenExpress (http://www.ebi.ac.uk/microarray-as/ae/). The downloaded array data were then analyzed and fold-change (log2) values for each experiment calculated. (See example in **Figure A2**, Appendix A).

3.5.10 Outcomes

Interestingly and overwhelmingly, our research approach managed to confirm AC activity in all of the studied 14 AC candidates as is detailed in **Table 2** below. In the table, it can be seen that 6 of the candidates have already been published by our group while 3 are currently under review, again from our group. Therefore, in order to avoid duplications in this chapter, we only presented results for the last outstanding 5 candidates (**Figure 2**) for publication consideration.

4. Impact of our study

It is worth mentioning that our breakthrough in the identification of ACs in *A. thaliana* had an impetus effect of prompting us and other plant scientists across the globe to also search for similar proteins in Arabidopsis and other plants, using either our own motif search approach or other different methods such as omics analysis and homologous cloning. Through the motif search approach, 6 additional ACs were identified in *A. thaliana* (AtKUP7, AtKUP5, AtNCED3, AtTIR1, AtAFB1 and AtAFB5) by two groups [75–79], 4 in the basal plant *Physcomitrella patens*

Protein name	Gene ID	Annotated or confirmed function(s)	Publication status as AC
AtTTM 1	At1g73980	Leaf senescence [70]	Unpublished
AtTTM 2	At1g26190	Pathogen resistance [71]	Unpublished
AtTTM 3	At2g11890	Root development [72]	Under Review
AtLRR	At3g14460	Pathogen defense [73]	Published [51]
AtENTH	At1g25240	Endocytosis, pollen germination and pollen tube growth	Under Review
AtPPR	At1g62590	Chloroplast biogenesis and restoration of cytoplasmic male sterility	Published [50]
AtClAP	At1g68110	Endocytosis and plant defense	Published [69]
AtMEE	At2g34780	Embryogenesis and response to abiotic stress	Published [52]
AtMTA	At3g02930	Regulation of microtubule assembly and growth	Unpublished
AtAC	At3g21465	Response to biotic stress	Published [53]
AtTIR-NBS-LRR	At3g04220	Disease resistance	Unpublished
AtLHL	At3g18035	Regulation of developmental and reproductive processes, and response to abiotic stress	Published [74]
AtFb	At3g28223	Regulation of the cell cycle	Unpublished
AtKRFb	At4g39756	Regulation of the cell cycle	Under Review

Table 2.
Current research status of the 14 Arabidopsis proteins harboring the AC search motif: [RK][YFW][DE][VIL][FV]X(8)[KR]X(1,3)[DE] [33].

(PpAFB1, PpAFB2, PpAFB3 and PpAFB4) by one of the two mentioned groups [78, 79], 1 in *Brassica napus* (BnFolD) by another totally new group [80], and 1 (GmAC) by our group in soybean (*Glycine max*) [81]. AtKUP7 and AtKUP5 are permeases responsible for K^+ ion flux [75, 76] while AtNCED3 participates in the biosynthesis of the stress-related hormone, abscisic acid (ABA) [77]. AtTIR1, AtAFB1, AtAFB5, PpAFB1, PpAFB2, PpAFB3 and PpAFB4 are all auxin receptors responsible for the auxin-mediated regulation of root growth [78, 79]. BnFolD regulates the metabolism of folate [80] while GmAC is involved in early plant development and stress response [81].

Using the other methods, other scientists identified a total of 13 additional ACs of which 1 was identified through the omics analysis method and 12 through homologous cloning. ZmRPP13-LK3, which participates in the ABA-mediated resistance of *Z. mays* plants to heat stress, was identified through omics analysis [82] while NbAC from *Nicotiana benthamiana* [83]; HpAC1 from *Hippeastrum hybridum* [84]; MpAC from *Marchantia polymorpha* [85]; AtDK4 from *A. thaliana* [12]; BdTTM3 and BdGUCD1 from *Brachypodium distachyon* [86, 87]; MdTTM1 and MdTTM2 from *Malus domestica* [88]; and ZjAC1, ZjAC2 and ZjAC3 from *Ziziphus jujuba* [89] were identified through homologous cloning. NbAC plays a role in tabtoxinine-β-lactam-induced cell death during the development of wildfire disease [83] while HpAC1 is involved in responses to infection by the plant fungal pathogen, *Phoma narcissi*, and also to injuries through mechanical damage [84]. MpAC is involved in male organ and cell development [85] while AtDK4 is responsible for the nitric oxide (NO) dependent pollen tube guidance and fertilization [12]. BdTTM3 is responsible for responses to mechanical wounding [86] while BdGUCD1 is involved in jasmonic acid (JA) mediated responses to *Fusarium pseudograminearum* infection [87]. MdTTM1 and MdTTM2 currently do not have any known function(s) [88] while ZjAC1, ZjAC2 and ZjAC3 are involved in the significant acceleration of seed germination, root growth, and flowering respectively [89]. Notably, ZmPSiP, responsible for the polarized growth and re-orientation of pollen tubes in *Z. mays* and identified through the homologous cloning method [11], is so far the only plant AC to have been identified before our own ACs in Arabidopsis. Intriguingly, of the 13 additional plant ACs discovered through either the omics analysis or homologous cloning method, 4 (ZmPSiP, NbAC, AtDK4 and ZmRPP13-LK3) coincidentally harbor the same rationally designed AC search motif, commonly found in the AC protein molecules studied by our group.

All in all, a total of 30 plant proteins are currently known to harbor the rationally designed AC search motif among which 22 have been experimentally confirmed as functional ACs while 8 are still yet to be confirmed [11, 12, 33, 50–53, 69, 73–83]. Apparently, when this whole lot of 30 plant proteins harboring the rationally designed AC search motif was clustered together phylogenetically, using ClustalO, a total of 10 distinct groups emerged (**Figure 3**). Interestingly and when closely analyzed, the AC search motif in each of those 10 generated groups, tended to have a certain pattern, whereby its key catalytic residues are somewhat uniformly conserved per group, proposing that the motif could perhaps be utilized to predict and/or assign a group to any newly identified plant ACs harboring the very same motif. This is quiet tempting and persuading considering that currently, plant ACs have not yet been systematically classified as compared to animal and microorganismal ACs that have six established and well-known classes [90]. Thus, in future, the AC search motif could possibly be of vital essence to both the identification and classification of more/new plant ACs.

5. Brief discussion and self-reflection

It is evident that the advent of high-throughput technologies in the twenty-first century ultimately revolutionized science, allowing the uncovering of most of the previously obscured areas such as the identification of ACs in plants in this case. New technologies such as omics analysis, homologous cloning and motif search approach proved very useful and key in the identification of ACs in plants. For instance, of the total (31) currently confirmed ACs in plants, the omics analysis method identified 1 AC, the homologous cloning method identified 12 while the motif search approach identified 18. More so, this whole AC identification spectrum also covered a wide range of plant species (eleven to be specific), which among them are seven herbaceous plants (*A. thaliana*, *N. benthamiana*, *Z. mays*, *H. hybridum*, *B. distachyon*, *B. napus* and *G. max*), two woody plants (*M. domestica* and *Z. jujuba*) and two basal plants (*M. polymorpha* and *P. patens*). In addition to this, there are also six dicots (*A. thaliana*, *N. benthamiana*, *M. domestica G. max*, *B. napus* and *Z. jujuba*) and three monocots (*Z. mays*, *H. hybridum* and *B. distachyon*). Ideally, the wide-spread distribution of ACs in plants noted here, clearly shows the importance and significance of this group of proteins in plants as key enzymes for essential cellular processes such as growth, development and survival.

Symbolically, the fact that our own research method (motif search approach) has overall identified up to 18 plant ACs, i.e., 12 in *A. thaliana* and 6 in other plants, strongly demonstrates its strength among other methods currently known and/or used. In addition, this also lends lots of confidence into this method for searching and identifying ACs in plants, and perhaps even in going further to classify such newly identified ACs into specific and/or particular groups. Thus, all in all, the motif search method can be confidently widened further to cover as many species as possible across the whole spectrum of the plant kingdom.

6. Conclusion

The report given here has three very key/important messages. Firstly, it proves the strength and usefulness of the method (motif search approach) we used in our research group to search and identify ACs in *A. thaliana* and therefore, the method can too be applied, confidently so, to other plants in the plant kingdom to bring back valid results. Secondly, it proposes the possibility of using the AC search motif to determine and classify newly identified ACs into specific and recognizable groups. Lastly, it proved the relevance and usefulness of *A. thaliana* as a model organism for plant research, particularly when it comes to the designing of experiments and modeling of concepts that are relevant and applicable to the rest of the plant kingdom.

7. Future prospects and recommendations

Considering that ACs and their functions have also now been identified in a number of commercially important plants such as maize, tobacco, soybean, apple, cabbage and jujube, more work underlying the mechanisms of action of these important proteins should be pursued and interrogated. Such an approach will then help elucidate or reveal the exact functions or roles of these enzymes and/or their cAMP-linked signaling systems in plants.

Author contributions

OR conceived the idea; PC, AS-M, TD, EB, KS, GM, NM, MT and DK did the experiments. OR wrote the manuscript and PC edited it. All authors read and approved the manuscript.

Funding

The project was funded by the National Research Foundation (NRF) of South Africa (Grant Numbers: CSUR78843 & CSUR93635) and the Organization for Women in Science for the Developing World (OWSD).

Conflict of interest

The authors declare no conflict of interest.

Appendix A: Illustration of some of the results obtained during our process of searching and identifying ACs in *A. thaliana*

(A)

Amino acid sequence

```
1                                                               65
MRISISSVVSSTTSRIVHRDLQGKGNPRIAPSSIDLCGMCYWGRAFSSGSGDYREILRNGLHDMK
LDDAIGLFGGMVKSRPLPSIVEFNKLLSAIAKMKKFDVVISLGEKMQRLEIVHGLYTYNILINCF
CRRSQISLALALLGKMMKLGYEPSIVTLSSLLNGYCHGKRISDAVALVDQMVEMGYRPDTITFTT
LIHGLFLHNKASEAVALVDRMVQRGCQPNLVTYGVVVNGLCKRGDTDLALNLLNKMEAAKIEADV
VIFNTIIDSLCKYRHVDDALNLFKEMETKGIRPNVVTYSSLISCLCSYGRWSDASQLLSDMIEKK
INPNLVTFNALIDAFVKEGKFVEAEKLYDDMIKRSIDPDIFTYNSLVNGFCMHDRLDKAKQMFEF
MVSKDCFPDVVTYNTLIKGFCKSKRVEDGTELFREMSHRGLVGDTVTYTTLIQGLFHDGDCDNAQ
KVFKQMVSDGVPPDIMTYSILLDGLCNNGKLEKALEVFDYMQKSEIKLDIYIYTTMIEGMCKAGK
VDDGWDLFCSLSLKGVKPNVVTYNTMISGLCSKRLLQEAYALLKKMKEDGPLPNSGTYNTLIRAH
LRDGDKAASAELIREMRSCRFVGDASTIGLVANMLHDGRLDKSFLDMLS-
```

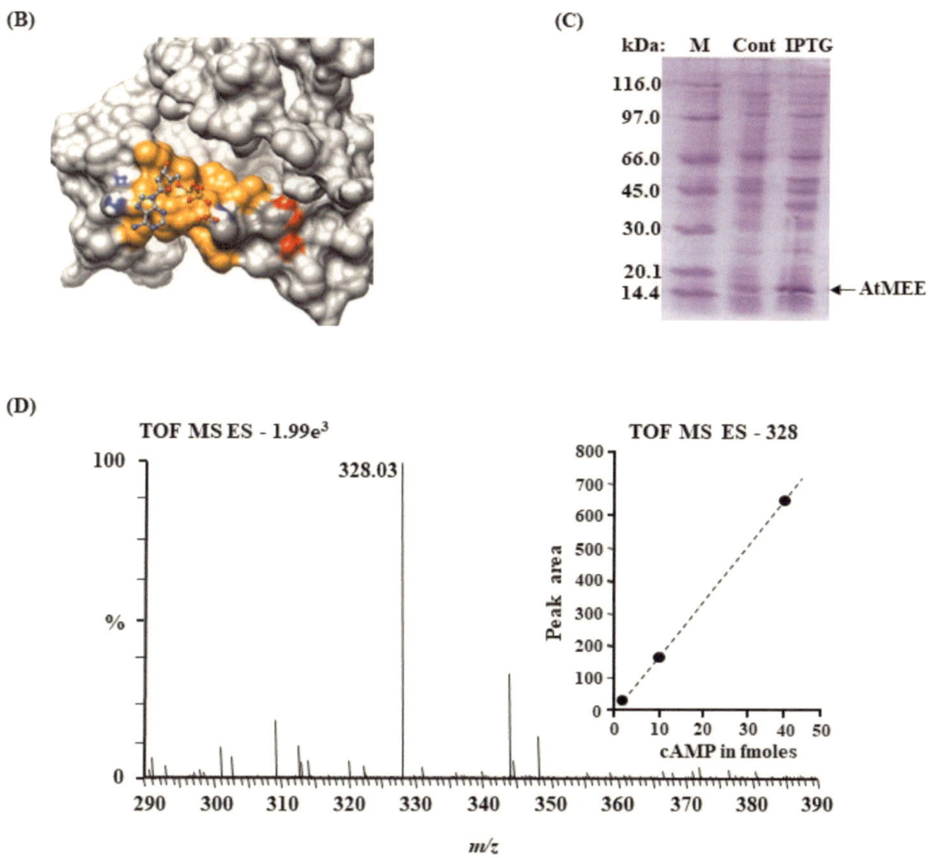

Figure A1.
(A) Amino acid sequence of AtPPR showing presence of the AC search motif in the cloned and recombinantly expressed protein fragment (inverted red triangles) [50]. (B) 3D structure of AtLRR showing the AC search motif (gold) in a solvent-exposed position that is amenable for an unimpeded ATP binding and ultimately catalysis [51]. (C) Expression profile of the recombinant AtMEE protein [52]. (D) Mass spectrometric analysis of the in vitro *AC activity of AtAC [53].*

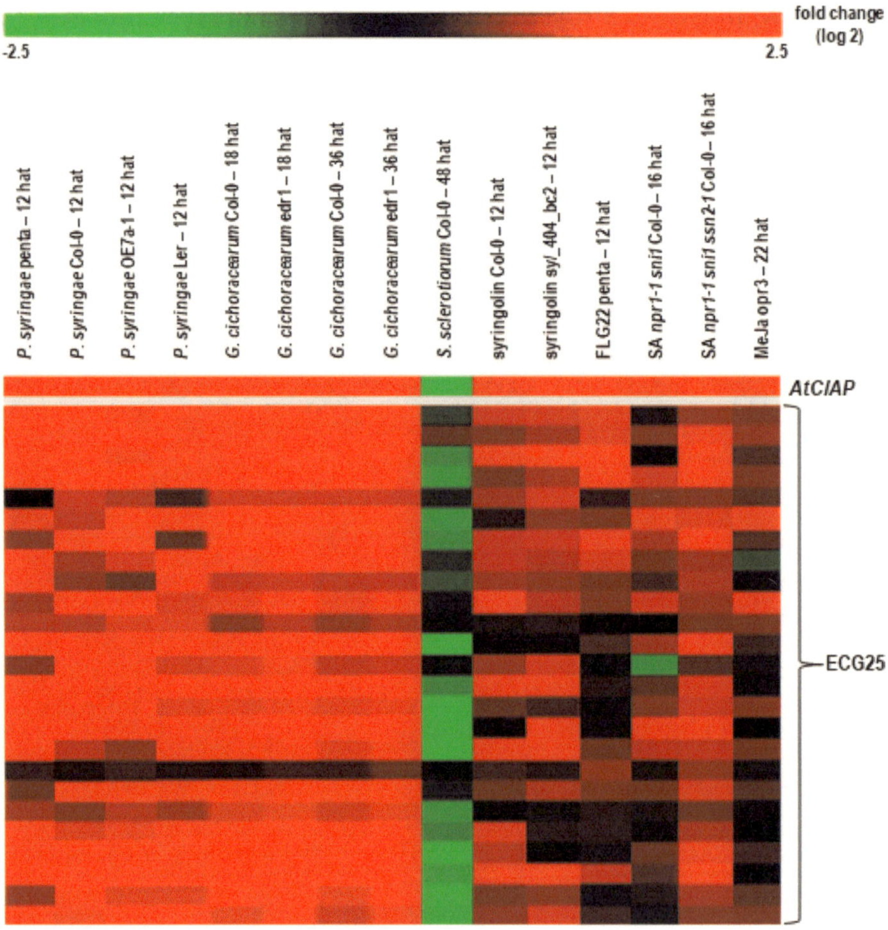

Figure A2.
Heat map of AtClAP showing its correlation and differential expression patterns as an AC [69].

Appendix B: Cloning of the Arabidopsis proteins harboring the AC search motif

AC molecule	Gene ID	Primer set
AtTTM 1*	At1g73980	Fwd: 5′-TCACCCAGAATAACTTTTGAAGTTAGTGTT-3′
		Rev.: 5′-AGATGCGGATGCGGAGAATGATAAGACAGA-3′
AtTTM 2*	At1g26190	Fwd: 5′-GGTCAAGACAGCAATGGAATTGAGTTTCAT-3′
		Rev.: 5′-GTCATAGTCCGTTAACCGTGGATCATCAAA-3′
AtTTM 3	At2g11890	Fwd: 5′-GAAGTCGAAGTCAAGCTCCGTCTCCTAACC-3′
		Rev.: 5′-AGGAAGTTTTCCTGACCGGAAAACAGCAAA-3′
AtLRR	At3g14460	Fwd: 5′-CCAGGGGTTGGAAAGACTACCTTGACAGAG-3′
		Rev.: 5′-TACTAATTCCTCCCTATCGAAAACATGACC-3′

AC molecule	Gene ID	Primer set
AtENTH	At1g25240	Fwd: 5′-GAATTTGGGGTCTCAAACGCGCACGACATT-3′
		Rev.: 5′-GAAGGTAATCAAATCTGGTAACGTGTGAGT-3′
AtPPR	At1g62590	Fwd: 5′-CGGAGCCAAATCGAGAAGATGAGGATCTCG-3′
		Rev.: 5′-CATTTCCACCATTTGATCAACCAAAGCTAC-3′
AtClAP	At1g68110	Fwd: 5′-GAATTCTGCAAAGGTTTCGGTGTCTCGAAC-3′
		Rev.: 5′-GAATGTAATCAAATCTGGCATTGTATAAGT-3′
AtMEE	At2g34780	Fwd: 5′-GCCCGGAAGGATCCAATGTCGGAGTTGGAG-3′
		Rev.: 5′-GCGCCGGAATTCCGAGACTAATTGCGCTTC-3′
AtMTA*	At3g02930	Fwd: 5′-ACTGATAAGAGGTCCCCCAAAGCTCCAACC-3′
		Rev.: 5′-CAAAGCCTTCAACCGTATTAACTCTGACGA-3′
AtAC	At3g21465	Fwd: 5′-GCTGCCAAAAGAGGAGACACAGAGTCGTTA-3′
		Rev.: 5′-GCTAAGAAGAGCTTCATTCTTGTTTAACTC-3′
AtTIR-NBS-LRR*	At3g04220	Fwd: 5′-GATTCTTCTTTTTTACTCGAAACTGTTGCT-3′
		Rev.: 5′-TGGATCCACTTTGTAGAAAATGACTATCAC-3′
AtLHL	At3g18035	Fwd: 5′-GGAAGGCCTAGGAGAGTTGTTGACCCTAGC-3′
		Rev.: 5′-GAACAGAGCTTCTTGCATTGCCTCTGCTTC-3′
AtFb*	At3g28223	Fwd: 5′-AACAACTATCGTGATCACCTTGTTGTATCC-3′
		Rev.: 5′-AAGCTTGTCGGTTGGAGGCGAGGAAGGGAT-3′
AtKRFb	At4g39756	Fwd: 5′-GGAATTCCCATGGCTACTGGTACGGAATCT-3′
		Rev.: 5′-GACTCGAGCGTAGCAGCCAATGCGGGAGAG-3′

Each primer set was designed in such a way that the two annealing regions kept the AC search motif in between so that each resultant gene fragment would be containing the motif. Asterisks represent candidates presented herein in this chapter for publication.

Table B1.
Primers used for amplification of the gene fragments of the Arabidopsis protein candidates harboring the AC search motif [RK][YFW][DE][VIL][FV]X(8)[KR]X(1,3)[DE] [33].

Author details

Patience Chatukuta, Angela Sibanda-Makuvise, Tshegofatso Dikobe, Enetia Bobo, Katlego Sehlabane, Grace Mabadahanye, Neo Mametja, Mutsa Takundwa, David Kawadza and Oziniel Ruzvidzo*
Department of Botany, School of Biological Sciences, North-West University, Mmabatho, South Africa

*Address all correspondence to: oziniel.ruzvidzo@nwu.ac.za

References

[1] Robinson GA, Butcher RW, Sutherland EW. Cyclic AMP. Annual Review of Biochemistry. 1968;**37**(1): 149-174. DOI: 10.1146/annurev.bi.37.070168.001053

[2] Goodman DBP, Rasmussen H, DiBella F, Guthrow CE. Cyclic adenosine 3′:5′-monophosphate-stimulated phosphorylation of isolated neurotubule subunits. Proceedings of the National Academy of Sciences of the United States of America. 1970;**67**(2):652-659. DOI: 10.1073/pnas.67.2.652

[3] Gerisch G, Fromm H, Huesgen A, Wick U. Control of cell-contact sites by cyclic AMP pulses in differentiating Dictyostelium cells. Nature. 1975; **255**(5509):547-549. DOI: 10.1038/255547a0

[4] Ehsan H, Reichheld JP, Roef L. Effect of indomethacin on cell cycle dependent cyclic AMP fluxes in tobacco BY-2 cells. FEBS Letters. 1998;**422**(2):165-169. DOI: 10.1016/S0014-5793(97)01610-4

[5] Bolwell PG. A role for phosphorylation in the down-regulation of phenylalanine ammonia-lyase in suspension-cultured cells of French bean. Phytochemistry. 1992;**31**(12): 4081-4086. DOI: 10.1016/0031-9422(92)80418-E

[6] Pietrowska-Borek M, Nuc K. Both cyclic-AMP and cyclic-GMP can act as regulators of the phenylpropanoid pathway in *Arabidopsis thaliana* seedlings. Plant Physiology and Biochemistry. 2013;**70**:142-149. DOI: 10.1016/j.plaphy.2013.05.029

[7] Jin XC, Wu WH. Involvement of cyclic AMP in ABA- and Ca^{+2}-mediated signal transduction of stomatal regulation in *Vicia faba*. Plant & Cell Physiology. 1999;**40**(11):1127-1133. DOI: 10.1093/oxfordjournals.pcp.a029497

[8] Maathuis FJ, Sanders D. Sodium uptake in Arabidopsis roots is regulated by cyclic nucleotides. Plant Physiology. 2001;**127**(4):1617-1625

[9] Tezuka T, Hiratsuka S, Takahashi SY. Promotion of the growth of self-incompatible pollen tubes in lily by cAMP. Plant & Cell Physiology. 1993; **34**(6):955-958. DOI: 10.1093/oxfordjournals.pcp.a078508

[10] Malho R. Signalling pathways in pollen tube growth and reorientation. Annals of Botany. 2000;**85**:59-68. DOI: 10.1006/anbo.1999.0991

[11] Moutinho A, Hussey PJ, Trewavas AJ, Malhó R. cAMP acts as a second messenger in pollen tube growth and reorientation. Proceedings of the National Academy of Sciences of the United States of America. 2001;**98**(18): 10481-10486. DOI: 10.1073/pnas.171104598

[12] Vaz Dias F, Serrazina S, Vitorino M, Marchese D, Heilmann I, Godinho M, et al. A role for diacylglycerol kinase 4 in signalling crosstalk during Arabidopsis pollen tube growth. The New Phytologist. 2019;**222**(3):1434-1446. DOI: 10.1111/nph.15674

[13] Yang WQ, Braun C, Plattner H, Purvee J, Van Houten JL. Cyclic nucleotides in glutamate chemosensory signal transduction of paramecium. Journal of Cell Science. 1997;**110**(20): 2567-2572. DOI: 10.1242/jcs.110.20.2567

[14] Ma W, Smigel A, Verma R, Berkowitz GA. Cyclic nucleotide gated

channels and related signaling components in plant innate immunity. Plant Signaling & Behavior. 2009;**4**(4): 277-282. DOI: 10.4161/psb.4.4.8103

[15] Sabetta W, Vandelle E, Locato V. Genetic buffering of cyclic AMP in *Arabidopsis thaliana* compromises the plant immune response triggered by an avirulent strain of *pseudomonas syringae* pv. *Tomato*. The Plant Journal. 2019;**98**(4):590-606. DOI: 10.1111/tpj.14275

[16] Blanco E, Fortunato S, Viggiano L, de Pinto MC. Cyclic AMP: A polyhedral signalling molecule in plants. International Journal of Molecular Sciences. 2020;**21**(14):4862. DOI: 10.3390/ijms21144862

[17] Meinke DW, Cherry JM, Dean C, Rounsley SD, Koornneef M. *Arabidopsis thaliana*: A model plant for genome analysis. Science. 1998;**282**(5389): 662-682. DOI: 10.1126/science.282.5389.662

[18] David MM, Coelho D, Barrote I, Correia MJ. Leaf age effects on photosynthetic activity and sugar accumulation in droughted and rewatered *Lupinus albus* plants. Functional Plant Biology. 1998;**25**(3): 299. DOI: 10.1071/PP97142

[19] Koncz C, Chua NH, Schell J. Methods in Arabidopsis Research. Singapore: World Scientific; 1992. DOI: 10.1142/1602

[20] Gepstein S, Horwitz BA. The impact of Arabidopsis research on plant biotechnology. Biotechnology Advances. 1995;**13**(3):403-414. DOI: 10.1016/0734-9750(95)02003-L

[21] Sutherland EW, Rall TW. The properties of an adenine ribonucleotide produced with cellular particles, ATP, Mg^{+2}, and epinephrine or glucagon. Journal of the American Chemical Society. 1957;**79**(13):3608-3608. DOI: 10.1021/ja01570a087

[22] Tesmer JJG, Sunahara RK, Gilman AG, Sprang SR. Crystal structure of the catalytic domains of adenylyl cyclase in a complex with $G_{s\alpha}$·GTPγS. Science. 1997;**278**(5345):1907-1916. DOI: 10.1126/science.278.5345.1907

[23] Guo YL, Kurz U, Schultz A. Interaction of Rv1625c, a mycobacterial class IIIa adenylyl cyclase, with a mammalian congener: Mycobacterial class IIIa adenylyl cyclase Rv1625c. Molecular Microbiology. 2005;**57**(3): 667-677. DOI: 10.1111/j.1365-2958.2005.04675.x

[24] Steegborn C, Litvin TN, Levin LR, Buck J, Wu H. Bicarbonate activation of adenylyl cyclase via promotion of catalytic active site closure and metal recruitment. Nature Structural & Molecular Biology. 2005;**12**(1):32-37. DOI: 10.1038/nsmb880

[25] Hintermann R, Parish RW. Determination of adenylate cyclase activity in a variety of organisms: Evidence against the occurrence of the enzyme in higher plants. Planta. 1979; **146**(4):459-461. DOI: 10.1007/BF00380860

[26] Newton RP, Roef L, Witters E, Van Onckelen H. Cyclic nucleotides in higher plants: The enduring paradox. The New Phytologist. 1999;**143**(3):427-455. DOI: 10.1046/j.1469-8137.1999.00478.x

[27] Bârzu O, Danchin A. Adenylyl Cyclases: A heterogeneous class of ATP-utilizing enzymes. Progress in Nucleic Acid Research and Molecular Biology. 1994;**49**:241-283. DOI: 10.1016/S0079-6603(08)60052-5

[28] Gerisch G, Hülser D, Malchow D, Wick U. Cell communication by periodic cyclic-AMP pulses. Philosophical Transactions of the Royal Society of London. Series B. 1975;**272**(915):181-192. DOI: 10.1098/rstb.1975.0080

[29] Wiegant VM. Cyclic nucleotides in nervous tissue. Brain Research Bulletin. 1978;**3**(6):611-622. DOI: 10.1016/0361-9230(78)90007-2

[30] Assmann SM. Cyclic AMP as a second messenger in higher plants (status and future prospects). Plant Physiology. 1995;**108**(3):885-889. DOI: 10.1104/pp.108.3.885

[31] Amrhein N. The current status of cyclic AMP in higher plants. Annual Review of Plant Physiology. 1977;**28**(1): 123-132. DOI: 10.1146/annurev.pp.28.060177.001011

[32] Martinez-Atienza J, Van Ingelgem C, Roef L, Maathuis FJM. Plant cyclic nucleotide signalling: Facts and fiction. Plant Signaling & Behavior. 2007;**2**(6): 540-543. DOI: 10.4161/psb.2.6.4789

[33] Gehring C. Adenyl cyclases and cAMP in plant signaling – Past and present. Cell Communication and Signaling: CCS. 2010;**8**(1):15. DOI: 10.1186/1478-811X-8-15

[34] Takahashi N, Ejiri SI, Katsumata T. Changes of cyclic AMP and cyclic GMP levels during germination of pine pollen. Agricultural and Biological Chemistry. 1978;**42**(8): 1605-1606. DOI: 10.1080/00021369.1978.10863215

[35] Gomperts BD, Kramer IJM, Tatham PER. Effector enzymes coupled to GTP binding proteins: Adenylyl cyclase and phospholipase C. In: Kramer IM, editor. Signal Transduction. Netherlands: Elsevier; 2009. pp. 131-157. DOI: 10.1016/B978-0-12-369441-6.00005-2

[36] Linder JU. Class III adenylyl cyclases: Molecular mechanisms of catalysis and regulation. Cellular and Molecular Life Sciences. 2006;**63**(15):1736-1751. DOI: 10.1007/s00018-006-6072-0

[37] Kurosaki F, Nishi A. Stimulation of calcium influx and calcium cascade by cyclic AMP in cultured carrot cells. Archives of Biochemistry and Biophysics. 1993;**302**(1):144-151. DOI: 10.1006/abbi.1993.1192

[38] Kurosaki F, Kaburaki H, Nishi A. Synthesis and degradation of cyclic AMP in cultured carrot cells treated with forskolin. Archives of Biochemistry and Biophysics. 1993;**303**(1):177-179. DOI: 10.1006/abbi.1993.1270

[39] Carricarte VC, Bianchini GM, Muschietti JP. Adenylate cyclase activity in a higher plant, alfalfa (*Medicago sativa*). The Biochemical Journal. 1988; **249**(3):807-811. DOI: 10.1042/bj2490807

[40] Li W, Luan S, Schreiber SL, Assmann SM. Cyclic AMP stimulates K^+ channel activity in mesophyll cells of *Vicia faba* L. Plant Physiology. 1994; **106**(3):957-961. DOI: 10.1104/pp.106.3.957

[41] Lemtiri-Chlieh F, Berkowitz GA. Cyclic adenosine monophosphate regulates calcium channels in the plasma membrane of Arabidopsis leaf guard and mesophyll cells. The Journal of Biological Chemistry. 2004;**279**(34):35306-35312. DOI: 10.1074/jbc.M400311200

[42] Wong A, Gehring C. Computational identification of candidate nucleotide cyclases in higher plants. In: Gehring C, editor. Cyclic Nucleotide Signaling in Plants. Methods in Molecular Biology.

Vol. 1016. New York: Humana Press; 2013. pp. 195-205. DOI: 10.1007/978-1-62703-441-8_13

[43] Kwezi L, Meier S, Mungur L, Ruzvidzo O, Irving H, Gehring C. The *Arabidopsis thaliana* brassinosteroid receptor (AtBRI1) contains a domain that functions as a guanylyl cyclase *in vitro*. PLoS One. 2007;**2**(5):e449. DOI: 10.1371/journal.pone.0000449

[44] Meier S, Ruzvidzo O, Morse M, Donaldson L, Kwezi L, Gehring C. The Arabidopsis wall associated kinase-like 10 gene encodes a functional guanylyl cyclase and is co-expressed with pathogen defense related genes. PLoS One. 2010;**5**(1):e8904. DOI: 10.1371/journal.pone.0008904

[45] Qi Z, Verma R, Gehring C. Ca^{+2} signaling by plant *Arabidopsis thaliana* pep peptides depends on AtPepR1, a receptor with guanylyl cyclase activity, and cGMP-activated Ca^{2+} channels. Proceedings of the National Academy of Sciences of the United States of America. 2010;**107**(49):21193-21198. DOI: 10.1073/pnas.1000191107

[46] Kwezi L, Ruzvidzo O, Wheeler JI. The phytosulfokine (PSK) receptor is capable of guanylate cyclase activity and enabling cyclic GMP-dependent signaling in plants. The Journal of Biological Chemistry. 2011;**286**(25): 22580-22588. DOI: 10.1074/jbc.M110.168823

[47] Mulaudzi T, Ludidi N, Ruzvidzo O. Identification of a novel *Arabidopsis thaliana* nitric oxide-binding molecule with guanylate cyclase activity *in vitro*. FEBS Letters. 2011;**585**(17):2693-2697. DOI: 10.1016/j.febslet.2011.07.023

[48] Ludidi N, Gehring C. Identification of a novel protein with guanylyl cyclase activity in *Arabidopsis thaliana*. The Journal of Biological Chemistry. 2003; **278**(8):6490-6494. DOI: 10.1074/jbc.M210983200

[49] Xu N, Zhang C, Lim LL, Wong A. Bioinformatic analysis of nucleotide cyclase functional centers and development of ACPred webserver. In: Proceedings of the 2018 ACM International Conference on Bioinformatics. Vol. 17. New York: Computational Biology and Health Informatics; 2018. pp. 122-129. DOI: 10.1145/3233547.3233549

[50] Ruzvidzo O, Dikobe BT, Kawadza DT, Mabadahanye GH, Chatukuta P, Kwezi L. Recombinant expression and functional testing of candidate adenylate cyclase domains. In: Gehring C, editor. Cyclic Nucleotide Signaling in Plants. Methods in Molecular Biology. Vol. 1016. New York: Humana Press; 2013. pp. 13-25. DOI: 10.1007/978-1-62703-441-8-2

[51] Ruzvidzo O, Gehring C, Wong A. New perspectives on plant adenylyl cyclases. Frontiers in Molecular Biosciences. 2019;**6**:136. DOI: 10.3389/fmolb.2019.00136

[52] Kawadza D, Dikobe T, Chatukuta P. An Arabidopsis maternal effect embryo arrest protein is an adenylyl cyclase with predicted roles in embryo development and response to abiotic stress. The Open Biotechnology Journal. 2023;**17**:1-10. DOI: 10.2174/18740707-v16-e221206-2022-10

[53] Sehlabane KS, Chatukuta P, Dikobe TB. A putative protein with no known function in *Arabidopsis thaliana* harbors a domain with adenylyl cyclase activity. Asian Journal of Pharmaceutical Sciences. 2022;**13**(07):943-959. DOI: 10.4236/ajps.2022.137062

[54] Varadi M, Anyango S, Deshpande M. AlphaFold protein structure database: Massively expanding the structural coverage of protein-sequence space with high-accuracy models. Nucleic Acids Research. 2022;**50**(D1):D439-D444. DOI: 10.1093/nar/gkab1061

[55] Pettersen EF, Goddard TD, Huang CC. UCSF ChimeraX: Structure visualization for researchers, educators, and developers. Protein Science. 2021; **30**(1):70-82. DOI: 10.1002/pro.3943

[56] Gastreich M, Lilienthal M, Briem H, Claussen H. Ultrafast *de novo* docking combining pharmacophores and combinatorics. Journal of Computer-Aided Molecular Design. 2007;**20**(12): 717-734. DOI: 10.1007/s10822-006-9091-x

[57] Trott O, Olson AJ. AutoDock Vina: Improving the speed and accuracy of docking with a new scoring function, efficient optimization, and multithreading. Journal of Computational Chemistry. 2009;**31**(2): 455-461. DOI: 10.1002/jcc.21334

[58] Bradford MM. A rapid and sensitive method for the quantitation of microgram quantities of protein utilizing the principle of protein-dye binding. Analytical Biochemistry. 1976;**72**(1–2): 248-254. DOI: 10.1016/0003-2697(76)90527-3

[59] Shah S, Peterkofsky A. Characterization and generation of *Escherichia coli* adenylate cyclase deletion mutants. Journal of Bacteriology. 1991; **173**(10):3238-3242. DOI: 10.1128/jb.173.10.3238-3242.1991

[60] Ullman A, Danchin A. Role of cyclic AMP in regulatory mechanisms in bacteria. Trends in Biochemical Sciences. 1980;**5**(4):95-96. DOI: 10.1016/0968-0004(80)90257-1

[61] Sweetlove LJ, Fernie AR. Regulation of metabolic networks: Understanding metabolic complexity in the systems biology era. The New Phytologist. 2005; **168**(1):9-24. DOI: 10.1111/j.1469-8137.2005.01513.x

[62] Jansen R, Embden JD, Gaastra W, Schouls LM. Identification of genes that are associated with DNA repeats in prokaryotes. Molecular Microbiology. 2002;**43**(6):1565-1575. DOI: 10.1046/j.1365-2958.2002.02839.x

[63] Allocco DJ, Kohane IS, Butte AJ. Quantifying the relationship between co-expression, co-regulation and gene function. BMC Bioinformatics. 2004; **5**(1):18. DOI: 10.1186/1471-2105-5-18

[64] Lee HK, Hsu AK, Sajdak J, Qin J, Pavlidis P. Co-expression analysis of human genes across many microarray data sets. Genome Research. 2004;**14**(6): 1085-1094. DOI: 10.1101/gr.1910904

[65] Manfield IW, Jen CH, Pinney JW. Arabidopsis Co-expression tool (ACT): Web server tools for microarray-based gene expression analysis. Nucleic Acids Research. 2006;**34**:W504-W509. DOI: 10.1093/nar/gkl204

[66] Zimmermann P, Hirsch-Hoffmann M, Hennig L, Genevestigator GW. Arabidopsis microarray database and analysis toolbox. Plant Physiology. 2004; **136**(1):2621-2632. DOI: 10.1104/pp.104.046367

[67] Barrett T, Troup DB, Wilhite SE. NCBI GEO: Archive for high-throughput functional genomic data. Nucleic Acids Research. 2009;**37**:D885-D890. DOI: 10.1093/nar/gkn764

[68] Craigon DJ. NASCArrays: A repository for microarray data generated by NASC's transcriptomics service.

Nucleic Acids Research. 2004;**32**(90001): 575D-5577D. DOI: 10.1093/nar/gkh133

[69] Chatukuta P, Dikobe T, Kawadza D. An Arabidopsis clathrin assembly protein with a predicted role in plant defense can function as an adenylate cyclase. Biomolecules. 2018;**8**(2):15. DOI: 10.3390/biom8020015

[70] Ung H, Karia P, Ebine K, Ueda T, Yoshioka K, Moeder W. Triphosphate tunnel metalloenzyme function in senescence highlights a biological diversification of this protein superfamily. Plant Physiology. 2017; **175**(1):473-485. DOI: 10.1104/pp.17.00700

[71] Ung H, Moeder W, Yoshioka K. Arabidopsis triphosphate tunnel metalloenzyme2 is a negative regulator of the salicylic acid-mediated feedback amplification loop for defense responses. Plant Physiology. 2014;**166**(2): 1009-1021. DOI: 10.1104/pp.114.248757

[72] Moeder W, Garcia-Petit C, Ung H. Crystal structure and biochemical analyses reveal that the Arabidopsis triphosphate tunnel metalloenzyme AtTTM3 is a tripolyphosphatase involved in root development. The Plant Journal. 2013;**76**(4):615-626. DOI: 10.1111/tpj.12325

[73] Bianchet C, Wong A, Quaglia M. An *Arabidopsis thaliana* leucine-rich repeat protein harbors an adenylyl cyclase catalytic center and affects responses to pathogens. Journal of Plant Physiology. 2019;**232**:12-22. DOI: 10.1016/j.jplph.2018.10.025

[74] Ruzvidzo O, Chatukuta P. An Arabidopsis linker histone-like protein harbours a domain with adenylyl cyclase activity. Plant Molecular Biology Reporter. 2023;**19**:1-15. DOI: 10.1007/s11105-023-01392-8

[75] Al-Younis I, Wong A, Gehring C. The *Arabidopsis thaliana* K^+-uptake permease 7 (AtKUP7) contains a functional cytosolic adenylate cyclase catalytic Centre. FEBS Letters. 2015;**589**: 3848-3852. DOI: 10.1016/j.febslet.2015.11.038

[76] Al-Younis I, Wong A, Lemtiri-Chlieh F. The *Arabidopsis thaliana* K^+-uptake permease 5 (AtKUP5) contains a functional cytosolic adenylate cyclase essential for K^+ transport. Frontiers in Plant Science. 2018;**9**:1645. DOI: 10.3389/fpls.2018.01645

[77] Al-Younis I, Moosa B, Kwiatkowski M, Jaworski K, Wong A, Gehring C. Functional crypto-adenylate cyclases operate in complex plant proteins. Frontiers in Plant Science. 2021;**12**:711749. DOI: 10.3389/fpls.2021.711749

[78] Qi L, Kwiatkowski M, Chen H. Adenylate cyclase activity of TIR1/AFB auxin receptors in plants. Nature. 2022; **611**(7934):133-138. DOI: 10.1038/s41586-022-05369-7

[79] Wong A, Tian X, Yang Y, Gehring C. Adenylate cyclase activity of TIR1/AFB links cAMP to auxin-dependent responses. Molecular Plant. 2022;**15**:1838-1840. DOI: 10.1016/j.molp.2022.11.012

[80] Kwiatkowski M, Wong A, Bi C, Gehring C, Jaworski K. Twin cyclic mononucleotide cyclase and phosphodiesterase domain architecture as a common feature in complex plant proteins. Plant Science. 2022;**325**:111493. DOI: 10.1016/j.plantsci.2022.111493

[81] Bobo DE, Sehlabane SK, Dikobe BT, Takundwa MM, Kawadza TD, Ruzvidzo O. Identification and characterization of a soybean protein

with adenylyl cyclase activity. Communications in Plant Sciences. 2022; **12**:50-59. DOI: 10.26814/cps2022007

[82] Yang H, Zhao Y, Chen N. A new adenylyl cyclase, putative disease-resistance RPP13-like protein 3, participates in abscisic acid-mediated resistance to heat stress in maize. Journal of Experimental Botany. 2021;**72**(2): 283-301. DOI: 10.1093/jxb/eraa431

[83] Ito M, Takahashi H, Sawasaki T, Ohnishi K, Hikichi Y, Kiba A. Novel type of adenylyl cyclase participates in tabtoxinine-β-lactam-induced cell death and occurrence of wildfire disease in *Nicotiana benthamiana*. Plant Signaling & Behavior. 2014;**9**(1):e27420. DOI: 10.4161/psb.27420

[84] Świeżawska B, Jaworski K, Pawełek A, Grzegorzewska W, Szewczuk P, Szmidt-Jaworska A. Molecular cloning and characterization of a novel adenylyl cyclase gene, HpAC1, involved in stress signaling in *Hippeastrum hybridum*. Plant Physiology and Biochemistry. 2014;**80**:41-52. DOI: 10.1016/j.plaphy.2014.03.010

[85] Kasahara M, Suetsugu N, Urano Y. An adenylyl cyclase with a phosphodiesterase domain in basal plants with a motile sperm system. Scientific Reports. 2016;**6**(1):39232. DOI: 10.1038/srep39232

[86] Świeżawska B, Duszyn M, Kwiatkowski M, Jaworski K, Pawełek A, Szmidt-Jaworska A. *Brachypodium distachyon* triphosphate tunnel metalloenzyme 3 is both a triphosphatase and an adenylyl cyclase upregulated by mechanical wounding. FEBS Letters. 2020;**594**(6):1101-1111. DOI: 10.1002/1873-3468.13701

[87] Duszyn M, Świeżawska-Boniecka B, Skorupa M, Jaworski K, Szmidt-Jaworska A. BdGUCD1 and cyclic GMP are required for responses of *Brachypodium distachyon* to *fusarium pseudograminearum* in the mechanism involving jasmonate. International Journal of Molecular Sciences. 2022; **23**(5):2674. DOI: 10.3390/ijms23052674

[88] Yuan Y, Liu Z, Wang L. Two triphosphate tunnel metalloenzymes from apple exhibit adenylyl cyclase activity. Frontiers in Plant Science. 2022; **13**:992488. DOI: 10.3389/fpls.2022.992488

[89] Liu Z, Yuan Y, Wang L. Three novel adenylate cyclase genes show significant biological functions in plant. Journal of Agricultural and Food Chemistry. 2023; **71**(2):1149-1161. DOI: 10.1021/acs.jafc.2c07683

[90] Danchin A. Phylogeny of adenylyl cyclases. Advances in Second Messenger and Phosphoprotein Research. 1993;**27**: 109